Methods in Molecular Biology

Series Editor
John M. Walker
School of Life and Medical Sciences
University of Hertfordshire
Hatfield, Hertfordshire, UK

For further volumes:
http://www.springer.com/series/7651

For over 35 years, biological scientists have come to rely on the research protocols and methodologies in the critically acclaimed *Methods in Molecular Biology* series. The series was the first to introduce the step-by-step protocols approach that has become the standard in all biomedical protocol publishing. Each protocol is provided in readily-reproducible step-by-step fashion, opening with an introductory overview, a list of the materials and reagents needed to complete the experiment, and followed by a detailed procedure that is supported with a helpful notes section offering tips and tricks of the trade as well as troubleshooting advice. These hallmark features were introduced by series editor Dr. John Walker and constitute the key ingredient in each and every volume of the *Methods in Molecular Biology* series. Tested and trusted, comprehensive and reliable, all protocols from the series are indexed in PubMed.

Intestinal Stem Cells

Methods and Protocols

Edited by

Paloma Ordóñez-Morán

Division of Cancer and Stem Cells, School of Medicine, Biodiscovery Institute, Centre for Cancer Sciences, University of Nottingham, Nottingham, UK

Humana Press

Editor
Paloma Ordóñez-Morán
Division of Cancer and Stem Cells
School of Medicine
Biodiscovery Institute
Centre for Cancer Sciences
University of Nottingham
Nottingham, UK

ISSN 1064-3745 ISSN 1940-6029 (electronic)
Methods in Molecular Biology
ISBN 978-1-0716-0746-6 ISBN 978-1-0716-0747-3 (eBook)
https://doi.org/10.1007/978-1-0716-0747-3

This Humana imprint is published by the registered company Springer Science+Business Media, LLC, part of Springer Nature.
The registered company address is: 1 New York Plaza, New York, NY 10004, U.S.A.

Preface

The intestinal epithelium is one of the most rapidly renewing types of tissue in the body, where intestinal stem cells are responsible for fueling the turnover of the tissue. A precise balance between self-renewal and differentiation of stem cells is essential to maintain homeostasis. Loss of this balance tends to lead to uncontrolled cell growth or prematuration and thus results in tumors, cancers, or tissue defects. During recent years, many researchers have undertaken great efforts to understand how the intestine replaces and repairs itself through the identification of the different intestinal stem cell populations and by defining its role in the continual renewal of the epithelial layer.

The goal of this book is to englobe the most up-to-date methods of the intestinal stem cell field. We provide here step-by-step guidance to a variety of techniques for studying intestinal stem cells properties. We aim to provide comprehensive and easy-to-follow protocols that are designed to be helpful to both seasoned researchers and newcomers to the field. The protocols included in this volume are separated into four different parts. Part I (Chapters 1–7) describes in vitro techniques to study different aspects of the intestinal stem cell functions by innovative imaging and functional assays. We have put particular emphasis on approaches to study the metabolism and niche of intestinal stem cells. Part II (Chapters 8 and 9) outlines the power of the single-cell transcriptional profiling method. In these recent years, the knowledge of intestinal stem cell heterogeneity has quickly advanced thanks to the development of this emerging technology. Part III (Chapters 10–17) presents protocols for the isolation of intestinal crypts to generate and establish 3D organoids to study stem cells. Functional analysis of stem cells and their environment can currently be performed by using innovative in vitro 3D technology that allows long-term culture and maintains basic crypt-villus physiology. This method allows a level of accessibility and tractability that is impossible to achieve in vivo and reduces animal experimentation. Furthermore, we also present protocols that use these 3D organoids as a tool to study intestinal stem cell properties. Finally, Part IV (Chapters 18–23) describes different animal models of gastrointestinal cancer and also presents examples of the use of in vivo state-of-the-art methods for studying intestinal tumor-initiating cells or cancer stem cells.

I would like to thank all of the contributors for sharing their expertise and for carefully guiding readers through all the details of their respective techniques. I am very grateful to the series editor, Dr. John Walker, for his help during the editing process.

Nottingham, UK *Paloma Ordóñez-Morán*

Contents

Part III Organoids and Applications

Part IV In Vivo Models

Contributors

JUDITH AGUDO • *Department of Cancer Immunology and Virology, Dana-Farber Cancer Institute, Boston, MA, USA; Department of Immunology, Harvard Medical School, Boston, MA, USA*

STEVEN J. ALTSCHULER • *Department of Pharmaceutical Chemistry, University of California, San Francisco, San Francisco, CA, USA*

JUMPEI ASANO • *Department of Biodefense Research, Medical Research Institute, Tokyo Medical and Dental University (TMDU), Tokyo, Japan; Seirei Women's Junior College, Akita, Japan*

DURGA ATTILI • *Department of Cell and Developmental Biology, University of Michigan, Ann Arbor, MI, USA*

KEVIN BARBER • *Eli and Edythe Broad Center of Regeneration Medicine and Stem Cell Research, University of California, San Francisco, San Francisco, CA, USA; Department of Biochemistry and Biophysics, University of California, San Francisco, San Francisco, CA, USA*

JAKE M. BIEBER • *Department of Pharmaceutical Chemistry, University of California, San Francisco, San Francisco, CA, USA; Graduate Program in Bioengineering, University of California, Berkeley, Berkeley, CA, USA*

COSTANZA BORRELLI • *Institute of Molecular Life Sciences, University of Zurich, Zurich, Switzerland; Institute of Molecular Cancer Research, University of Zurich, Zurich, Switzerland*

ERIN C. BUSH • *Department of Systems Biology, Columbia University Irving Medical Center, New York, NY, USA; Department of Biochemistry & Molecular Biophysics, Columbia University Irving Medical Center, New York, NY, USA*

RUBEN I. CALDERON • *Division of Digestive and Liver Diseases, Department of Medicine, Columbia Center for Human Development, Columbia Stem Cell Initiative, Columbia University Irving Medical Center, New York, NY, USA; Department of Genetics and Development, Columbia University Irving Medical Center, New York, NY, USA*

JUAN ANTONIO CÁMARA • *Preclinical Imaging Platform, Vall d'Hebron Institute of Research (VHIR), Barcelona, Spain*

CLAUDIA CAPDEVILA • *Division of Digestive and Liver Diseases, Department of Medicine, Columbia Center for Human Development, Columbia Stem Cell Initiative, Columbia University Irving Medical Center, New York, NY, USA; Department of Genetics and Development, Columbia University Irving Medical Center, New York, NY, USA*

INA W. CHEN • *Department of Pharmaceutical Chemistry, University of California, San Francisco, San Francisco, CA, USA*

CHIA-WEI CHENG • *The David H. Koch Institute for Integrative Cancer Research at MIT, Cambridge, MA, USA; Department of Biology, MIT, Cambridge, MA, USA*

IRENE CHICOTE • *Stem Cells and Cancer Laboratory, Vall d'Hebron Institute of Oncology (VHIO), Barcelona, Spain*

HANS CLEVERS • *Hubrecht Institute, Royal Netherlands Academy of Arts and Sciences (KNAW), Utrecht, The Netherlands; Oncode Institute, Utrecht, The Netherlands; University Medical Center Utrecht, Cancer Genomics Netherlands, Utrecht, The Netherlands; Princess Maxima Center for Pediatric Oncology, Utrecht, The Netherlands*

Justin A. Colacino • *Department of Environmental Health Sciences, University of Michigan School of Public Health, Ann Arbor, MI, USA; Department of Nutritional Sciences, University of Michigan School of Public Health, Ann Arbor, MI, USA; Center for Computational Medicine and Bioinformatics, University of Michigan, Ann Arbor, MI, USA*

Caroline Dafflon • *Disease Area Oncology, Novartis Institutes for BioMedical Research, Basel, Switzerland*

Michael K. Dame • *Division of Gastroenterology, Department of Internal Medicine, University of Michigan Medical School, Ann Arbor, MI, USA*

Pierre Dessen • *Swiss Institute for Experimental Cancer Research, École Polytechnique Fédé rale de Lausanne (EPFL), Lausanne, Switzerland*

Ruslan I. Dmitriev • *School of Biochemistry and Cell Biology, University College Cork, Cork, Ireland*

Faranak Fattahi • *Eli and Edythe Broad Center of Regeneration Medicine and Stem Cell Research, University of California, San Francisco, San Francisco, CA, USA; Department of Biochemistry and Biophysics, University of California, San Francisco, San Francisco, CA, USA*

Jared M. Fischer • *Cancer Early Detection Advanced Research Center, Knight Cancer Institute, Oregon Health & Science University, Portland, OR, USA; Department of Molecular and Medical Genetics, Oregon Health & Science University, Portland, OR, USA*

Masayuki Fujii • *Department of Organoid Medicine, Keio University School of Medicine, Tokyo, Japan*

Hartmut Geiger • *Division of Experimental Hematology, Cancer Biology & Stem Cell Biology, Cincinnati Children's Hospital Medical Center, Cincinnati, OH, USA*

David L. Hacker • *Protein Production and Structure Core Facility (PPSCF), School of Life Sciences, École Polytechnique Fédérale de Lausanne (EPFL), Lausanne, Switzerland*

Jonathan M. Harnoss • *Cancer Immunology, Genentech, Inc., South San Francisco, CA, USA*

Michael A. Helmrath • *Department of Pediatric General and Thoracic Surgery, Cincinnati Children's Hospital Medical Center, Cincinnati, OH, USA; Department of Pediatrics, University of Cincinnati, Cincinnati, OH, USA*

Sha Huang • *Division of Gastroenterology, Department of Internal Medicine, University of Michigan Medical School, Ann Arbor, MI, USA; Department of Cell and Developmental Biology, University of Michigan, Ann Arbor, MI, USA*

Joerg Huelsken • *Swiss Institute for Experimental Cancer Research, École Polytechnique Fé dérale de Lausanne (EPFL), Lausanne, Switzerland*

Masamichi Imajo • *Institute for Chemical Reaction Design and Discovery (WPI-ICReDD), Hokkaido University, Sapporo, Japan*

Nicholas Jinks • *Cancer Genetics & Stem Cell Group, BioDiscovery Institute, Division of Cancer and Stem Cells, School of Medicine, University of Nottingham, Nottingham, UK*

Hossein Kashfi • *Cancer Genetics & Stem Cell Group, BioDiscovery Institute, Division of Cancer and Stem Cells, School of Medicine, University of Nottingham, Nottingham, UK*

Taryn E. Kawashima • *Cancer Early Detection Advanced Research Center, Knight Cancer Institute, Oregon Health & Science University, Portland, OR, USA*

OPHIR D. KLEIN • *Department of Pediatrics, University of California, San Francisco, San Francisco, CA, USA; Program in Craniofacial Biology and Department of Orofacial Sciences, University of California, San Francisco, San Francisco, CA, USA; Department of Pediatrics and Institute for Human Genetics, University of California, San Francisco, San Francisco, CA, USA; Klein Lab, UCSF, San Francisco, CA, USA*

NOELYN KLJAVIN • *Molecular Oncology, Genentech, Inc., South San Francisco, CA, USA*

KEVIN G. LEONG • *Discovery Biology, Exelixis Inc., Alameda, CA, USA; Discover Oncology, Genentech, Inc., South San Francisco, CA, USA*

MAXIME M. MAHE • *Department of Pediatric General and Thoracic Surgery, Cincinnati Children's Hospital Medical Center, Cincinnati, OH, USA; Department of Pediatrics, University of Cincinnati, Cincinnati, OH, USA; Université de Nantes, INSERM, TENS, The Enteric Nervous System in Gut and Brain Diseases, IMAD, Nantes, France*

ELISE C. MANALO • *Cancer Early Detection Advanced Research Center, Knight Cancer Institute, Oregon Health & Science University, Portland, OR, USA*

MARIA M. MIHAYLOVA • *Department of Biological Chemistry and Pharmacology, The Ohio State University, Columbus, OH, USA; The Ohio State University Comprehensive Cancer Center, Columbus, OH, USA*

SHIZUKA MIURA • *Division of Organogenesis and Regeneration, Medical Institute of Bioregulation, Kyushu University, Fukuoka, Japan*

TOMOHIRO MIZUTANI • *Hubrecht Institute, Royal Netherlands Academy of Arts and Sciences (KNAW), Utrecht, The Netherlands; Oncode Institute, Utrecht, The Netherlands*

ANDREAS E. MOOR • *Institute of Molecular Cancer Research, University of Zurich, Zurich, Switzerland*

KODANDARAMIREDDY NALAPAREDDY • *Division of Experimental Hematology, Cancer Biology & Stem Cell Biology, Cincinnati Children's Hospital Medical Center, Cincinnati, OH, USA*

ABDOLRAHMAN S. NATERI • *Cancer Genetics & Stem Cell Group, BioDiscovery Institute, Division of Cancer and Stem Cells, School of Medicine, University of Nottingham, Nottingham, UK*

THERESA N. NGUYEN • *Cancer Early Detection Advanced Research Center, Knight Cancer Institute, Oregon Health & Science University, Portland, OR, USA*

MAXIM NORKIN • *Swiss Institute for Experimental Cancer Research, École Polytechnique Fédé rale de Lausanne, Lausanne, Switzerland*

TOSHIAKI OHTEKI • *Department of Biodefense Research, Medical Research Institute, Tokyo Medical and Dental University (TMDU), Tokyo, Japan*

IRINA A. OKKELMAN • *School of Biochemistry and Cell Biology, University College Cork, Cork, Ireland*

PALOMA ORDÓÑEZ-MORÁN • *Division of Cancer and Stem Cells, School of Medicine, Biodiscovery Institute, Centre for Cancer Sciences, University of Nottingham, Nottingham, UK*

HÉCTOR G. PALMER • *Stem Cells and Cancer Laboratory, Vall d'Hebron Institute of Oncology (VHIO), Barcelona, Spain*

DMITRI B. PAPKOVSKY • *School of Biochemistry and Cell Biology, University College Cork, Cork, Ireland*

HOLLY M. POLING • *Department of Pediatric General and Thoracic Surgery, Cincinnati Children's Hospital Medical Center, Cincinnati, OH, USA*

JENS PUSCHHOF • *Hubrecht Institute, Royal Netherlands Academy of Arts and Sciences (KNAW), Utrecht, The Netherlands; Oncode Institute, Utrecht, The Netherlands*

CHATHRUCKAN RAJENDRA • *Department of Pediatrics, University of California, San Francisco, San Francisco, CA, USA; Program in Craniofacial Biology and Department of Orofacial Sciences, University of California, San Francisco, San Francisco, CA, USA*
LAURA E. SANMAN • *Department of Pharmaceutical Chemistry, University of California, San Francisco, San Francisco, CA, USA*
ALBERT SANTAMARÍA-MARTÍNEZ • *Tumor Ecology Lab, Department of Oncology, Microbiology and Immunology, Faculty of Science and Medicine, University of Fribourg, Fribourg, Switzerland*
TAKU SATO • *Department of Biodefense Research, Medical Research Institute, Tokyo Medical and Dental University (TMDU), Tokyo, Japan*
TOSHIRO SATO • *Department of Organoid Medicine, Keio University School of Medicine, Tokyo, Japan; Department of Gastroenterology, Keio University School of Medicine, Tokyo, Japan*
FREDERIC J. DE SAUVAGE • *Molecular Oncology, Genentech, Inc., South San Francisco, CA, USA*
RYAN SCOTT • *Research Biology, Genentech, Inc., South San Francisco, CA, USA*
KISMET SHELDON-COLLINS • *Columbia Center for Human Development, Columbia Stem Cell Initiative, Division of Digestive & Liver Diseases, Department of Medicine, Columbia University Irving Medical Center, New York, NY, USA; Department of Genetics & Development, Columbia University Irving Medical Center, New York, NY, USA*
STEPHEN E. SHERMAN • *Department of Cancer Immunology and Virology, Dana-Farber Cancer Institute, Boston, MA, USA*
PETER A. SIMS • *Department of Systems Biology, Columbia University Irving Medical Center, New York, NY, USA; Department of Biochemistry & Molecular Biophysics, Columbia University Irving Medical Center, New York, NY, USA*
CATHERINE SOHN • *Research Biology, Genentech, Inc., South San Francisco, CA, USA*
FELIPE DE SOUSA E MELO • *Discovery Oncology, Genentech, Inc., South San Francisco, CA, USA*
JASON R. SPENCE • *Division of Gastroenterology, Department of Internal Medicine, University of Michigan Medical School, Ann Arbor, MI, USA; Department of Cell and Developmental Biology, University of Michigan, Ann Arbor, MI, USA; Department of Biomedical Engineering, College of Engineering, University of Michigan, Ann Arbor, MI, USA; Center for Organogenesis, University of Michigan Medical School, Ann Arbor, MI, USA*
TIANHONG SU • *Division of Digestive and Liver Diseases, Department of Medicine, Columbia Center for Human Development, Columbia Stem Cell Initiative, New York, NY, USA; Department of Genetics and Development, Columbia University Irving Medical Center, New York, NY, USA*
SHINYA SUGIMOTO • *Department of Organoid Medicine, Keio University School of Medicine, Tokyo, Japan; Department of Gastroenterology, Keio University School of Medicine, Tokyo, Japan*
NAMBIRAJAN SUNDARAM • *Department of Pediatric General and Thoracic Surgery, Cincinnati Children's Hospital Medical Center, Cincinnati, OH, USA*
ATSUSHI SUZUKI • *Division of Organogenesis and Regeneration, Medical Institute of Bioregulation, Kyushu University, Fukuoka, Japan*
CURTIS A. THORNE • *Department of Cellular and Molecular Medicine, University of Arizona Cancer Center, University of Arizona, Tucson, AZ, USA*

MARIA TRIFAS • *Division of Digestive and Liver Diseases, Department of Medicine, Columbia Center for Human Development, Columbia Stem Cell Initiative, New York, NY, USA; Department of Genetics and Development, Columbia University Irving Medical Center, New York, NY, USA*

SIMON VALES • *Department of Pediatric General and Thoracic Surgery, Cincinnati Children's Hospital Medical Center, Cincinnati, OH, USA*

TOMAS WALD • *Program in Craniofacial Biology and Department of Orofacial Sciences, University of California, San Francisco, San Francisco, CA, USA*

LANI F. WU • *Department of Pharmaceutical Chemistry, University of California, San Francisco, San Francisco, CA, USA*

KELLEY S. YAN • *Division of Digestive and Liver Diseases, Department of Medicine, Columbia Center for Human Development, Columbia Stem Cell Initiative, Columbia University Irving Medical Center, New York, NY, USA; Department of Genetics and Development, Columbia University Irving Medical Center, New York, NY, USA*

OMER H. YILMAZ • *The David H. Koch Institute for Integrative Cancer Research at MIT, Cambridge, MA, USA; Department of Biology, MIT, Cambridge, MA, USA; Broad Institute of Harvard and MIT, Cambridge, MA, USA; Department of Pathology, Massachusetts General Hospital and Harvard Medical School, Cambridge, MA, USA*

Part I

Characterization, Imaging, and Functional Assays

Chapter 1

Identification and Isolation of Human LGR5+ Cells Using an Antibody-Based Strategy

Michael K. Dame, Sha Huang, Durga Attili, Jason R. Spence, and Justin A. Colacino

Abstract

Leucine-rich repeat-containing G protein-coupled receptor 5 (Lgr5) has been identified as a marker of stem cells across multiple tissues. Lgr5-expressing cells are also regulators of tissue homeostasis and wound repair, and drivers of carcinogenic progression. The majority of information about Lgr5-expressing cells derives from genetically engineered mouse models. Human studies have been limited by a lack of specific reagents and experimental procedures for the purification of these cells. We recently demonstrated that antibody-based purification can be used to obtain viable LGR5-expressing cells from human primary tissues and patient derived organoids. Here, we provide detailed methods for the purification of these cells from colonic epithelial organoids generated from patient-derived tissues, from induced pluripotent stem cell (iPSC) derived intestinal organoids, and from freshly isolated patient tissue intestinal crypts. These methods will facilitate experimental analysis of human LGR5-expressing cells in development, wound healing, and cancer.

Key words Lgr5, Organoid, Colonoid, Enteroid, Antibody, MACS, Colon, Intestine, Stem cell, Human

1 Introduction

Leucine-rich repeat-containing G protein-coupled receptor 5 (Lgr5) has been characterized as a stem cell marker across multiple organ sites [1–4]. The LGR family of proteins act as receptors for R-spondin proteins, which function to potentiate Wnt/β-catenin signaling [5–11]. Lgr5 was first identified as a stem cell marker in the mouse intestine, where it marked a select subpopulation of cells at the base of the crypt [1]. Lineage tracing experiments have confirmed that Lgr5+ cells are actively cycling stem cells at the crypt base and give rise to differentiated cell lineages along the gastrointestinal tract. Lgr5 expressing cells are essential for tissue homeostasis, injury repair and regeneration in a number of tissues, including the pancreas, small intestine, stomach, and hair follicles

Paloma Ordóñez-Morán (ed.), *Intestinal Stem Cells: Methods and Protocols*, Methods in Molecular Biology, vol. 2171, https://doi.org/10.1007/978-1-0716-0747-3_1, © Springer Science+Business Media, LLC, part of Springer Nature 2020

[12–15]. Lineage tracing studies have also shown that Lgr5 marks a population of tumor initiating cells in precancerous adenoma lesions, which precede invasive cancer development [16]. Lgr5-expressing cells are implicated as the drivers of metastasis in colon cancer [17]. Due to their established role in cancer initiation and metastasis, Lgr5-expressing cells are currently under intense investigation as a target for chemotherapeutics [18, 19].

Many fundamental discoveries about stem cell biology have been made using mouse models genetically engineered to express an Lgr5 reporter [1]. Characterizing the biology of Lgr5 expressing cells in normal and tumor tissues from humans has been more challenging, due to a lack of methods for the accurate identification and purification of Lgr5-expressing cells, specifically due to unsuccessful efforts to generate effective LGR5-targeting antibodies [20]. RNA in situ hybridization has been utilized to detect *LGR5*-expressing cells in human tissues [21, 22]; however, this approach does not allow for the isolation of live cell populations. More recently, LGR5-expression reporter human organoids lines have been developed using gene editing techniques [23]; however, these methods are not broadly applicable to unmodified primary human tissues. We describe here the procedures for the dissociation of human organoids or fresh tissue crypts into viable single cells, labeling of cells with a magnetic bead-bound anti-LGR5 antibody, magnetic separation, and flow cytometry analysis (Fig. 1). Each aspect of this procedure required rigorous optimization, and we detail the culmination of that optimization process here. We have applied this procedure to human colonoids (normal and adenoma derived epithelial organoids) [24], human pluripotent stem cell (hPSC)-derived intestinal organoids, and freshly prepared normal colon tissue crypts [24]. The initiation and maintenance of colonoid or hPSC-derived organoid cultures is well established and previously described [25–28]. Primary colon organoids can be grown in conditions that are highly enriched in stem cells [1] or that recapitulate the tissue differentiation hierarchy [29]. Thus, patient-derived organoids provide a robust experimental platform for the interrogation of human LGR5 expressing cells. This protocol enables the antibody-based purification of viable LGR5-expressing cells, which will facilitate the experimental analysis of these cells in development, tissue homeostasis, wound repair, and carcinogenesis.

2 Materials

All plasticware that comes into contact with tissue, organoids, or cells throughout the procedure is coated with 0.1% bovine serum albumen to minimize adherence of cells to plastic surfaces (*see* **Note 1**). Where noted, BSA-coated pipette tips are cut to

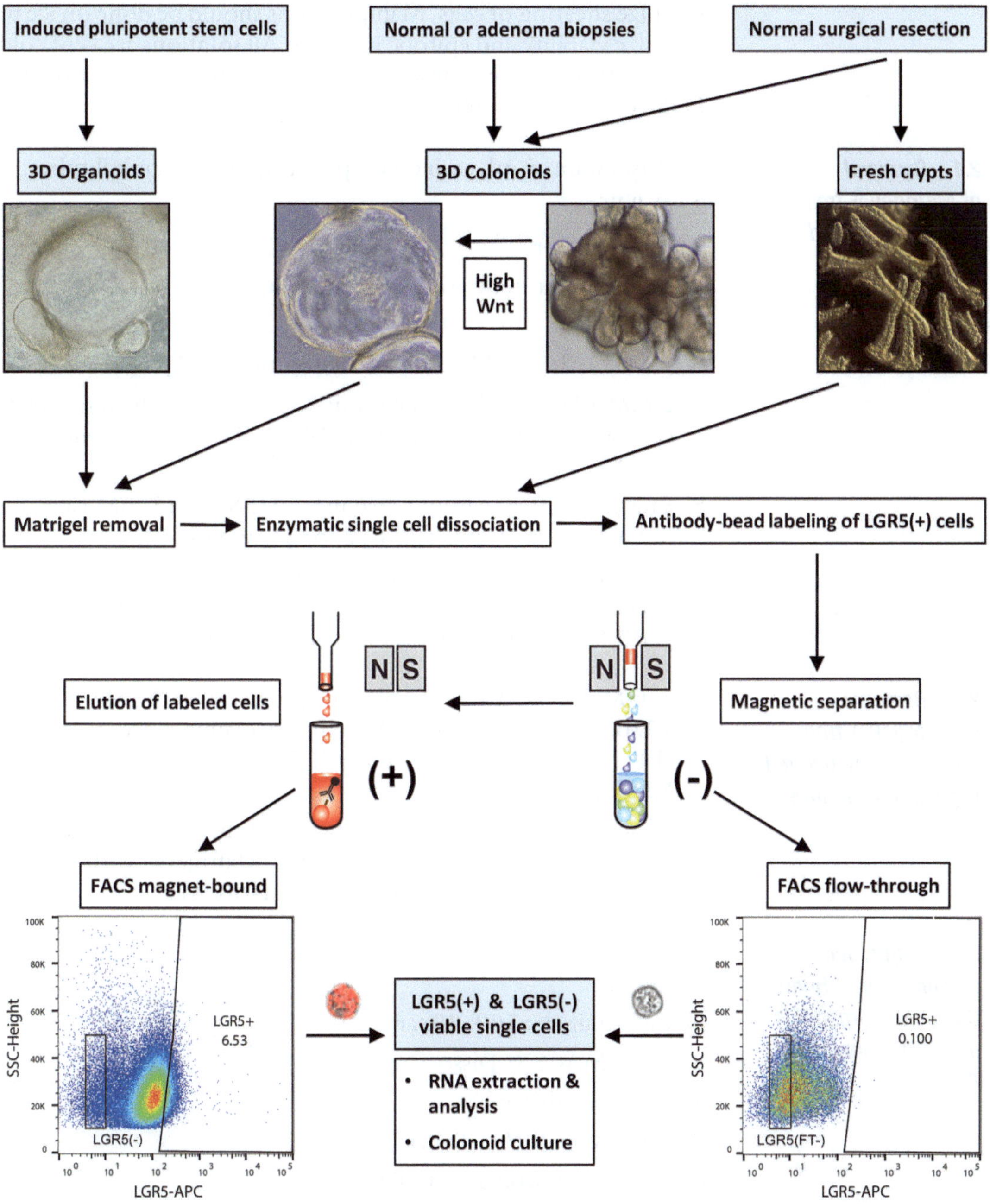

Fig. 1 Graphical illustration of the strategy for the isolation of LGR5(+) cells. Outlined here are methods to isolate LGR5(+) cells from cultured human colonoids (normal and adenoma-derived), iPSC-derived organoids (composed of epithelium and mesenchyme), and from freshly isolated colonic tissue crypts. High-Wnt3a containing L-WRN medium is used to drive a thin-walled cystic morphology in the colonoid culture, enriching for the stem cell component. Matrigel is first removed, followed by single cell dissociation with a gentle enzyme preparation. Cells are labeled with anti-human LGR5 antibody-magnetic bead conjugate, followed by an anti-bead allophycocyanin (APC) stain. Where mesenchyme is present or contaminating, the epithelial marker EpCAM is used to discriminate epithelium. Cells are passed through magnetic columns to enrich for

minimize shearing of cells. Manipulations should be done on ice to maximize viability and epitope integrity. All solutions are kept cold in ice, with the exception of the 37 °C Tumor Dissociation Kit (TDK) enzyme preparation.

2.1 Removal of Matrigel™ from Cultured Colonoid Structures

1. Plasticware: 5 mL serological pipettes, 15 mL conical tubes.
2. Cell lifter.
3. Tube rotator.
4. Swinging bucket refrigerated centrifuge.
5. DPBS.
6. Y27632: 2.5 mM Y27632 stock solution in H_2O (*see* **Note 2**).
7. 2 mM EDTA-Y27632 solution: 2 mM ethylenediaminetetraacetic acid (EDTA) solution in DPBS, pH 7.3–7.4, containing 5 μM Y27632.
8. Enzyme Buffer (without enzymes): HBSS, 0.13 mM calcium, 0.9 mM magnesium, 5 μM Y27632 (*see* **Note 3**).
9. Human colonoids: cultures embedded in 10–50 μL-sized adherent droplets of 8 mg/mL Matrigel™ with 250 μL Matrigel™/well of a 6-well dish (*see* **Notes 4** and **5**).

2.2 Removal of Matrigel™ from Cultured iPSC-Derived Organoid Structures

1. 4 mM EDTA-Y27632 solution (in place of 2 mM EDTA-Y27632 solution): 4 mM ethylenediaminetetraacetic acid (EDTA) solution in DPBS, pH 7.3–7.4, containing 5 μM Y27632.
2. Human iPSC organoids: cultures are embedded in 50–100 μL sized adherent droplets of 8 mg/mL Matrigel™ with 400 μL Matrigel™/well of a 6-well dish (*see* **Note 5**).

2.3 Single Cell Dissociation of Human Colonoids, Organoids or Fresh Colonic Crypts

1. Plasticware: 5 mL serological pipettes, 15 and 50 mL conical tubes.
2. Cell strainers: 100, 40, and 20 μm.
3. gentleMACS™ Dissociator.
4. 37 °C incubator.
5. gentleMACS™ C Tubes.
6. Countess Automated Cell Counter.
7. Tumor Dissociation Kit (TDK) enzyme preparation: 400 μL enzyme H, 200 μL enzyme R, 50 μL enzyme A, add up to 10 mL total Enzyme Buffer including cell pellet volume (*see* **Note 6**).

Fig. 1 (continued) the LGR5-magnetic bead fraction, and FACS sorted for DAPI(−) and LGR5-APC(+) cells. Representative scatterplots are for cells isolated from an adenoma-derived colonoid. Final LGR5(+) and LGR5 (−) single cells flow-imaged with the Amnis ImageStreamX (60× magnification; brightfield and APC fluorescence). (Content adapted from Dame et al., 2018 [24] with permission from *Development*)

8. Labeling Buffer: 2 mM EDTA-Y27632, 0.5% BSA.
9. Trypan Blue solution.

2.4 Antibody Labeling

2.4.1 LGR5 Antibody Labeling

1. Plasticware: 2 mL V-bottom Eppendorf tubes, cut-down P200 pipette tips, uncut P200 pipette tips.
2. MACS™ LS Column.
3. FcR Blocking Reagent.
4. Anti-human Lgr5-Microbead: microbead conjugated to monoclonal anti-human LGR5, rat IgG2b, clone 22H2.8.
5. Labeling Check Reagent-allophycocyanin (APC): protect from light.
6. Labeling Buffer: 2 mM EDTA-Y27632, 0.5% BSA.
7. Column Buffer: Enzyme Buffer, 0.5% BSA, 200 Kunitz units/ mL DNase.
8. Flow Buffer: 2 mM EDTA, DPBS, 0.1% BSA, 10 μM Y27632.

2.4.2 EpCAM Labeling of iPSC-Derived Organoid or Freshly Isolated Crypt Cells

1. Plasticware: 2 mL V-bottom Eppendorf tubes, cut-down P200 pipette tips, uncut P200 pipette tips.
2. EpCAM antibody: phycoerythrin (PE)-conjugated anti-human EpCAM, mouse IgG2b κ, clone 9C4, protect from light.
3. EpCAM isotype control antibody: PE-mouse IgG2b κ, protect from light.
4. Labeling Buffer: 2 mM EDTA-Y27632, 0.5% BSA.
5. Flow Buffer: 2 mM EDTA, DPBS, 0.1% BSA, 10 μM Y27632.
6. Column Buffer: Enzyme Buffer, 0.5% BSA, 200 Kunitz units/ mL DNase.

2.5 Magnetic Activated Cell Separation (MACS™) of LGR5(+) and (−) Fractions

1. Plasticware: 15 mL conical tubes, cut-down P200 pipette tips, uncut P200 and P1000 pipette tips.
2. MACS™ Separator permanent magnet.
3. 20 μm cell strainer.
4. MACS™ LS Column: prepped from above Subheading 2.3.
5. Flow Buffer: 2 mM EDTA, DPBS, 0.1% BSA, 10 μM Y27632.
6. Column Buffer: Enzyme Buffer, 0.5% BSA, 200 Kunitz units/ mL DNase.

2.6 Flow Analysis and FACS of LGR5(+) and (−) Fractions

1. Plasticware: uncut P200 and P1000 pipette tips, 2 mL V-bottom Eppendorf tubes.
2. Flow Buffer: 2 mM EDTA, DPBS, 0.1% BSA, 10 μM Y27632.
3. DAPI working solution: 100 μM 4′,6-diamidino-2-phenylindole dilactate in H_2O (*see* **Note 7**).

4. RLT Lysis Buffer.
5. RNeasy Micro Kit with on-column DNase digestion.
6. Serum-free L-WRN conditioned medium (SF-LWRN): L-WRN 24-h-conditioned media in Advanced DMEM/F-12, 2 mM GlutaMax, 10 mM HEPES, 1× N-2 media supplement, 1× B-27 supplement minus vitamin A, 1 mM *N*-Acetyl-L-cysteine, 50 μg/mL Primocin, 5 ng/mL human EGF, and 5 ng/mL FGF-2.
7. Single Cell Culture Medium (SCC Medium) [23, 30–33]: Advanced DMEM/F-12, 50% SF-LWRN, 2 mM GlutaMax, 10 mM HEPES, 1× N-2 media supplement, 1× B-27 supplement minus vitamin A, 1 mM *N*-Acetyl-L-cysteine, 100 ng human EGF/mL, 10 mM Nicotinamide, 10 nM PGE2, 10 nM Gastrin, 10 μM Y27632, and 100 μg/mL Primocin. For normal colonoid cultures, supplement with 3 μM SB202190 and 500 nM A83-01. 5 μM CHIR99021, and 2 μM Jagged-1 are added for the first 2 days of single cell culture.
8. FACS Live Cell Collection Medium: SCC Medium with added 0.25 mg/mL Matrigel™.

2.7 Single Cell Colonoid-Forming Culture

1. 12-well tissue culture plate: 37 °C-warmed.
2. L-WRN growth medium (*see* **Notes 4** and **8**).

3 Methods

Detailed in vitro culture methodologies are outlined for the tissue-derived epithelium-only colonoids/enteroids by Sugimoto and Sato [34], and for the iPSC-derived organoids [26, 35–37]. We further cultured adenoma-derived colonoids [28] under conditions which enriched for the LGR5(+) stem cell component [24] (*see* **Note 4**). Colonoids were established from tissue acquired by endoscopy, from surgical resection or from deceased donors, and according to protocols approved by the University of Michigan Institutional Review Board. Procedures for patient-derived epithelial colonoids begin at Subheading 3.1.1, for iPSC-derived organoids at Subheading 3.1.2, and for freshly isolated tissue colon crypts at Subheading 3.2. See schematic illustration for overview of entire procedure (Fig. 1).

3.1 Removal of Matrigel™

3.1.1 Removal of Matrigel™ from Cultured Colonoid Structures (See Note 9) (1+h)

1. Treat cultures with 10 μM Y27632 for at least 2.5 h prior to harvesting (*see* **Note 2**).
2. Remove culture medium from wells and wash with cold DPBS.
3. Transfer Matrigel™ droplets with cell lifter in cold 2 mM EDTA-Y27632 into 15 mL tube(s) (up to 1000 μL Matrigel™ per 15 mL). Triturate vigorously 10× in 10 mL with 5 mL serological pipette. Fill tube(s) to 15 mL with 2 mM EDTA-Y27632.
4. Incubate with slow rotation (approximately 15 rpm) for 15 min at 4 °C (*see* **Note 10**).
5. Triturate 20× with 5 mL serological pipette. Centrifuge at 250 × *g* for 3 min at 4 °C.
6. Aspirate supernatant (first wash) from the 15 mL tube(s) and combine pellet(s) by gently triturating 5× with a 5 mL serological pipette in 7 mL cold DPBS. Fill tube with DPBS and slow-spin at 100 × *g* to additionally remove dead single cells.
7. Aspirate supernatant (second wash). Add 7 mL cold DPBS, gently triturate 5× with a 5 mL serological pipette, and centrifuge at 250 × *g* at 4 °C.
8. Aspirate supernatant (third wash). Add 7 mL cold Enzyme Buffer (without enzymes), gently triturate 5× with a 5 mL serological pipette, and centrifuge at 250 × *g* at 4 °C.
9. Proceed to Subheading 3.2.

3.1.2 Removal of Matrigel™ from Cultured iPSC-Derived Organoid Structures (2+ h) (See Note 11)

1. Treat cultures with 10 μM Y27632 for at least 2.5 h prior to harvesting (*see* **Note 2**).
2. Remove culture medium from wells and wash with cold DPBS.
3. Transfer Matrigel™ droplets with cell lifter in cold 4 mM EDTA-Y27632 into 15 mL tube(s) (up to 1000 μL Matrigel™/15 mL). Triturate 10× in 10 mL with 5 mL serological pipette. Fill tube(s) to 15 mL with 4 mM EDTA-Y27632.
4. Incubate with slow rotation (approximately 15 rpm) for 30 min at 4 °C.
5. Triturate 30× with 5 mL serological pipette. Centrifuge at 300 × *g* for 5 min at 4 °C.
6. Aspirate supernatant (first wash). Wash pellet by triturating 20× with a 5 mL serological pipette in 7 mL cold DPBS. Fill tubes to 15 mL with DPBS. Centrifuge at 300 × *g* for 5 min at 4 °C.
7. Aspirate supernatant (second wash). Add 7 mL cold DPBS, gently triturate 10× with a 5 mL serological pipette, and centrifuge at 300 × *g*.

8. Aspirate supernatant (third wash). Add 7 mL cold Enzyme Buffer (without enzymes), gently triturate 5× with a 5 mL serological pipette. Slow-spin at 100 × *g* for 5 min at 4 °C to additionally remove dead single cells.
9. Proceed to Subheading 3.2.

3.2 Single Cell Dissociation of Human Colonoids, Organoids or Fresh Colonic Crypts (2+ h)

1. Aspirate supernatant and resuspend pellet in 37 °C warm TDK-enzyme Preparation to a final volume of 10 mL. Transfer to warm MACS™ C tube.
   (a) Colonoid/organoid cultures: up to a 0.5 mL pellet per 10 mL TDK-enzyme Preparation.

   or

   (b) Freshly prepared tissue crypts (*see* **Notes 5** and **12**): Prior to enzyme digestion, wash newly isolated crypts once with Enzyme Buffer (without enzymes) and slow-spin at 100 × *g* for 5 min at 4 °C, aspirate buffer, and then add 10 mL TDK-enzyme Preparation per 1 mL of a dense crypt pellet.
2. Mechanically assist enzyme dissociation with a gentleMACS™ Dissociator using the program *h_tumor_01*, once at the beginning, and then every 15 min for 1 h. Finish with two *h_tumor_01* program runs. Slowly rotate suspension at 37 °C between runs (*see* **Note 13**).
3. Place tube on ice. Triturate 30× with a 5 mL serological pipette and pipet over 100 μm strainer into a 50 mL tube containing 15 mL Labeling Buffer.
4. Triturate 5× with a 5 mL serological pipette and pipet over 40 μm strainer into a 50 mL tube containing 5 mL Labeling Buffer.
5. Centrifuge at 500 × *g* for 5 min at 4 °C.
6. Aspirate supernatant and resuspend pellet in 2 mL Labeling Buffer by triturating 30× with a P1000. Add 28 mL of Labeling Buffer and pipet over 20 μm strainer(s) into two 15 mL tubes (first wash).
7. Centrifuge at 500 × *g* for 5 min at 4 °C.
8. Aspirate supernatant and resuspend pellet in 2 mL Labeling Buffer by triturating 30× with a P1000. Add 13 mL of Labeling Buffer. Estimate cell count and viability by trypan blue exclusion (second wash) (*see* **Note 14**).
9. Centrifuge at 500 × *g* for 5 min at 4 °C.
10. Aspirate supernatant and resuspend in 10 mL Labeling Buffer by triturating 5–10× with a 5 mL serological pipette. Aliquot the appropriate number of cells to control and sort 15 mL tubes and bring up to 10 mL per tube (third wash):

(a) Control cell 15 mL tube: colonoid cells: 0.2–1.0 × 10^6 cells (cells for 1 FACS control) and iPSC organoid or fresh crypt cells: 0.6–3.0 × 10^6 cells (cells for 3 FACS controls).

and

(b) Sort cell 15 mL tube: add remaining cells.

11. Pellet both tubes at 500 × *g* for 5 min at 4 °C.

3.3 LGR5 Antibody Labeling (1–1.5 h)

All manipulations are conducted on ice, while incubations are at 4 °C. During incubations, at 5 min intervals, gently tap tubes to disperse cells. Resuspend or mix cells with a cut-down or wide-bore P200 pipette tip to minimize sheering of cells. Volumes listed below are for ≤10^7 cells; adjust volumes 2× for 1–2 × 10^7; 3× for 2–3 × 10^7 and so on.

1. Add 10 μL FcR Blocking reagent to empty 2 mL Eppendorf tubes (*see* **Note 15**).
   (a) Colonoid cells: two Eppendorf tubes (one "Sort" and one "No Stain" control).

   or

   (b) iPSC organoid or crypt cells: four Eppendorf tubes (one "Sort" and three control tubes: "No Stain," "EpCAM only," and "EpCAM isotype").
2. Aspirate supernatant from tubes in Subheading 3.2, **step 11** and resuspend cell pellets in the following volumes of cold Labeling Buffer.
   (a) Colonoid cells: 70 μL (sort 15 mL tube) and 70 μL (control 15 mL tube).

   or

   (b) iPSC organoid or crypt cells: 70 μL (sort 15 mL tube) and 210 μL (control 15 mL tube).
3. Add 70 μL of cells in Labeling Buffer to FcR-Eppendorf tubes (total volume now 80 μL per Eppendorf tube).
4. Incubate for 10 min at 4 °C. Set aside control Eppendorf tubes at 4 °C; periodically tap tubes to disperse cells.
5. During incubation prepare the LS column(s) (≤1 × 10^7 cells per column) for Subheading 3.4 (at least 45 min before use). Precoat the LS column by applying 2.5 mL of Column Buffer to column. After buffer begins to drip from column, stop end with syringe cap wrapped in Parafilm. Place at 4 °C.
6. Add 20 μL Lgr5-Microbeads to Eppendorf sort tube and gently mix (total volume now 100 μL).
7. Incubate for 15 min at 4 °C.
8. Add 1.7 mL Labeling Buffer to Eppendorf "Sort" tube (total volume now 1.8 mL).

9. Centrifuge Eppendorf "Sort" tube at 500 × *g* for 5 min at 4 °C.
10. Aspirate supernatant and resuspend pellet in 90 μL of Labeling Buffer.
11. Add 10 μL APC Labeling Check Reagent to Eppendorf "Sort" tube and gently mix (total volume now 100 μL). Protect cells from light henceforward.
12. Incubate for 10 min at 4 °C.
13. Centrifuge Eppendorf "Sort" tube at 500 × *g* for 5 min at 4 °C.
14. Aspirate supernatant and resuspend Eppendorf "Sort" tube and Eppendorf "No Stain" control tube with up to 1.8 mL Labeling Buffer (first wash).
15. Centrifuge at 500 × *g* for 5 min at 4 °C.

3.3.1 Colonoid Cells: No EpCAM Labeling Required

1. Aspirate supernatant and resuspend Eppendorf "No Stain" control tube in 1 mL Flow Buffer and place on ice.
2. Aspirate supernatant and resuspend Eppendorf "Sort" tube in 1.8 mL of Labeling Buffer (second wash).
3. Centrifuge at 500 × *g* for 5 min at 4 °C.
4. Aspirate supernatant and resuspend Eppendorf "Sort" tube in 1 mL Column Buffer by triturating 30× with an uncut P200 pipette tip and proceed with magnetic separation (MACS™).
5. Proceed to Subheading 3.4.

3.3.2 iPSC-Derived Organoid or Freshly Isolated Crypt Cells: EpCAM Labeling (See Note 16)

1. Aspirate supernatant and resuspend Eppendorf "No Stain" control tube in 1 mL Flow Buffer and place on ice.
2. Aspirate supernatant and resuspend the Eppendorf "EpCAM only" and "EpCAM isotype" tubes in 100 μL Labeling Buffer.
3. Add 1 μL EpCAM isotype antibody to "EpCAM isotype" Eppendorf tube, and mix well.
4. Add 4 μL EpCAM antibody to the "EpCAM only" Eppendorf tube and mix well (*see* **Note 17**).
5. Aspirate supernatant and resuspend the Eppendorf "Sort" tube in 400 μL Labeling Buffer per 1×10^7 cells.
6. Add 16 μL EpCAM antibody to the Eppendorf "Sort" tube and mix well (*see* **Note 18**).
7. Incubate tubes for 10 min at 4 °C.
8. Resuspend all Eppendorf tubes up to 1.8 mL of Labeling Buffer (first EpCAM wash).
9. Centrifuge at 500 × *g* for 5 min at 4 °C.
10. Repeat wash and centrifugation (second EpCAM wash).

11. Aspirate supernatant and resuspend the "EpCAM Only" and "EpCAM isotype" control Eppendorf tubes in 1 mL Flow Buffer and place on ice.
12. Resuspend Eppendorf "Sort" tube in 1 mL Column Buffer by triturating 30× with an uncut P200 pipette tip and proceed with magnetic separation (MACS™).

3.4 Magnetic Activated Cell Separation (MACS™) of LGR5(+) and (−) Fractions (1 h)

1. Snap column to the magnet. Remove column end cap to drain coating buffer. Position 15 mL flow-through collection tube on ice.
2. Place 20 μm strainer above column. If cells have settled since resuspension at Subheading 3.3.1, **step 4** or 3.3.2, **step 12**, triturate cells 30× immediately with an uncut P200 pipette tip before applying to column to ensure that they are in single cell suspension. Apply 1 mL cell suspension with a P1000 uncut pipette tip (*see* **Note 19**).
3. After the entire cell volume has passed through, apply 3 mL cold Column Buffer (*see* **Note 20**).
4. Repeat 3 mL cold Column Buffer wash twice more, continuing to collect flow-through unbound fractions on ice. Perform a cell count and viability assessment of the flow-through fraction by trypan blue exclusion.
5. Remove column from magnet and place over 15 mL collection tube on ice. Apply 2.5 mL Column Buffer and vigorously flush out magnet-bound cells by firmly pushing plunger into column. Perform a cell count and viability assessment of the magnet-bound positive fraction by trypan blue exclusion.
6. Centrifuge both the flow-through and magnet-bound fractions at 500 × *g* for 10 min at 4 °C.
7. Resuspend both the flow-through and magnet-bound fractions in cold Flow Buffer with an P200 uncut pipette tip at concentrations approximately 3–5 × 10^6 cells/mL, depending on the specifications of your FACS instrument.

3.5 Flow Analysis and FACS of LGR5(+) and (−) Fractions (See Note 21)

Stain with 1 μM DAPI for 1 min prior to analysis for viability assessment: add 10 μL 100 μM DAPI working solution per mL cells.

1. Triturate cells 20× before analysis/sorting with P1000 uncut pipette tip to ensure cells are in a single cell suspension prior to analysis.
2. With unstained control cells, set forward and side scatter gating strategy to exclude debris and doublets.
3. Add DAPI to unstained control cells. Gate and exclude nonviable cells, including intermediate dying cells (*see* **Note 22**). Proceed to **step 5** for colonoid-derived cells (no EpCAM).

4. For iPSC-derived organoid cells and fresh tissue crypt cells (EpCAM gating).
   (a) Add DAPI to EpCAM isotype-PE-stained cells to set gating threshold for EpCAM at 0.1% events based on EpCAM isotype-PE staining in DAPI(−) cells.

   and

   (b) Add DAPI to EpCAM-PE-stained cells and. Analyze for DAPI(−) EpCAM(+) cells (*see* **Note 23**). Further LGR5 analyses should be gated for DAPI(−)EpCAM(+) cells.
5. Add DAPI to the MACS™ flow-through negative fraction. Set the threshold APC-positive gating at 0.01–0.1% APC (*see* **Note 24**).
6. FACS sort viable LGR5(−) cells from both flow-through negative and magnet-bound positive fractions, while collecting LGR5(+) cells from the magnet-bound positive fraction (*see* **Note 25**). See representative Flow scatter plots of LGR5(+) cells isolated from iPSC-derived organoids (Fig. 2).
7. FACS cell collection options:
   (a) RNA analyses: FACS collect into 100 μL RLT Lysis Buffer for ≤5000 cells; 350 μL for 5000–50,000 cells. For ≥50,000 sorted cells, add 100 μL additional lysis buffer immediately after sort. Vortex for 1 min and place on ice for RNA processing (*see* **Note 26**). We extract RNA with RNeasy Micro Kit with on-column DNase digestion according to manufacturer's instructions. We have used RNA extracted from these cells for RNA sequencing of LGR5(+) and LGR5(−) cells from patient-derived colonoids (Fig. 3).

   or

   (b) Live-cell collection for analysis or culture: FACS collect approximately 2000 cells into 13 mL cold FACS Live Cell Collection Medium into a 15 mL tube. Proceed to Subheading 3.6 for culture of FACS single cells from colonoid cultures.

3.6 Single Cell Colonoid-Forming Culture

1. Pellet FACS sorted cells at 500 × *g* for 10 min at 4 °C.
2. Resuspend pellet in 22 μL SCC Medium (including cell pellet-residual supernatant volume) and mix with 88 μL ice-cold Matrigel™ to a final concentration of 8 mg/mL Matrigel™ (assuming Matrigel™ stock is 10 mg/mL) for a total of 110 μL Matrigel™-cell mixture.
3. Pipet 10–12 raised 10 μL Matrigel-cell drops (approximately 200 cells/10 μL Matrigel™) onto the well surface of a prewarmed 12-well plate placed on a warm pack.

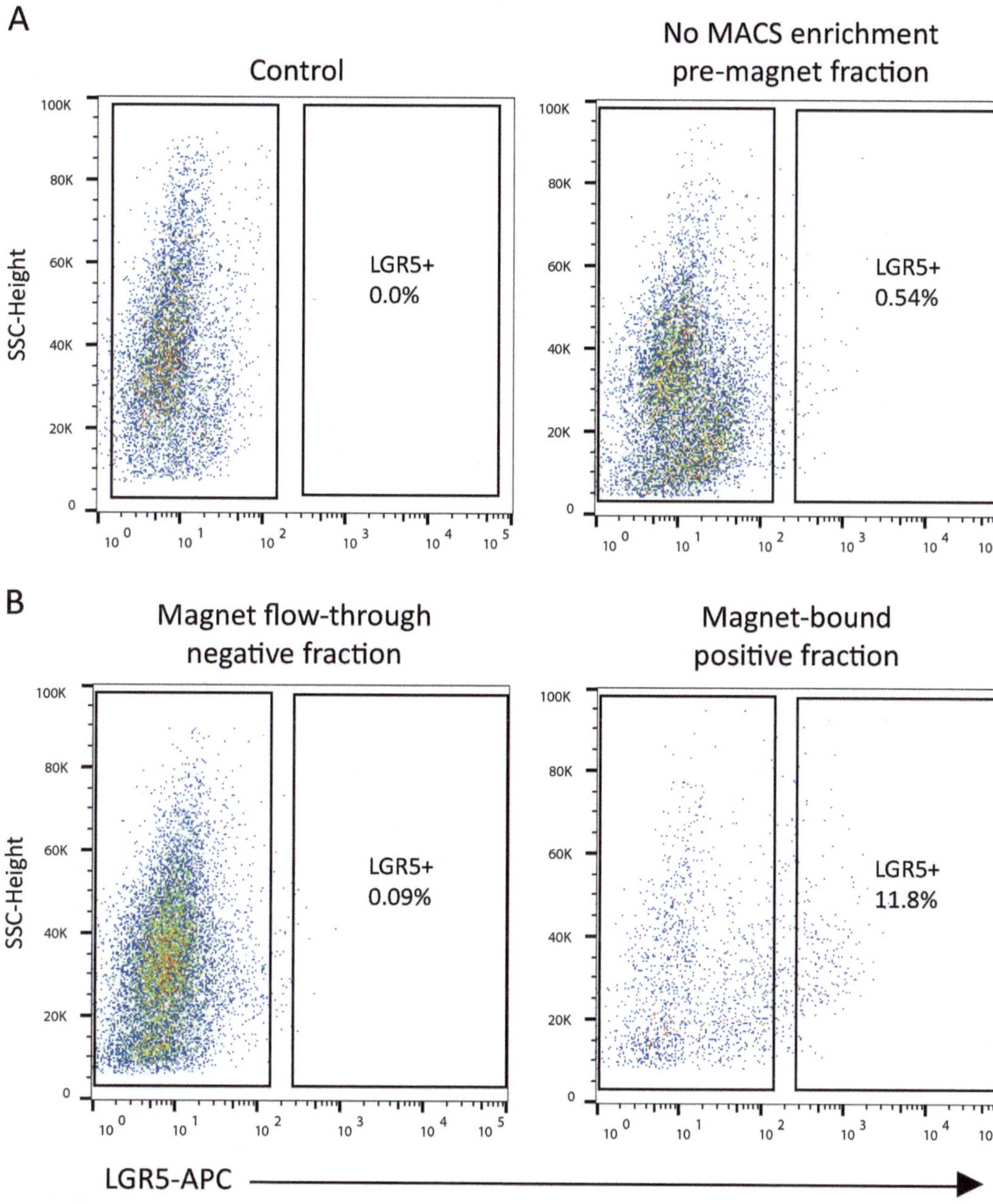

Fig. 2 Isolation of LGR5(+) cells from human iPSC-derived organoids. Cells were first isolated on the live DAPI (−), EpCAM-PE(+) epithelial cell markers to discriminate epithelial cells from the associated mesenchymal iPSC cell lineage (scatter plots not shown). (**a**) Control FMO-stained cells are DAPI(−), and EpCAM-PE(+), minus stain for APC. Representative scatterplots of LGR5-APC events are shown before and (**b**) after magnetic bead separation (MACS). The flow-through effluent is depleted of LGR5-APC(+) cells, while the magnetic bead-bound fraction is enriched 20-fold over the pre-MACS fraction for LGR5-APC(+) cells

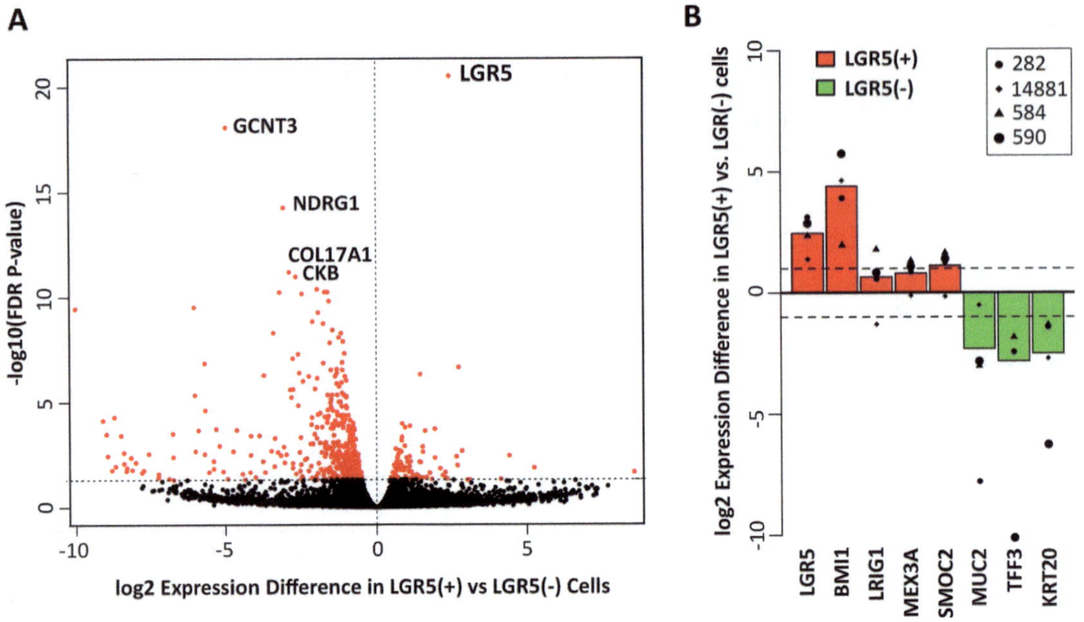

Fig. 3 Transcriptomic analysis of isolated LGR5(+) vs LGR5(−) cells. Colonoid cultures were established from four patient-derived, genetically diverse, tubular adenomas (patient identifiers 14881, 282, 584, and 590). LGR5(+) cells were isolated and compared to LGR5(−) cells for differential gene expression across these four specimens. (**a**) FDR volcano plot of the log(2) ratio of gene expression between the LGR5(+) and LGR5(−) cells, based on the top 500 most variable genes. *LGR5* had the highest level of statistical enrichment in the LGR5(+) cells (FDR, 3.8E−21) and was expressed an average of 5.5-fold higher compared with in LGR5(−) cells. (**b**) Log(2) fold change in gene expression between LGR5(+) and LGR5(−) cells for known markers of colon stem (red) and differentiated (green) cells. Stem cell markers associated with the colon, as well as other tissue-specific stem cell markers, including BMI1, MEX3A, and SMOC2, were upregulated in LGR5(+) cells, whereas known markers of colonic differentiation, including MUC2, TFF3, and KRT20, were downregulated. (Content adapted from Dame et al., 2018 [24] with permission from *Development*)

4. After 10 min, add 800 μL of warm SCC Medium.
5. Change medium every 2 days.
6. When small cyst-like structures are evident (6–8 days), change to standard LWRN medium, or continue in SCC Medium (*see* **Note 8**).

4 Notes

1. All plastic surfaces are coated in 0.1% bovine serum albumen in DPBS (tissue culture grade BSA). Prepare a 10% BSA stock solution in DPBS and sterilize with a low-protein binding syringe-filter. Dilute in sterile DPBS to prepare 0.1% BSA working solution. Store solutions at 4 °C for up to 6 months; can be reused multiple times. Coat by filling plastic components at room temperature for below time periods or overnight at 4 °C. Polypropylene surfaces (conical tubes, Eppendorf

tubes, MACS™ C-tubes, P200 and P1000 pipette tips, cell strainers, and columns) require at least 20 min of coating time. Serological pipettes require 3 min of coating time. Coated plasticware can be stored at 4 °C for at least 2 weeks. Do not let surfaces dry. After coating, rinse once in DBPS.

2. Cultures are pretreated with the Rho-associated protein kinase (ROCK) inhibitor, Y27632, to reduce anoikis of dissociated single cells. Cultures are also routinely treated with Y27632 at passaging, while some colonoid cultures are maintained with 10 μM Y27632 in their regular growth medium. 2.5 mM stock solutions are prepared in sterile H_2O and stored as per manufacturer's instructions.

3. HBSS-containing standard calcium/magnesium is reduced 10:1 with HBSS that is calcium/magnesium-free. These cations are reduced to minimize stem cell differentiation but at sufficient levels to facilitate enzymatic activity.

4. To enrich cultures for LGR5(+) stem cells, we have transitioned adenoma colonoid cultures [28] from KGM-Gold to L-WRN conditioned medium, which is high in Wnt3a, R-spondin-3, and Noggin [38]. L-WRN can promote a shift from a differentiated budding phenotype to a stem-cell enriched cystic phenotype [24, 39–43]. These cystic-driven cultures can also serve as robust positive controls for the LGR5(+) cell isolation procedure. Cultures are gradually transitioned from KGM-Gold to L-WRN: 3 days post-passage switch for 5 days to 50/50 L-WRN, then 5 days 75/25, to finally 100% L-WRN. The complete L-WRN medium contains Advanced DMEM/F-12, 2 mM GlutaMax, 10 mM HEPES, 1× N-2 media supplement, 1× B-27 supplement minus vitamin A, 1 mM *N*-Acetyl-L-cysteine, 100 ng/mL EGF, 10 mM nicotinamide, and 100 μg/mL Primocin. For normal colonoid cultures, supplement with 10 μM SB202190, 500 nM A83-01, plus/minus 10 μM Y27632.

5. To isolate sufficient human LGR5(+) cells for experimental analysis, we recommend starting with at least 2 wells of a 6-well plate of dense cystic colonoids (500 μL Matrigel; 10E6 total starting cells), 15 wells of a 6-well plate of iPSC-derived organoids (3750 μL Matrigel; 15E6 total starting cells including mesenchymal cells), or a 9 cm^2-sized colon resection of freshly isolated crypts (20E6 total starting cells). We have successfully scaled this procedure up to 30 wells of colonoids or to 25 cm^2 of freshly resected colon tissue.

6. We found that both trypsin and the Neural Tissue Dissociation Kit were too aggressive for this application and removed some relevant cell surface epitopes. We employed a gentle enzyme preparation, Tumor Dissociation Kit. TDK enzymes H and R

were reconstituted with warm Enzyme Solution described here, and enzyme A was reconstituted with Buffer A supplied with the kit. Enzymes were aliquoted in working volumes and stored at −80 °C according to manufacturer's instructions.

7. 10 mM DAPI dilactate stock solution prepared in H_2O. Store at −20 °C for up to 9 months. Day of use prepare 100 μM working solution in H_2O. Protect from light.
8. The serum-free Single Cell Culture Medium (SCC Medium) is employed for early initiation of cystic colonoids from FACS single cells. Once small cysts are evident (≥50 μm in size, after approximately 5–8 days), we have employed a serum-containing commercially available colonoid medium (IntestiCult™ Organoid Growth Medium), plus 10 μM Y27632, to enhance the establishment of colonoids from the starter cysts.
9. We have found that EDTA is not required for routine passaging of cultures. Instead, to passage colonoids: (1) 30× triturate the Matrigel™ pads with a P1000 pipette in cold DPBS (2 mL per 250 μL Matrigel™), (2) fill tube to 5 mL, (3) pellet at 300 × *g* 4 °C for 3 min, (4) 30× triturate pellet with a P200 pipette in the small volume of medium required to dilute the stock Matrigel™ to 8 mg/mL final, and (5) 15× triturate pellet with a P200 pipette in Matrigel™ before plating as 10–50 μL drops.
10. Incubate in 2 mM EDTA to remove Matrigel™ to prepare for enzymatic dissociation. Extend the EDTA incubation time period when colonoids/organoids have been in culture for periods more than 7 days (before next subculture); add 3 min for every subsequent day of culture.
11. iPSC-derived organoids require more vigorous methods to remove associated Matrigel™, owing to the mesenchymal contribution (30 min 4 mM EDTA). During the initial wash steps many Matrigel™ fragments with organoids will remain suspended in the supernatant after pelleting. Recover these fragments along with the pellet by permitting them to settle on ice (4 min) after centrifugations, before discarding the supernatant.
12. Fresh human intestinal crypts are prepared from surgical tissue resections as described by Jung et al. [27]. After crypts have been released from the tissue, keep on ice to preserve viability.
13. A customized program has been designed for this specific protocol which integrates program run steps (*h_tumor_01*) and incubations steps for use with the heated gentleMACS Dissociator.
14. We counted and estimated viability via trypan blue exclusion with an Invitrogen Countess Automated Cell Counter. Mix cell sample 1:1 with trypan blue solution and inject into

automated hemocytometer slide. If necessary, adjust cell dilution to recommended instrument cell number range of accuracy. Use this process to also visually note single cell efficiency of digestion.

15. Eppendorf tubes containing FcR blocking reagent are prepared earlier during breaks in steps and stored at 4 °C.
16. LGR5 has been identified in both epithelial and stromal tissue by our laboratory [24] and others [44, 45]. iPSC-derived organoids have both an epithelial and mesenchymal component, and freshly isolated tissue crypts will have some percentage of contaminating stroma. EpCAM is used as a marker to separate epithelial cells from the stroma.
17. The EpCAM IgG2b κ isotype control antibody concentration was according to manufacturer's instruction. The EpCAM antibody concentration, although ½ the concentration of the isotype control, was determined based on a concentration which would efficiently delineate a positive population without shifting the entire negative population of cells.
18. For cell numbers greater than 1×10^7 cells, adjust volume of buffer and EpCAM antibody 2× for 1–2 $\times 10^7$ cells (i.e., 800 μL Labeling buffer/32 μL antibody), 3× for 2–3 $\times 10^7$ cells, and so on.
19. Avoid bubbles in Column Solution which may cause blockage. If at any point the column appears to stop, place thumb over the top of the column and push down to pressurize restart.
20. Drain column completely before applying next volume.
21. Maintain cells on ice before analysis/FACS. A refrigerated sorting instrument is highly recommended.
22. When the viable cell population shifts with sort time, we readjust gating to exclude dying cells.
23. EpCAM gating also resulted in enhanced selection of viable cells. We have included EpCAM staining even with epithelium-only colonoids when added assurances of long-term viability are of high priority (i.e., FACS for subsequent single cell colonoid-forming efficiency).
24. The APC Check Reagent is highly specific to the microbead. The unbound column flow-through cells provide a robust negative control for the LGR5-microbead-APC conjugate. Rigorous FMO (fluorescence-minus-one) controls are suitable for samples that are not MACS-processed [DAPI(−), EpCAM (+)].
25. We observe an approximately tenfold plus enrichment of LGR5 (+) cells by MACS, which results in a significantly abbreviated FACS time. We believe this reduced processing time likely

translates into higher stem cell viability and possible preservation of the LGR5-bound moiety.

26. We extract RNA from lysed sorted cells with the RNeasy Micro Kit (with on-column DNase digestion) according to manufacturer instructions. We observe a significant enhancement in quantity and quality of RNA when extracted soon after FACS, as opposed to freezing for later extraction.

Acknowledgments

This work was supported by National Institute of Environmental Health Sciences of the National Institutes of Health (R01ES028802 and P30 ES017885 to J.A.C.); the University of Michigan Rogel Comprehensive Cancer Center Research Grant Fund (P30CA046592); the University of Michigan Center for Gastrointestinal Research (UMCGR) (5P30DK034933); the University of Michigan MCubed Program (to J.A.C.); the Ravitz Family Foundation (to J.A.C.); the Intestinal Stem Cell Consortium (U01DK103141 to J.R.S.), a collaborative research project funded by the National Institute of Diabetes and Digestive and Kidney Diseases (NIDDK); and the National Institute of Allergy and Infectious Diseases (NIAID); and the NIAID Novel Alternative Model Systems for Enteric Diseases (NAMSED) consortium (U19AI116482 to J.R.S.). We thank Maliha Berner, Erica Katz, Caroline McCarthy, Gina Newsome, and Angeline Wu of the Translational Tissue Modeling Laboratory (TTML) of the Department of Internal Medicine, University of Michigan; Michael Dellheim of the University of Michigan BRCF Flow Cytometry Core; and Robin Kunkel of the Department of Pathology, University of Michigan, for graphic design.

References

1. Barker N, van Es JH, Kuipers J, Kujala P, van den Born M, Cozijnsen M, Haegebarth A, Korving J, Begthel H, Peters PJ, Clevers H (2007) Identification of stem cells in small intestine and colon by marker gene Lgr5. Nature 449(7165):1003–1007. https://doi.org/10.1038/nature06196. nature06196 [pii]
2. de Visser KE, Ciampricotti M, Michalak EM, Tan DW, Speksnijder EN, Hau CS, Clevers H, Barker N, Jonkers J (2012) Developmental stage-specific contribution of LGR5(+) cells to basal and luminal epithelial lineages in the postnatal mammary gland. J Pathol 228 (3):300–309. https://doi.org/10.1002/path.4096
3. Jaks V, Barker N, Kasper M, van Es JH, Snippert HJ, Clevers H, Toftgard R (2008) Lgr5 marks cycling, yet long-lived, hair follicle stem cells. Nat Genet 40(11):1291–1299. https://doi.org/10.1038/ng.239
4. Barker N, Rookmaaker MB, Kujala P, Ng A, Leushacke M, Snippert H, van de Wetering M, Tan S, Van Es JH, Huch M, Poulsom R, Verhaar MC, Peters PJ, Clevers H (2012) Lgr5 (+ve) stem/progenitor cells contribute to nephron formation during kidney development. Cell Rep 2(3):540–552. https://doi.org/10.1016/j.celrep.2012.08.018
5. Koo BK, Spit M, Jordens I, Low TY, Stange DE, van de Wetering M, van Es JH, Mohammed S, Heck AJ, Maurice MM, Clevers

H (2012) Tumour suppressor RNF43 is a stem-cell E3 ligase that induces endocytosis of Wnt receptors. Nature 488(7413):665–669. https://doi.org/10.1038/nature11308

6. de Lau W, Barker N, Low TY, Koo BK, Li VS, Teunissen H, Kujala P, Haegebarth A, Peters PJ, van de Wetering M, Stange DE, van Es JE, Guardavaccaro D, Schasfoort RB, Mohri Y, Nishimori K, Mohammed S, Heck AJ, Clevers H (2011) Lgr5 homologues associate with Wnt receptors and mediate R-spondin signalling. Nature 476(7360):293–297. https://doi.org/10.1038/nature10337
7. Fafilek B, Krausova M, Vojtechova M, Pospichalova V, Tumova L, Sloncova E, Huranova M, Stancikova J, Hlavata A, Svec J, Sedlacek R, Luksan O, Oliverius M, Voska L, Jirsa M, Paces J, Kolar M, Krivjanska M, Klimesova K, Tlaskalova-Hogenova H, Korinek V (2013) Troy, a tumor necrosis factor receptor family member, interacts with lgr5 to inhibit wnt signaling in intestinal stem cells. Gastroenterology 144(2):381–391. https://doi.org/10.1053/j.gastro.2012.10.048
8. Ruffner H, Sprunger J, Charlat O, Leighton-Davies J, Grosshans B, Salathe A, Zietzling S, Beck V, Therier M, Isken A, Xie Y, Zhang Y, Hao H, Shi X, Liu D, Song Q, Clay I, Hintzen G, Tchorz J, Bouchez LC, Michaud G, Finan P, Myer VE, Bouwmeester T, Porter J, Hild M, Bassilana F, Parker CN, Cong F (2012) R-Spondin potentiates Wnt/beta-catenin signaling through orphan receptors LGR4 and LGR5. PLoS One 7(7):e40976. https://doi.org/10.1371/journal.pone.0040976
9. Carmon KS, Lin Q, Gong X, Thomas A, Liu Q (2012) LGR5 interacts and cointernalizes with Wnt receptors to modulate Wnt/beta-catenin signaling. Mol Cell Biol 32(11):2054–2064. https://doi.org/10.1128/mcb.00272-12
10. Glinka A, Dolde C, Kirsch N, Huang YL, Kazanskaya O, Ingelfinger D, Boutros M, Cruciat CM, Niehrs C (2011) LGR4 and LGR5 are R-spondin receptors mediating Wnt/beta-catenin and Wnt/PCP signalling. EMBO Rep 12(10):1055–1061. https://doi.org/10.1038/embor.2011.175
11. Carmon KS, Gong X, Lin Q, Thomas A, Liu Q (2011) R-spondins function as ligands of the orphan receptors LGR4 and LGR5 to regulate Wnt/beta-catenin signaling. Proc Natl Acad Sci U S A 108(28):11452–11457. https://doi.org/10.1073/pnas.1106083108
12. Huch M, Bonfanti P, Boj SF, Sato T, Loomans CJ, van de Wetering M, Sojoodi M, Li VS, Schuijers J, Gracanin A, Ringnalda F, Begthel H, Hamer K, Mulder J, van Es JH, de Koning E, Vries RG, Heimberg H, Clevers H (2013) Unlimited in vitro expansion of adult bi-potent pancreas progenitors through the Lgr5/R-spondin axis. EMBO J 32(20):2708–2721. https://doi.org/10.1038/emboj.2013.204
13. van Es JH, Sato T, van de Wetering M, Lyubimova A, Yee Nee AN, Gregorieff A, Sasaki N, Zeinstra L, van den Born M, Korving J, Martens ACM, Barker N, van Oudenaarden A, Clevers H (2012) Dll1+ secretory progenitor cells revert to stem cells upon crypt damage. Nat Cell Biol 14(10):1099–1104. https://doi.org/10.1038/ncb2581
14. Sigal M, Logan CY, Kapalczynska M, Mollenkopf HJ, Berger H, Wiedenmann B, Nusse R, Amieva MR, Meyer TF (2017) Stromal R-spondin orchestrates gastric epithelial stem cells and gland homeostasis. Nature 548(7668):451–455. https://doi.org/10.1038/nature23642
15. Hoeck JD, Biehs B, Kurtova AV, Kljavin NM, de Sousa EMF, Alicke B, Koeppen H, Modrusan Z, Piskol R, de Sauvage FJ (2017) Stem cell plasticity enables hair regeneration following Lgr5(+) cell loss. Nat Cell Biol 19(6):666–676. https://doi.org/10.1038/ncb3535
16. Schepers AG, Snippert HJ, Stange DE, van den Born M, van Es JH, van de Wetering M, Clevers H (2012) Lineage tracing reveals Lgr5+ stem cell activity in mouse intestinal adenomas. Science 337(6095):730–735. https://doi.org/10.1126/science.1224676
17. de Sousa e Melo F, Kurtova AV, Harnoss JM, Kljavin N, Hoeck JD, Hung J, Anderson JE, Storm EE, Modrusan Z, Koeppen H, Dijkgraaf GJ, Piskol R, de Sauvage FJ (2017) A distinct role for Lgr5(+) stem cells in primary and metastatic colon cancer. Nature 543(7647):676–680. https://doi.org/10.1038/nature21713
18. Junttila MR, Mao W, Wang X, Wang BE, Pham T, Flygare J, Yu SF, Yee S, Goldenberg D, Fields C, Eastham-Anderson J, Singh M, Vij R, Hongo JA, Firestein R, Schutten M, Flagella K, Polakis P, Polson AG (2015) Targeting LGR5+ cells with an antibody-drug conjugate for the treatment of colon cancer. Sci Transl Med 7(314):314ra186. https://doi.org/10.1126/scitranslmed.aac7433
19. Gong X, Azhdarinia A, Ghosh SC, Xiong W, An Z, Liu Q, Carmon KS (2016) LGR5-targeted antibody-drug conjugate eradicates gastrointestinal tumors and prevents recurrence. Mol Cancer Ther 15(7):1580–1590.

https://doi.org/10.1158/1535-7163.MCT-16-0114
20. Barker N (2014) Adult intestinal stem cells: critical drivers of epithelial homeostasis and regeneration. Nat Rev Mol Cell Biol 15 (1):19–33. https://doi.org/10.1038/nrm3721
21. Baker AM, Graham TA, Elia G, Wright NA, Rodriguez-Justo M (2015) Characterization of LGR5 stem cells in colorectal adenomas and carcinomas. Sci Rep 5:8654. https://doi.org/10.1038/srep08654
22. Jang BG, Lee BL, Kim WH (2013) Distribution of LGR5+ cells and associated implications during the early stage of gastric tumorigenesis. PLoS One 8(12):e82390. https://doi.org/10.1371/journal.pone.0082390
23. Shimokawa M, Ohta Y, Nishikori S, Matano M, Takano A, Fujii M, Date S, Sugimoto S, Kanai T, Sato T (2017) Visualization and targeting of LGR5(+) human colon cancer stem cells. Nature 545(7653):187–192. https://doi.org/10.1038/nature22081
24. Dame MK, Attili D, McClintock SD, Dedhia PH, Ouillette P, Hardt O, Chin AM, Xue X, Laliberte J, Katz EL, Newsome GM, Hill DR, Miller AJ, Tsai YH, Agorku D, Altheim CH, Bosio A, Simon B, Samuelson LC, Stoerker JA, Appelman HD, Varani J, Wicha MS, Brenner DE, Shah YM, Spence JR, Colacino JA (2018) Identification, isolation and characterization of human LGR5-positive colon adenoma cells. Development 145(6). https://doi.org/10.1242/dev.153049
25. Sato T, Stange DE, Ferrante M, Vries RG, Van Es JH, Van den Brink S, Van Houdt WJ, Pronk A, Van Gorp J, Siersema PD, Clevers H (2011) Long-term expansion of epithelial organoids from human colon, adenoma, adenocarcinoma, and Barrett's epithelium. Gastroenterology 141(5):1762–1772. https://doi.org/10.1053/j.gastro.2011.07.050. S0016-5085(11)01108-5 [pii]
26. Spence JR, Mayhew CN, Rankin SA, Kuhar MF, Vallance JE, Tolle K, Hoskins EE, Kalinichenko VV, Wells SI, Zorn AM, Shroyer NF, Wells JM (2011) Directed differentiation of human pluripotent stem cells into intestinal tissue in vitro. Nature 470(7332):105–109. https://doi.org/10.1038/nature09691
27. Jung P, Sato T, Merlos-Suarez A, Barriga FM, Iglesias M, Rossell D, Auer H, Gallardo M, Blasco MA, Sancho E, Clevers H, Batlle E (2011) Isolation and in vitro expansion of human colonic stem cells. Nat Med 17 (10):1225–1227. https://doi.org/10.1038/nm.2470. nm.2470 [pii]
28. Dame MK, Jiang Y, Appelman HD, Copley KD, McClintock SD, Aslam MN, Attili D, Elmunzer BJ, Brenner DE, Varani J, Turgeon DK (2014) Human colonic crypts in culture: segregation of immunochemical markers in normal versus adenoma-derived. Lab Investig 94(2):222–234. https://doi.org/10.1038/labinvest.2013.145
29. Fujii M, Matano M, Toshimitsu K, Takano A, Mikami Y, Nishikori S, Sugimoto S, Sato T (2018) Human intestinal organoids maintain self-renewal capacity and cellular diversity in Niche-inspired culture condition. Cell Stem Cell 23(6):787–793.e786. https://doi.org/10.1016/j.stem.2018.11.016
30. Yin X, Farin HF, van Es JH, Clevers H, Langer R, Karp JM (2014) Niche-independent high-purity cultures of Lgr5+ intestinal stem cells and their progeny. Nat Methods 11 (1):106–112. https://doi.org/10.1038/nmeth.2737
31. von Furstenberg RJ, Buczacki SJ, Smith BJ, Seiler KM, Winton DJ, Henning SJ (2014) Side population sorting separates subfractions of cycling and non-cycling intestinal stem cells. Stem Cell Res 12(2):364–375. https://doi.org/10.1016/j.scr.2013.10.012
32. Wang F, Scoville D, He XC, Mahe MM, Box A, Perry JM, Smith NR, Lei NY, Davies PS, Fuller MK, Haug JS, McClain M, Gracz AD, Ding S, Stelzner M, Dunn JCY, Magness ST, Wong MH, Martin MG, Helmrath M, Li L (2013) Isolation and characterization of intestinal stem cells based on surface marker combinations and colony-formation assay. Gastroenterology 145 (2):383–395.e321. https://doi.org/10.1053/j.gastro.2013.04.050
33. Jung P, Sommer C, Barriga FM, Buczacki SJ, Hernando-Momblona X, Sevillano M, Duran-Frigola M, Aloy P, Selbach M, Winton DJ, Batlle E (2015) Isolation of human colon stem cells using surface expression of PTK7. Stem Cell Rep 5(6):979–987. https://doi.org/10.1016/j.stemcr.2015.10.003
34. Sugimoto S, Sato T (2017) Establishment of 3D intestinal organoid cultures from intestinal stem cells. In: Koledova Z (ed) 3D Cell culture: methods and protocols. Springer, New York, pp 97–105. https://doi.org/10.1007/978-1-4939-7021-6_7
35. McCracken KW, Howell JC, Wells JM, Spence JR (2011) Generating human intestinal tissue from pluripotent stem cells in vitro. Nat Protoc 6(12):1920–1928. https://doi.org/10.1038/nprot.2011.410
36. Capeling MM, Czerwinski M, Huang S, Tsai YH, Wu A, Nagy MS, Juliar B, Sundaram N, Song Y, Han WM, Takayama S, Alsberg E,

Garcia AJ, Helmrath M, Putnam AJ, Spence JR (2019) Nonadhesive alginate hydrogels support growth of pluripotent stem cell-derived intestinal organoids. Stem Cell Rep 12 (2):381–394. https://doi.org/10.1016/j.stemcr.2018.12.001

37. Finkbeiner SR, Hill DR, Altheim CH, Dedhia PH, Taylor MJ, Tsai YH, Chin AM, Mahe MM, Watson CL, Freeman JJ, Nattiv R, Thomson M, Klein OD, Shroyer NF, Helmrath MA, Teitelbaum DH, Dempsey PJ, Spence JR (2015) Transcriptome-wide analysis reveals hallmarks of human intestine development and maturation in vitro and in vivo. Stem Cell Rep. https://doi.org/10.1016/j.stemcr.2015.04.010

38. Miyoshi H, Stappenbeck TS (2013) In vitro expansion and genetic modification of gastrointestinal stem cells in spheroid culture. Nat Protoc 8(12):2471–2482. https://doi.org/10.1038/nprot.2013.153. http://www.nature.com/nprot/journal/v8/n12/abs/nprot.2013.153.html#supplementary-information

39. Onuma K, Ochiai M, Orihashi K, Takahashi M, Imai T, Nakagama H, Hippo Y (2013) Genetic reconstitution of tumorigenesis in primary intestinal cells. Proc Natl Acad Sci U S A 110 (27):11127–11132. https://doi.org/10.1073/pnas.1221926110

40. Drost J, van Jaarsveld RH, Ponsioen B, Zimberlin C, van Boxtel R, Buijs A, Sachs N, Overmeer RM, Offerhaus GJ, Begthel H, Korving J, van de Wetering M, Schwank G, Logtenberg M, Cuppen E, Snippert HJ, Medema JP, Kops GJ, Clevers H (2015) Sequential cancer mutations in cultured human intestinal stem cells. Nature 521 (7550):43–47. https://doi.org/10.1038/nature14415

41. Sato T, van Es JH, Snippert HJ, Stange DE, Vries RG, van den Born M, Barker N, Shroyer NF, van de Wetering M, Clevers H (2011) Paneth cells constitute the niche for Lgr5 stem cells in intestinal crypts. Nature 469 (7330):415–418. https://doi.org/10.1038/nature09637. nature09637 [pii]

42. Farin HF, Van Es JH, Clevers H (2012) Redundant sources of Wnt regulate intestinal stem cells and promote formation of Paneth cells. Gastroenterology 143(6):1518–1529.e1517. https://doi.org/10.1053/j.gastro.2012.08.031

43. Matano M, Date S, Shimokawa M, Takano A, Fujii M, Ohta Y, Watanabe T, Kanai T, Sato T (2015) Modeling colorectal cancer using CRISPR-Cas9-mediated engineering of human intestinal organoids. Nat Med 21 (3):256–262. https://doi.org/10.1038/nm.3802

44. Boddupally K, Wang G, Chen Y, Kobielak A (2016) Lgr5 marks neural crest derived multipotent oral stromal stem cells. Stem Cells 34 (3):720–731. https://doi.org/10.1002/stem.2314

45. Lee J-H, Tammela T, Hofree M, Choi J, Marjanovic ND, Han S, Canner D, Wu K, Paschini M, Bhang DH, Jacks T, Regev A, Kim CF (2017) Anatomically and functionally distinct lung mesenchymal populations marked by Lgr5 and Lgr6. Cell 170(6):1149–1163.e1112. https://doi.org/10.1016/j.cell.2017.07.028

Chapter 2

Immune-Mediated Specific Depletion of Intestinal Stem Cells

Stephen E. Sherman and Judith Agudo

Abstract

Functional studies of specific stem cell populations often require depletion of tissue-specific stem cells in an in vivo model to allow for the interrogation of their contribution to the maintenance and/or regeneration of their home tissue. Depletion methods need an exquisite specificity to uniquely eliminate the target cell type. To achieve such specificity, a commonly used approach has been murine models with expression of the Diphtheria Toxin Receptor (DTR) in the cell of interest. The major caveat of using these DTR-expressing transgenic mice is the need to generate new DTR models for every new cell population of interest. While DTR-expressing models are limited, the number of available GFP-expressing mice is large. To take advantage of this plethora of cell type-specific GFP-reporter mice, we sought to exploit the body's own killer cells as a depletion tool. Thus, we generated a mouse model whose cytotoxic T cells recognize and kill GFP-expressing cells, called the Jedi (Agudo et al., Nat Biotechnol 33:1287–1292, 2015). Jedi T cells now enable the depletion of virtually almost any cell type by using a suitable GFP-expressing transgenic mouse (Agudo et al., Nat Biotechnol 33:1287–1292, 2015; Chen et al., J Clin Invest 128(8):3413–3424, 2018). Here, we explain in detail how to achieve depletion of Lgr5+ stem cells in the intestine with a single injection of Jedi T cells (Agudo et al., Immunity 48:271–285.e5, 2018) with a methodology that can be extrapolated to any other GFP-expressing cell.

Key words Green fluorescent protein, Cell depletion, T cells, Fluorescent reporters, Cell function, Intestinal stem cells, Cytotoxicity

1 Introduction

In order to dissect the role of distinct stem cell populations in vivo, the most commonly used approaches are cell fate-mapping and specific cell depletion. Specific depletion of a given cell type can be achieved in some instances by means of antibodies that recognize surface antigens. This method is extensively used to deplete immune cells such as T cells, NK cells and macrophages, but its use is limited for depletion of stem cells, as the existence of surface molecules that are stem cell type specific and the availability of antibodies that target these populations is restricted. Alternatively, transgenic mice expressing the Diphtheria Toxin Receptor (DTR)

Paloma Ordóñez-Morán (ed.), *Intestinal Stem Cells: Methods and Protocols*, Methods in Molecular Biology, vol. 2171, https://doi.org/10.1007/978-1-0716-0747-3_2,

in specific cell types have been developed. This system enables the use of tissue or cell-specific promoters that exquisitely drive expression of DTR only in the cells of interest [1–4]. The presence of DTR does not cause abnormalities in the bearing mouse and does not alter the function of the DTR-expressing cell population [1–4]. Then, depletion is achieved in a timely controlled manner when Diphtheria Toxin (DT) is systemically injected in the DTR-expressing transgenic mice [1–4]. Although its use has greatly helped to advance our understanding in stem cell biology and the function of many other cell types, it is limited by the fact that a new mouse model must be generated for every new population to study. Moreover, DT-mediated cell death may be inflammatory, and it is not well understood how much this type of cell death may influence the behavior of the surrounding tissue. To improve the accessibility for depletion models within the research community, we developed a tool that takes advantage of a more "natural" way to eliminate cells, exploiting preexisting transgenic mouse models to investigate virtually any tissue and/or cell type of interest. Our bodies (equivalent in mice) possess a remarkably specific and efficient CD8+ cytotoxic T lymphocyte population whose job is to recognize and eliminate specific cells. Each T cell expresses a T cell receptor (TCR), and each TCR recognizes a different peptide loaded–MHC class I complex. All nucleated cells possess the property of antigen presentation by MHC class I, and hence, are susceptible to T cell-mediated death when presenting the right antigen. Therefore, we created a mouse model whose CD8+ T cells possess a TCR that is specific for GFP that we called Just EGFP Death Inducing or JEDI mouse [5]. Now, Jedi T cells can be transferred to GFP-expressing mouse and they can recognize and kill the GFP-expressing cell population [6].

The discovery of GFP was awarded the Nobel Prize in Chemistry in 2008 as its use has shed light on countless biological processes. Thus, multiple engineered eukaryotic cell lines and hundreds of GFP-expressing transgenic mice have been developed and are widely used (in the GENSAT project and the Jackson Laboratories). Thus, by combining Jedi T cells and mice that express GFP in a given cell population, depletion of the GFP+ cells can be efficiently achieved without the need of developing new models. Here we describe how to utilize the Jedi technology to deplete Lgr5+ intestinal stem cells in Lgr5-EGFP-IRES-CreERT2 mice developed by Hans Clevers [7]. This method faithfully recapitulated previous observations by the Klein and Sauvage labs, showing that Lgr5+ cells are dispensable during steady state [3], but we showed they are necessary for proper tissue homeostasis after irradiation-mediated damage [8]. Beyond its use for specific depletion of Lgr5+ cells in the intestine, this chapter provides a methodology that can be used with any other GFP-expressing transgenic mouse to deplete other cells of interest in the intestine or in any other epithelia.

2 Materials

2.1 Mouse Models

1. CD45.1 B10D2xB6 Jedi mice that have CD8+ T cells that recognize $EGFP_{200-208}$ peptide presented on the MHC class I allele H-2K^{d} (Jackson Laboratories cat#028062). These mice are in a mixed B10D2 C57Bl/6 background and are homozygous for the H2K^{d} allele. Moreover, they were bred with SJL mice to incorporate the CD45 allele CD45.1 (as homozygous) to allow their identification upon adoptive transfer into CD45.2 mice.
2. Lgr5-EGFP-IRES-CreERT2 (aka Lgr5-EGFP mice) or your GFP-reporter mouse of interest × B10D2 (F_1): Lgr5-EGFP-IRES-CreERT2 mice are in a C57BL/6J. This background has the H2K^{b} allele, which is not compatible with Jedi T cells. In order to gain the H2K^{d} allele while retaining a C57 background, mice are bred with B10D2.

2.2 CD8 T Cell Isolation, Injection, and Flow Cytometry Analysis

1. FACS buffer: 0.5% Bovine Serum Albumin (BSA) 2 mM EDTA PBS (pH 7.2), stored at 4 °C and kept on ice during processing of tissue. This same buffer is used supplemented with 20 mM EDTA for preparation of single cell suspensions from intestinal epithelium for flow cytometry.
2. Cell strainers, both 100 μm and 70 μm sizes.
3. 50 mL conical falcon tubes.
4. Surgical scissors and tweezers, tissue papers, or Kimwipes.
5. Mouse CD8+ negative isolation kit or any other kit for negative mouse CD8 T cell isolation.
6. Trypan blue, hemocytometer and optical inverted microscope for counting live cells.
7. 50 U insulin syringe for injection of isolated T cells.
8. Mouse strainer and a heat lamp.
9. 1× Red Blood Lysis (RBC) buffer prepared from 10× RBC lysis buffer diluted into deionized water.
10. 1× Accutase.
11. Regular 5 mL polystyrene round-bottom flow cytometry tube and similar polystyrene 5 mL flow cytometry tubes with a 70 μm cell strainer.
12. 4,6-Diamidino-2-phenylindole, dihydrochloride (DAPI) solution. DAPI stock is prepared by dissolving DAPI powder in deionized water at 100 μM and stored in aliquots at −20 °C. Working solution is prepared by dissolving the stock at 1:10,000 in FACS buffer and stored at 4 °C protected from light.

13. Antibodies: anti-mouse CD45.1-PE; anti-mouse CD45.2-FITC; anti-mouse CD8a-APC; anti-mouse CD45-APC; and anti-mouse CD16/32 (for Fc blocking, 1 mg/mL).
14. Fortessa flow cytometer.

2.3 Immunofluorescence Analysis of GFP

1. 20% sucrose w/v 4% paraformaldehyde (PFA): 8 g of sucrose are diluted in 40 mL of 4% PFA. 4% PFA is prepared from powder and pH is adjusted to 7. Aliquots of 40 mL are kept frozen at −20 °C. The same day that tissue must be harvested, 4% PFA is thawed and sucrose is added. Leftovers are discarded.
2. Glass slides and coverslips.
3. Cryostat.
4. DAPI stock diluted in PBS (1:10,000).
5. Vectashield Antifade Mounting Medium.

3 Methods

3.1 Breeding and Generation of Suitable Mouse Lines

1. Generation of mice expressing GFP in Lgr5 cells (or the cell of interest) with $H2K^d$ allele: because Jedi T cells only recognize $EGFP_{200–208}$ when is presented on $H\text{-}2K^d$, all mouse experiments must be done with the progeny of the EGFP-expressing mouse of interest crossed with B10D2. B10D2 is a pure background and hence homozygous for $H2K^d$. One copy of $H2K^d$ is sufficient to enable T cell recognition and killing, and the use of the F_1 prevents the need to genotype for the H2-Kl alleles, but subsequent generations can be used as long as they express both GFP and $H2K^d$. Either a male Lgr5-EGFP-IRES-CreERT2 can be bred with multiple B10D2 females or a B10D2 male with several Lgr5-EGFP-IRES-CreERT2 females. Lgr5-EGFP-IRES-CreERT2 is heterozygous; hence, litters from these mice must be genotyped following the protocol established from the Jackson Laboratories.
2. Breeding of Jedi mice: Jedi mice were generated by somatic cell nuclear transfer from a GFP-specific T cell; thus, Jedi mice carry the rearranged alpha and beta chains for the TCR that recognizes GFP in all their cells (these chains are Vα1-J30 and Vβ4-D1-J1.6-C1). Thus, genotyping can be performed with DNA extracted from tail, ear or blood, as these cells also have the rearranged TCR. Importantly, this is not a transgenic mouse as no foreign DNA was inserted. Since each rearranged chain is in its endogenous locus, each chain must be genotyped separately, as they are located in different chromosomes and transmitted independently to the progeny. Jedi males and females can be bred together to ensure maintenance of both

the $H2K^d$ and the CD45.1 alleles. Genotyping of Jedi mice must be performed following the protocol established from the Jackson Laboratories.

3.2 Isolation and Adoptive Transfer of Jedi CD8+ T Cells

1. For efficient depletion of Lgr5+ cells in the intestine it is required to adoptively transfer 4–5 × 10^6 Jedi CD8+ T cells. Prior to isolation, calculate the necessary number of Jedi mice that you will need for your depletion experiment. Of note, not all of the CD8+ T cells from Jedi mice are GFP-specific, as further rearrangement of the alpha chain allows for development of polyclonal T cells. The percentage of GFP-specific T cells in Jedi mice varies from 30% to 60% [5], being this percentage the highest in young individuals and declining with age. Thus, higher percentage of GFP-specific T cells are present in young mice, but those have smaller spleens and lymph nodes (LN) and hence, lower number of overall T cells. For this reason, we recommend using Jedi mice that are 6–8 weeks old as they provide the optimal number of GFP-specific T cells. Depending on the efficiency in LN collection and the performance of the isolation kit, the number of CD8+ T cells that can be obtained from one Jedi mouse varies from 8 to 16 million T cells. If only the spleen is used (which is easier to harvest), the number will be 6–10 million. Thus, one Jedi mouse should suffice to obtain enough T cells for depletion in 2 Lgr5-GFP mice.
2. Euthanize the number of Jedi mice that will be required by CO_2 inhalation following the approved procedures from your IACUC animal protocol. Immediately after euthanasia, open the peritoneal cavity and harvest the spleen and cranial, axillar, brachial, inguinal, and mesenteric LNs. Place them in cold FACS buffer on ice.
3. Obtain a single cell suspension from spleen and LNs by placing the spleen on the 70 μm cell strainer on a 50 mL falcon tube, add 1–2 mL of cold FACS buffer to get the strainer wet and use the plunger of a 1 mL or a 5 mL syringe to gently smash the spleen against the mesh of the strainer with a circular motion. Cells must be forced to pass through the mesh by washing with FACS buffer.
4. Repeat the same process with all the harvested LNs using the same cell strainer and 50 mL tube.
5. Repeat with spleens and LNs from further Jedi mice if high numbers or Jedi T cells are required. Wash the cell strainer to make all cells go through the mesh into the 50 mL tube.
6. Once all lymph nodes and spleen are processed into a single cell suspension, count total number of live cells by using trypan blue and a hemocytometer.

7. When using the CD8+ T cells negative isolation kit from Stem Cell Technologies, no red blood cell lysis is necessary before isolation. Follow the manufacturer's instructions for CD8+ T cell isolation (*see* **Note 1**).
8. Count the total live isolated cells at the microscope with trypan blue again to know the number of isolated T cells. Centrifuge at $300 \times g$ for 10 min, remove supernatant and resuspend in the adequate volume of injection, taking into account that T cells are injected in 100 μL of sterile PBS or saline solution per mouse and each mouse will receive 4–5 $\times 10^6$ T cells. Thus, if three mice need to be injected, T cells will be resuspended in ~350 μL to account for the lost volume when loading syringes. If more T cells than necessary are obtained, we recommend to just divide and inject as no adverse events have been observed when injecting even ten million T cells.
9. Keep the isolated cells in ice until the moment of injection.

3.3 Injection of Jedi T Cells and LV.GFP

Jedi T cells are naïve when isolated from Jedi mice [5], as they have never experienced their cognate antigen EGFP or GFP, not expressed in Jedi mice. Naïve T cells lack the capacity to undergo killing unless properly educated by professional antigen presenting cells (APCs). Hence, Jedi T cells require "vaccination" of the recipient mice against GFP in order to be properly activated by APCs. Moreover, Lgr5-GFP and any other GFP-expressing reporter mice express GFP as a self-antigen and hence, develop tolerance to it. In order to (1) activate the Jedi T cells and (2) break tolerance in the recipient Lgr5-EGFP mice, mice must be "vaccinated" against GFP. To this end, in parallel to adoptive transfer of isolated Jedi T cells, Lgr5-EGFP mice receive intravenous injection of $\sim 3 \times 10^8$ transducing units (TU) of a vesicular stomatitis virus (VSV)-pseudotyped lentiviral vector (LV) encoding GFP under the control of a ubiquitous promoter, such as phosphoglycerate kinase 1 (PGK) herein referred as LV.GFP (*see* **Note 2**). The production of this LV.GFP has been extensively described in another MiMB chapter from Baccarini et al. [9]. Systemic injection of LV results in efficient transduction of APCs in spleen (Fig. 1) and liver, concomitantly to a high and transient type I interferon secretion (IFN-I) [10]. GFP presentation by APCs along with acute high IFN-I production constitute a very efficient approach to activate Jedi T cells.

1. Take LV.GFP aliquots out of the −80 °C, where LV must be stored, and thaw in ice. Always work in a BSL-2 cell culture hood when manipulating LV and follow the biosafety instructions from your institution to work with LV.
2. Resuspend LV with sterile PBS in the appropriate volume for injection. For proper activation, each mouse must receive $\sim 3 \times 10^8$ TU by tail vein. We recommend injecting this

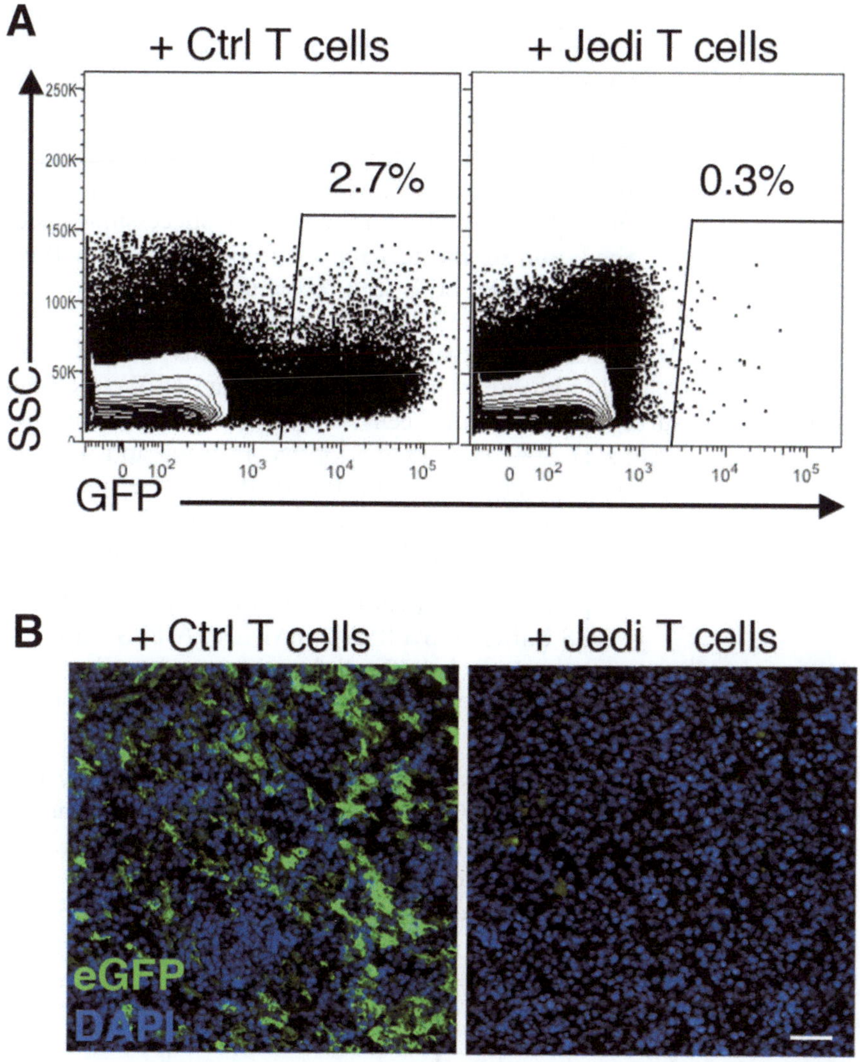

Fig. 1 Analysis of GFP expression after tail vein injection of LV.GFP. Mice were intravenously injected with a lentivirus expressing GFP (LV.GFP) at a dose of 3×10^8 TU and 2 days later control (Ctrl) or Jedi T cells were injected intravenously. Flow cytometry analysis was performed to measure the frequency of GFP-expressing cells in the spleen 5 days after transfer of Jedi T cells. (**a**) Shown here is a representative flow cytometry plot ($n = 4$ mice/group). (**b**) Fluorescent microscopy analysis of the spleen. Representative images are shown. White bar represents 100 μm

amount of LV in 100 μL, but a bigger volume can be used if this makes the injection easier (up to 200 μL). Keep LV.GFP diluted in PBS in ice until the moment of injection.

3. Place the mouse in the mouse restrainer with the tail out and place the heat lamp close enough to warm the tail but importantly ensuring the well-being of the mouse and avoiding overheating. We recommend having the researcher's hand close to the tail of the mouse to feel the intensity of the warmth.

4. During this warming time, load one 50 U syringe with the 100 μL of isolated T cells and another with 100 μL of LV.GFP to inject ~5 million T cells and ~3 × 10^8 TU, respectively. Remove bubbles.
5. Once the veins are clearly visible (the two vessels on the sides of the tail), wipe the tail with 70% alcohol and inject the T cells in one side and the LV on the other. If successfully inside the vein, the liquid should feel almost no pressure. After the injection a little bit of blood is seen to come out.

3.4 Monitoring Jedi T Cell Activation and Expansion

Activation and expansion of Jedi T cells can be monitored by analyzing blood by flow cytometry (Fig. 2). We recommend measuring CD45.1 Jedi T cells in the blood of recipient mice (that are CD45.2) 3–4 days after adoptive transfer and 1–2 days before euthanasia (e.g., day 3 and day 6). If adoptive transfer and activation of Jedi T cells was performed correctly, an increase in the percentage of CD45.1 Jedi T cells should be observed (*see* **Note 3**).

1. Prepare clean Eppendorf tubes containing 500 μL of sterile FACS buffer (0.5% BSA 2 mM EDTA PBS), use one tube per mouse where Jedi T cell function needs to be assessed. Keep them closed and in a clean container to transport to the mouse facility.
2. Utilize clean sterile scissors or a clean sterile razor to nick the tip of the tail. Gently massage the tail from the base to the tip to get a drop of blood. Let the drop fall inside the FACS buffer and immediately mix by gently inverting a few times. Repeat once more and mix the second drop with FACS buffer. Drops

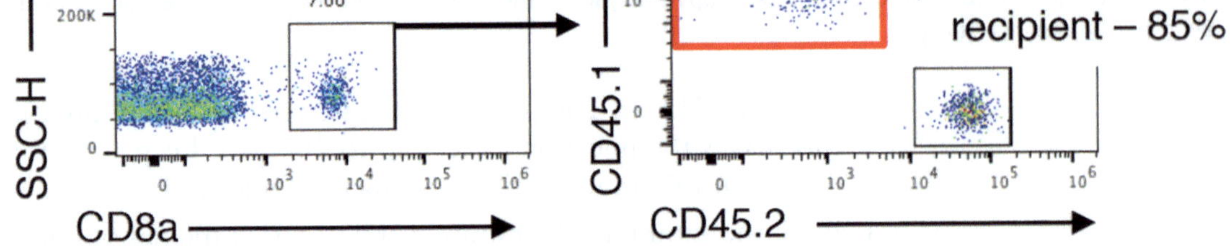

Fig. 2 Jedi T cells can be quantified in the blood of recipient mice. Flow cytometry plots shows the frequency of CD45.1 Jedi CD8+ T cells in the blood of a Lgr5-EGFP-IRES-CreERT mouse 3 days after T cell adoptive transfer. Anti-CD8a-APC antibody was used to label CD8 T cells. CD45.2-FITC was used to label hematopoietic cells from the recipient mouse and CD45.1-PE was used to label Jedi T cells, highlighted in red

are usually ~20 μL, so two drops represent ~40 μL of blood, which is enough to detect Jedi T cells' expansion. Keep the FACS buffer containing blood on ice until processing (blood can be kept in the fridge for ~12 h with no detectable cell loss).

3. Centrifuge the FACS buffer containing blood at 600 × *g* for 5 min at 4 °C and remove the supernatant. Resuspend in 500 μL of RBC lysis buffer and leave the tubes for 5 min at room temperature. After the 5 min are complete, add 500 μL of FACS buffer, mix gently by pipetting and centrifuge at 600 × *g* for 5 min at 4 °C.
4. Staining with master mix to label Jedi T cells: after centrifugation in **step 3**, remove supernatant and resuspend in 50 μL of staining master mix. For the preparation of this master mix, calculate 50 μL of FACS buffer per sample (and one extra sample to account for pipetting error). Pipet the adequate volume of FACS buffer in an Eppendorf tube and add the necessary antibodies. The volume of antibodies per sample are: 0.5 μL of anti-CD45.1-PE, 1 μL of anti-CD45.2-FITC and 0.5 μL of anti-CD8a-APC and 0.5 μL of anti-CD16/32 Fc blocking.
5. Stain for 15 min on ice.
6. Wash by adding 1 mL of cold FACS buffer and gently pipetting. Centrifuge at 600 × *g* for 5 min at 4 °C.
7. Remove the supernatant and resuspend in 200–300 μL of FACS buffer and transfer to 5 mL FACS tubes.
8. Use Fortessa or any other suitable flow cytometer to analyze the presence of Jedi T cells (Fig. 2).

3.5 Analysis of Lgr5 + Cells After Jedi T Cell Adoptive Transfer by Flow Cytometry Analysis

Depletion of GFP+ intestinal cells can be quantified by flow cytometry analysis. To this end, intestines from Lgr5-GFP mice (or your GFP-expressing mouse model of choice) that have received Jedi T cells are harvested and processed to obtain a single cell suspension and GFP is quantified by flow cytometry (Fig. 3).

1. After proper euthanasia, the peritoneal cavity is opened, and intestine is harvested. Then the intestine is opened with small scissors to expose all the lumen. Big pieces of feces can be very gently removed with the scissors or wet tissue paper.
2. The intestine is placed in a 10 cm petri dish containing 8–10 mL of cold PBS. By using tweezers and shaking the tissue inside the PBS, the tissue is rinsed to remove smaller pieces of feces and mucus. From here, it is transferred to a second petri dish that contains cold PBS for a second rinse, then to a third for a final wash.

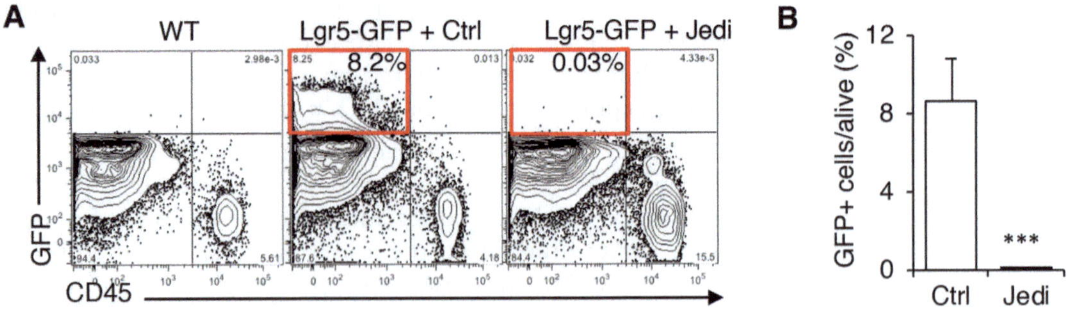

Fig. 3 Lgr5-GFP+ cells depletion can be visualized by flow cytometry analysis. Lgr5-GFP mice were injected with Jedi or control CD8+ T cells and vaccinated with LV.GFP. (**a**) Flow cytometry analysis of the frequency of GFP+ cells in the gut 1 week after T cell transfer. Cells were stained with CD45 to mark hematopoietic cells. (**b**) Graph presents the mean ± S.D. of the frequency of GFP+ cells relative to the total live cells ($n = 7–9$ mice/group)

3. Once cleaned, intestines are transferred to 50 mL tubes containing 10 mL of cold 0.5% BSA PBS supplemented with 20 mM EDTA and cut into small pieces. Tubes are then vigorously shaken and placed on ice. Tubes are incubated for ~30 min with vigorous agitation every ~5–10 min and placing it back on ice.
4. Intestinal crypts are enriched by filtration through a 100 μm cell strainer. Thus, after incubation with 20 mM EDTA and mechanical agitation, a 100 μm cell strainer is placed on a new falcon tube and the intestines are transferred using a 10 mL serological pipette. Cold FACS buffer can be used to collect remaining pieces of intestine that remain in the original 50 mL tube and to help wash crypts through the cell strainer into the new collection tube. Big pieces of intestine that do not go through the cell strainer are discarded.
5. Immediately after filtering through the 100 μm cell strainer, the disaggregated tissue is then filtered through a 70 μm cell strainer to remove mucus. This is done by placing a 70 μm cell strainer on a new falcon tube and transferring the tissue using a 25 mL serological pipette.
6. Centrifuge at 800 × *g* for 10 min at 4 °C and remove supernatant. Subsequently, the crypts are dissociated by digestion with 2 mL of 1× Accutase for 3 min at 37 °C.
7. Add 20 mL of 10% FBS supplemented FACS buffer to deactivate and wash off the Accutase. Mix gently and filter again through a 70 μm cell strainer as described in **step 5** (*see* **Note 4**). Centrifuge at 800 × *g* for 10 min at 4 °C and remove supernatant.
8. For red blood cell (RBC) lysis, resuspend the pellet in 1 mL of RBC lysis buffer and incubate for 3 min at room temperature. Add 10 mL of cold FACS buffer and mix gently. Filter again

through a 70 μm cell strainer into a new 50 mL tube to remove mucus. Centrifuge cells at 800 × *g* for 10 min at 4 °C and remove supernatant.

9. Stain cells with an antibody master mix containing anti-CD16/32 Fc blocking (1:100) and CD45-APC (1:300) in an appropriate volume (we recommend 100 μL for ~2 million cells). Use this step to transfer the sample to a 1.5 mL Eppendorf for the staining. Stain for 20–30 min on ice. Quantification of GFP + cells does not require staining of CD45 hematopoietic cells in the intestine, but it can help when performing the analysis and the gating strategy as GFP+ epithelial cells must be CD45-negative. Other antibodies of interest for the specific research can be used at this step.
10. Wash all samples with 1 mL of cold FACS buffer. Filter again into polystyrene FACS tubes using a 70 μm cell strainer immediately before analysis.

3.6 Analysis of Lgr5 + Cells After Jedi T Cell Adoptive Transfer by Immunofluorescence Analysis

Depletion of Lgr5-GFP+ intestinal stem cells or other GFP+ cells in the intestine can be determined by immunofluorescence (IF) analysis (Fig. 4). We have confirmed that depletion of Lgr5+ cells occurs as early as 5 days and is maintained at least until day 15 [1]. For IF analysis, GFP can be directly visualized in OCT-embedded frozen sections.

1. After euthanasia, the peritoneal cavity of the recipient mouse is opened using scissors and tweezers where 2–3 pieces (3–5 mm each) of the small intestine are plated in a 12-well plate (one well per mouse) containing 1 mL of cold 20% sucrose, 4% paraformaldehyde and placed in the fridge for 5–8 h.
2. Transfer the tissue to a new plate containing PBS to rinse the sucrose and paraformaldehyde for 3–5 min while keeping it protected from light.
3. Prior to embedding, use a paper tissue to gently wipe any excess liquid then transfer into a 1 cm × 1 cm plastic mold and pour OCT compound to cover the tissue. Use small tweezers to keep the pieces of intestine in a vertical position and place the mold on dry ice while keeping the tissue well-positioned. This results in tissue sections where the crypts and the lumen can be visualized (Fig. 4). Store the OCT-embedded frozen tissue at −80 °C until the moment of sectioning.
4. Take the molds with the frozen section out from the −80 °C freezer and place in a container with dry ice to keep frozen. Bring your samples to your cryostat machine and obtain 8 μm sections. Keep the sections protected from light in an opaque box while sectioning more tissue.

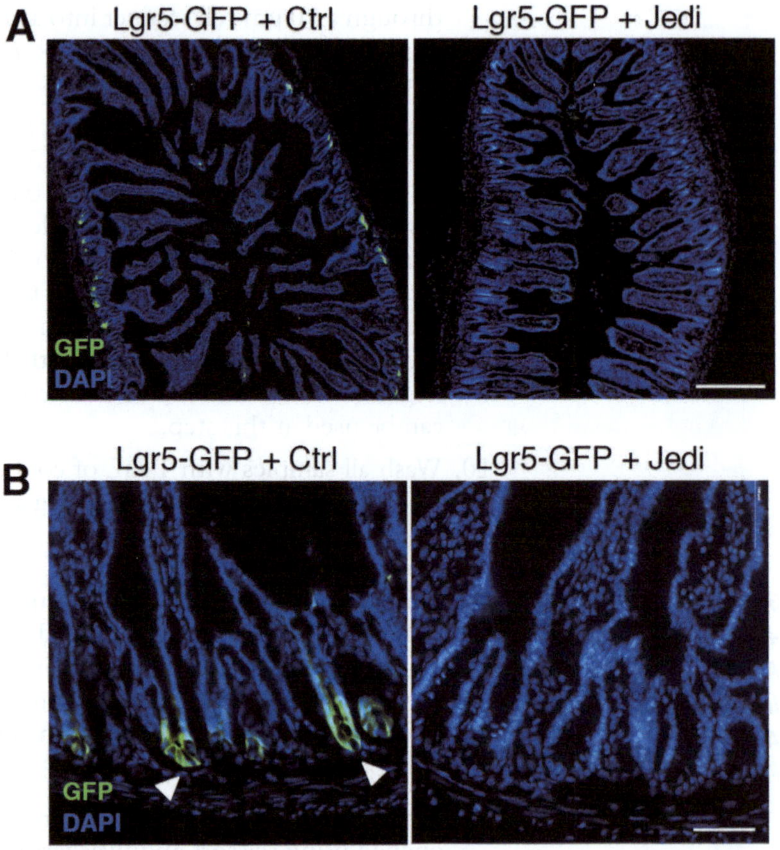

Fig. 4 Lgr5-GFP+ cell depletion can be visualized by immunofluorescence analysis. Lgr5-GFP mice were injected with Jedi or control CD8+ T cells and vaccinated with LV.GFP. Images show florescent microscopy analysis of the intestine 1 week after adoptive transfer. Representative images from 7 to 9 mice per group from three independent experiments are shown. (**a**) White bar represents 500 μm. (**b**) White bar represents 100 μm

5. If tissue sections are obtained within 2 days after harvesting and properly protected from light, GFP signal can be directly visualized without staining (*see* **Note 5**). Thus, after sectioning, slides can be dried for 15 min at 37 °C in the dark and then stained with DAPI for nuclei labeling. To this end, a solution of DAPI in PBS (1:10,000) is added on the slides for 5 min and washed immediately after.
6. After washing, use the mounting media and add a coverslip. Remove the excess of mounting media by using a paper tissue. Keep in the fridge until analysis.
7. Take pictures with upright wide-field microscope using the FITC channel for GFP and the DAPI channel for DAPI (Fig. 4). Pictures are taken and analyzed with the NisElements software.

4 Notes

1. Although we recommend a LV encoding for EGFP under the control of the PGK promoter, any other promoter capable of high expression in APCs could be used, such as other ubiquitous promoters (e.g., CAG or SFFVintron) or APC-specific promoters (e.g., CD11c). The use of intravenous injection of LV not only enables efficient activation of Jedi T cells but it also provides an unmatched internal control for killing. When Jedi T cells are properly activated and functional, all GFP+ splenocytes are killed, whereas splenocytes in control LV-treated mice should display GFP+ cells in the spleen (Fig. 1). Thus, if GFP+ splenocytes are observed after Jedi T cell transfer, this indicates there was a problem with the transfer or the genotyping of the donor Jedi mouse. If Jedi expansion is observed in the blood and GFP+ splenocytes are killed but the cell of interest is not depleted, that indicates that this cell type displays a mechanism of resistance to T cell killing, as observed in other stem cell populations [8].
2. If other kits such as eBiosciences or Miltenyi are used, red blood cell lysis prior to isolation is required. Prior to counting live cells to perform the isolation with the chosen kit, 1× RBC lysis buffer is used for 3 min at room temperature and washed with cold FACS buffer. Then proceed with the isolation following the manufacturer's instructions.
3. Jedi T cells expand in the recipient mouse after proper activation with LV.GFP. This expansion can easily be determined by quantifying the percentage of CD45.1 Jedi T cells into the recipient CD45.2 mouse. This percentage increases between days 3 and 10 after adoptive transfer with a peak around day 6. If CD45.1 T cells are not observed, it might be due to a problem with the staining. Then, we recommend using blood from a CD45.1 Jedi mouse as a positive control to ensure the antibody works well and also as an aid for the gating strategy with the flow cytometry analysis. If CD45.1+ cells are observed but they are in a very low proportion (<5%) and it is not clear whether this percentage is increasing over time, objective measurements of activation and expansion should be performed. We recommend staining blood with anti-CD45.1-PE along with anti-CD69-APC, which is an activation marker. Moreover, in order to troubleshoot the lack of expansion, we recommend repeating the experiment and staining isolated T cells with a lipophilic dye (e.g., CellTrace™ Violet Cell Proliferation Kit) prior to injection. Dilution of the dye provides an objective quantification of cell division and hence T cell activation and expansion. We do not expect this dye to alter the ability of T cells to kill their cognate target cells.

4. A frequent difficulty when preparing singe cell suspensions from intestines is the fact that cells do not form a pellet and stay floating after centrifugation. This is mostly due to the presence of mucus and can be solved by filtering through a 70 μm cell strainer right before centrifugation. Thus, we strongly recommend filtering at every step that requires centrifugation. If the problem persists, the sample can be filtered through a 40 μm cell strainer and/or centrifuged at higher speed (1000 × g for 10 min, but we do not recommend higher than that).
5. If GFP signal is not observed in the control-treated mice in IF analysis, there may be a problem with fixation and GFP degradation. The time between obtaining the sample and imaging should be minimized (within 1–2 days) where samples and slides must be always kept protected from light. If the problem persists, it can be resolved by staining GFP with an anti-GFP antibody. Moreover, if paraffin-embedded sections are preferred over frozen sections, this can be done by staining with an anti-GFP antibody. We recommend anti-GFP-Alexa Fluor 488 polyclonal antibody. Slides must be properly blocked prior to staining to minimize background signal.

Acknowledgments

These protocols are currently used in the Agudo lab, but they were developed in the Brown lab (Icahn School of Medicine at Mount Sinai, New York). Thus, we would like to thank past and current members of the Brown lab for their help in developing these methods, specially to Dr. Brian Brown, Navpreet Tung, and Dr. Albert Ruzo. J.A. is supported by a Footbridge grant from the MIT-Bridge Project, a Claudia Barr grant in Innovative Science, a Harvard Stem Cell Institute junior faculty award, and the Mary Kay foundation award.

References

1. Saito M, Iwawaki T, Taya C, Yonekawa H, Noda M, Inui Y et al (2001) Diphtheria toxin receptor-mediated conditional and targeted cell ablation in transgenic mice. Nat Biotechnol. https://doi.org/10.1038/90795
2. Buch T, Heppner FL, Tertilt C, Heinen TJAJ, Kremer M, Wunderlich FT et al (2005) A Cre-inducible diphtheria toxin receptor mediates cell lineage ablation after toxin administration. Nat Methods. https://doi.org/10.1038/nmeth762
3. Tian H, Biehs B, Warming S, Leong KG, Rangell L, Klein OD, de Sauvage FJ (2011) A reserve stem cell population in small intestine renders Lgr5-positive cells dispensable. Nature 478:255–259
4. Tittel AP, Heuser C, Ohliger C, Llanto C, Yona S, Hämmerling GJ et al (2012) Functionally relevant neutrophilia in CD11c diphtheria toxin receptor transgenic mice. Nat Methods. https://doi.org/10.1038/nmeth.1905
5. Agudo J, Ruzo A, Park ES, Sweeney R, Kana V, Wu M et al (2015) GFP-specific CD8 T cells

enable targeted cell depletion and visualization of T-cell interactions. Nat Biotechnol 33:1287–1292

6. Chen A, Lee K, D'Agati VD, Wei C, Fu J, Guan TJ et al (2018) Bowman's capsule provides a protective niche for podocytes from cytotoxic CD8+ T cells. J Clin Invest 128 (8):3413–3424. https://doi.org/10.1172/JCI97879
7. Barker N, van Es JH, Kuipers J, Kujala P, van den Born M, Cozijnsen M et al (2007) Identification of stem cells in small intestine and colon by marker gene Lgr5. Nature 449:1003–1007
8. Agudo J, Park ES, Rose SA, Alibo E, Sweeney R, Dhainaut M et al (2018) Quiescent tissue stem cells evade immune surveillance. Immunity 48:271–285.e5
9. Baccarini A, Brown BD (2010) Monitoring microRNA activity and validating microRNA targets by reporter-based approaches. Methods Mol Biol 667:215–233
10. Agudo J, Ruzo A, Kitur K, Sachidanandam R, Blander JM, Brown BD (2012) A TLR and non-TLR mediated innate response to lentiviruses restricts hepatocyte entry and can be ameliorated by pharmacological blockade. Mol Ther 20:2257–2267

Chapter 3

Analysis of Aged Dysfunctional Intestinal Stem Cells

Kodandaramireddy Nalapareddy and Hartmut Geiger

Abstract

Aging is a multifactorial process. Organ maintenance and tissue regeneration are impaired upon aging mainly due to loss of stem cell function in organs that depend on stem cell in the adult. Intestine is such an organ, and upon aging intestinal regeneration is impaired due to decline of intestinal stem cell function. To determine the aging status of intestine and intestinal stem cells, histological analyses; analyses of the level of proliferation markers in tissue by immunofluorescence and/or quantitative RT-PCR; and gene expression analysis for stemness related genes in isolated crypts, intestinal stem cells (ISC), and Paneth cells can be used. To analyze the level of regeneration in intestine and thus determine a decline in ISC function, techniques like in vitro organoid cultures and lineage tracing with BrdU, lineage tracing using transgenic mice and histological analyses of tissue regeneration after 3 and 5 days after two rounds of 10 Gy of radiation (a 10 + 10 Gy IR experiment) can be applied. In this chapter we will focus on protocols for lineage tracing, the 10 + 10 gy IR experiment and for organoid cultures from young and aged mouse intestine.

Lineage tracing experiments in intestine can be done in many ways. In this chapter we describe a protocol for lineage tracing upon BrdU incorporation and lineage tracing using the $Lgr5^{eGFPCreERT2}$ $Rosa26^{YFP}$ transgenic mouse. For BrdU based-lineage tracing BrdU is administrated via intraperitoneal injections into mice. Animals will be analyzed 3 days (72 h) after BrdU administration. For experiments involving $Lgr5^{eGFPCreERT2}$ $Rosa26^{YFP}$ mice, mice will be analyzed after tamoxifen injection that activates Cre in Lgr5 positive (ISC) cells, which will result in permanent YFP expression. This allows for tracing of YFP positive cells in the intestine. The time point for the analysis of the intestinal tissue will depend in this case on the underlying scientific question that will be addressed. For 10 + 10 Gy experiments, animals will be irradiated with a radiation dose of 10 Gy on 2 consecutive days. The intestinal tissue will be analyzed 3 and 5 days after the second dose of radiation. Quantitative analyses of crypt depth and determination of the rate of crypt fission upon histochemistry will provide an estimation on the in vivo regenerative potential of ISCs. For serial organoid culture experiments, crypts will be harvested from mouse intestine, initially plated at concentrations ranging from 500 to 1000 crypts per well in Matrigel and grown in conditional medium or ISC medium. ISC Medium is changed every 2 days. After 1 week in culture, the organoids will be disrupted via a syringe and replated in fresh Matrigel. As the ability to form multilobed organoids is considered to be a direct stem cell function, the frequency of organoid formation in serial replating experiments can serve as a quantitative measurement of ISC function. For example, we demonstrated a reduced frequency of organoid formation as well as a reduction in number of lobes formed per organoid after 4–5 replatings of intestinal organoids from aged compared to young mice. These three techniques are thus, in combination, able to quantify the regeneration potential of intestinal stem cells and thus determine the extent to which intestinal stem cell regenerative function is reduced upon aging.

Key words Aging, Regeneration, Irradiation, Mice, Organoids, Intestinal stem cells, Lgr5

Paloma Ordóñez-Morán (ed.), *Intestinal Stem Cells: Methods and Protocols*, Methods in Molecular Biology, vol. 2171, https://doi.org/10.1007/978-1-0716-0747-3_3, © Springer Science+Business Media, LLC, part of Springer Nature 2020

1 Introduction

Aging alters organ maintenance and tissue homeostasis. Intestine epithelium is an organ system that depends on stem cells called intestinal stem cells (ISCs) for tissue homeostasis and thus maintenance of epithelial cells within the tissue and the intestinal architecture [1, 2]. Similar to other tissues and organs that depend on stem cells for homeostasis like the hematopoietic system or muscle, also intestine ages [2, 3]. Intestine presents with phenotypic changes and a reduced regenerative potential upon aging [4–7]. Intestinal epithelium can be divided into two parts, the crypt and the villus. ISCs, which give rise to all other cell types in the crypt as well as the villus, reside at the base of the crypt. The villus hosts the absorptive function of the intestine [8]. ISCs are positive for Lgr5, a G protein coupled receptor, one of the Canonical Wnt signaling gene. ISCs are flanked by differentiated Paneth cells which serve as niches to ISCs. ISCs divide and give rise to transient amplifying progenitor cells which move along the crypt up to the tip of the crypt. When they reach the tip of the crypt, they will terminally differentiate to form the villus which is composed mostly of enterocytes, but also Goblet cells and enteroendocrine cells. A cell originated at the base of the crypt from a dividing ISC reaches the tip of the villus in 3–5 days and is then expelled into lumen of the intestine [9]. We use BrdU tracing experiments to determine this distance that cells travel from the crypt base toward the tip of the villus in a given time-frame (usually 3 days, [10]). To this end, BrdU is injected into animals and will be incorporated into dividing stem and transient amplifying cells at the crypt base. Intestine will be harvested 72 h later. The average distance travelled by BrdU positive cells into the villus from the base of the crypt will be quantified. Our experiments demonstrated that there is a reduction in the distance travelled by BrdU positive cells in the intestine of aged animals, implying reduced overall regeneration at the crypt base.

For another type of lineage tracing experiment, we use Lgr5^{eGFPCreERT2} Rosa26^{YFP} mice (JAX: 006148 R26R-EYFP). These mice contain an enhanced yellow fluorescent Protein gene (EYFP) inserted into the *Gt(ROSA)26Sor* locus. Expression of *EYFP* is blocked by an upstream *loxP*-flanked STOP sequence. Cre recombinase is then activated in Lgr5 positive cells by tamoxifen injection to delete the floxed stop codon region in the Rosa26 locus to activate YFP (YFP in this case, it could be any other color (CFP, RFP etc.) depending on the knock-in color used to generate the mouse strain). The YFP positive cells will travel/amplify out of the niche along the transient amplifying zone and move into the villus to reach the tip of villus in between 3 and 5 days in young mice. The number of YFP positive villi is a measurement of ISC function. There are more YFP positive crypts/villus in young compared to aged mice, indicating a defect in the function of Lgr5+ ISCs upon aging [8].

To quantify the extent of the regeneration potential in young and aged intestine in more detail, intestine is damaged with irradiation (IR) and the subsequent regeneration is measured 3 or 5 days after irradiation. To induce damage to the intestinal epithelium and ISCs themselves, 10 Gy of total body irradiation will be delivered on 2 consecutive days (at least 24 h apart), which was termed a 10 + 10 Gy IR treatment. Intestine will be harvested and analyzed both on the third day and the fifth day after the last IR treatment by histochemistry. The height of the now regenerative crypts (aka a longer crypt indicates more active regeneration) and the frequency of crypts that underwent fission (crypts divide by fission to cover the loss of crypts by IR) [5, 11] is determined as a quantitative measurement to determine the regenerative potential of intestinal epithelium and thus ISCs. Determination of crypt height and crypt fission in young or aged mouse intestine upon damage demonstrated a decline in intestinal regeneration in aged intestine compared to young. Thus, BrdU tracing experiments as well as $Lgr5^{eGFPCreERT2}$ $Rosa26^{YFP}$ tracing or determination of the regenerative index after an 10 + 10 Gy IR allow to determine the regenerative function of ISCs in vivo upon aging. Thus, all the three assays (BrdU, $Lgr5^{eGFPCreERT2}$ $Rosa26^{YFP}$ tracing and 10 + 10 Gy IR) in combination are useful tools to investigate changes in the in vivo regenerative potential of ISCs upon aging.

The ex vivo function of ISCs can be quantified by organoid assays [8]. For organoid cultures, intestinal crypts are harvested and plated in Matrigel and grown in conditioned medium with canonical Wnt proteins R-spondin1, growth factors such as EGF and some cytokines (*see* below in Subheading 2). It is established in the field that organoid formation and their subsequent growth in culture mainly depend on ISC function [8, 12]. Organoids can be serially replated to determine the self-renewal potential of ISCs ex vivo. In this assay, the frequency of organoids formed and the number of lobes per organoid serve as quantitative parameters for the function of ISCs. Intestinal cells from aged mice show a decline in frequency of organoids formed and also in the number of lobes per organoid upon serial transplantation when compared to young ISCs. The decline in function of aged ISCs is reverted by addition of the canonical Wnt Wnt3a to the ISC medium, indicating upon aging, the decline in the function of ISC might be reverted by enhancing canonical Wnt signaling in aged ISCs [6].

2 Materials

2.1 Mice

1. Young (2–4 months) male and female C57BL/6 mice and aged (18–22 months old).
2. *Lgr5*^{eGFPCreERT2} animals (C57BL/6.129/SvEv mice).
3. *Lgr5*^{eGFPCreERT2} mice are crossed with (R26R-EYFP) R26-loxp-stop-loxp-EYFP mutant mice, simply termed *Rosa26*^{YFP} throughout the protocol, to obtain *Lgr5*^{eGFPCreERT2} *Rosa26*^{YFP} mice. Animals might be aged for up to 2 years. All analyses will be performed on the proximal (8–9 cm from the start of the small intestine).

2.2 BrdU Lineage Tracing

1. 4% Paraformaldehyde.
2. Xylene.
3. 100% ethanol, 90% ethanol, and 70% ethanol.
4. Tissue processer for making paraffin blocks.
5. BrdU (5-bromo-2′ deoxyuridine).
6. 10 mM sodium citrate buffer.
7. PBS (Phosphate buffered saline).
8. Anti-BrdU primary antibody (1:100 dilution in PBS).
9. Anti-rat FITC-conjugated secondary antibody.
10. Mounting medium with DAPI.

2.3 Irradiation Experiment

1. 4% paraformaldehyde.
2. Xylene.
3. 100% ethanol, 90% ethanol, and 70% ethanol.
4. Methanol.
5. Concentrated HCl.
6. Tissue processer for making paraffin blocks.
7. Coplin Jar.
8. Slide tray.
9. 10 mM sodium citrate buffer.
10. PBS.
11. Ki67 primary antibody (SP6, 1:100 dilution in PBS).
12. Anti-rabbit Biotin-SP conjugated secondary antibody.
13. Vector stein PK6100 ABC kit.
14. DAB substrate (3,3′ diaminobenzidine tetrahydrochloride).
15. 0.1% sodium bicarbonate.
16. Hydrogen peroxide.
17. Cytoseal 60.

2.4 Intestinal Stem Cell Medium

1. DMEM/F12.
2. 2 mM GlutaMax.
3. 10 mM 4-(2-hydroxyethyl)-1-piperazineethanesulfonic acid (HEPES) buffer.
4. 0.5 U/mL penicillin/streptomycin.
5. N2 supplement.
6. B27 supplement.
7. 50 ng/mL mouse recombinant epithelial growth factor.
8. 100 ng/mL mouse recombinant Noggin.
9. 500 ng/mL human recombinant R-spondin1.
10. 100 ng/mL mouse recombinant Wnt3A is added to some cultures. For regular use though, the intestinal stem cell medium does not contain Wnt3a.
11. Matrigel is used to grow crypts for organoid cultures.

3 Methods

3.1 General Protocol for Harvesting Intestinal Tissue and Tracing of GFP/YFP

1. Euthanize animal.
2. Cut open the abdomen and unwind the intestine slowly by cutting the connective tissue using scissors.
3. Take the intestine out of the animal from the end of the stomach to the beginning of cecum.
4. The intestinal tube is flushed with ice-cold PBS and then with ice-cold 4% PFA.
5. Flushed intestine is cut into three parts, proximal, middle, and distal regions. We refer to the first 8–9 cm from the end of the stomach as duodenum, the last 4–5 cm above the cecum as ileum and the remaining middle part as jejunum.
6. Each region can be cut further into half inch pieces and stored separately in 5–10 mL of 4% PFA for further processing or each piece can be cut open longitudinally and positioned upward such a way that the mesenchymal layer is at the base and villi are pointing up, make sushi roles of the intestine with villus inside and mesenchyme outside of the sushi role and then store in 5 mL of 4% PFA.
7. After overnight fixation, 4% PFA is removed, the tissue washed twice with 10 mL of ice-cold PBS and processed further for paraffin blocks of the intestinal tissue. 5–7 μm sections are made on a microtome for further analyses.
8. For preparing the tissue for cryo-sections instead of paraffin sections (to preserve and analyze the endogenous GFP or YFP signal in Lgr5^{eGFPCreERT2} Rosa26^{YFP} mouse intestine), after

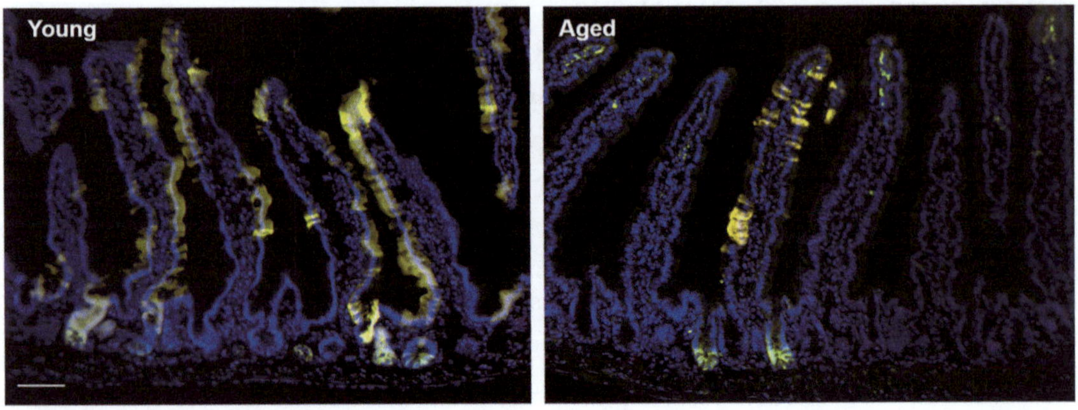

Fig. 1 Representative immunofluorescence pictures of the proximal part of the intestine from young and aged *Lgr5-EGFP-IRES-CreER*^{T2}:*Rosa26*^{YFP} animals. Pictures were taken 4 weeks after tamoxifen injection. Scale bar: 100 μm

washing 4% PFA with ice-cold PBS, further incubate them for 24–48 h in 5–6 mL of 30% sucrose solution at 4 °C.

9. Pieces are imbedded in Tissue-Tek O.C.T and blocks obtained by freezing the tissue on dry ice. Prepare 5–7 μm sections on a cryotome to further analysis.
10. Store the cryo sections at −20 °C in a dark box.
11. For further processing, take the sections out from freezer, leave them at room temperature in the Coplin glass staining jar for slides for 5–10 min.
12. Wash the sections twice for 5 min with PBS at room temperature (RT).
13. Fix the sections again 4% PFA for 5 min and wash again twice with PBS at RT on shaker.
14. Mount the sections with mounting medium with Dapi covered with cover slip on it and go for fluorescent microscopy (we use an Apotome microscope from Zeiss, 10× or 20× objective) (Fig. 1) [6].

3.2 BrdU Tracing

1. BrdU is injected at 100 mg/kg body weight (both for young and aged mice).
2. Harvest intestine 72 h after BrdU injection (*see* **Note 1**) (*see* Subheading 3.1).
3. After overnight fixation in 4% PFA, 4% PFA is removed, the tissue washed twice with ice-cold PBS and processed further for paraffin blocks of the intestinal tissue.
4. Prepare about 6 μm thick paraffin-embedded tissue sections (microtome), deparaffinize them with xylene (three times 5 min incubation), rehydrate them in 100%, 90%, and 70% ethanol series (consecutively 5 min in each).

5. Permeabilize the slides by heating in appropriate volume of (100–150 mL) of 10 mM sodium citrate buffer. For permeabilization, heat the sections in microwave at high (power level 100) for 4 min and medium (power level 50) for 4 min, make sure the sections are completely inside the sodium citrate buffer during the process.
6. After heating leave the sections at room temperature for 20 min.
7. Wash three times in PBS for 5 min each at RT on shaker.
8. Add BrdU primary antibody (1:100 dilution in PBS) and incubate overnight at 4 °C in humid chamber (slide box with PBS wet paper towels placed inside at the bottom of the slide box). Make sure to cover the entire section with antibody solution.
9. Wash three times in PBS for 5 min each on a shaker, incubate with anti-rat FITC-conjugated secondary antibody (1:200 dilution, for 1 h at RT).
10. Wash three times with PBS for 5 min each on shaker, dry slides, mount the dried slides on a coverslip using fluorescent mounting medium with DAPI.
11. Take pictures with a 10× objective (we use an Apotome microscope from Zeiss).
12. Measure the distance traveled by BrdU from the crypt base to the midpoint of all BrdU-positive cells in a villus using ImageJ software (Fig. 2) [6].

3.3 Irradiation Experiment (10 + 10 Gy Experiment)

1. Irradiate (both young and aged animals) with 10 Gy followed by a second 10 Gy dose 24 h later. We use a mark I-68A cesium 137 irradiator.
2. Harvest 3 days and/or 5 days after irradiation (Refer to general protocol for harvesting intestine).
3. Prepare 6-μm-thick paraffin-embedded tissue sections.
4. Deparaffinize, rehydrate with ethanol series, and permeabilize the tissue sections by heating in 10 mM sodium citrate buffer (see above).
5. After 15–20 min at RT wash with water twice 5 min each at RT.
6. Incubate in 0.5% hydrogen peroxide in methanol for 20 min to inhibit endogenous peroxidases (on the shaker). Wash with water.
7. Wash with three times with PBS for 5 min each on shaker.
8. Mark the section with DAKO pen (Water proof marker pen, will stop antibody solution to move out of the marked region from the tissue section) and add Ki67 primary antibody (1:100 dilution) to section, incubate at 4 °C overnight.

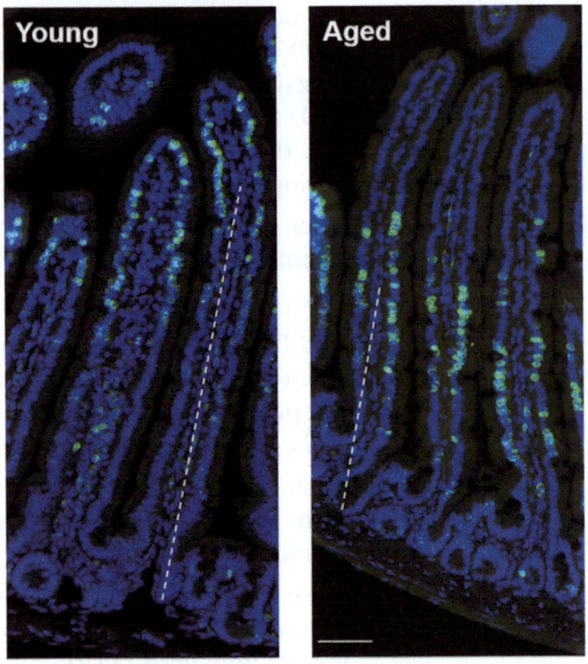

Fig. 2 Representative pictures of the BrdU signal (GFP stripe) in the proximal part of young and aged intestine 72 h after injection of BrdU. The dotted lines indicate the distance from the crypt base to the middle of the BrdU positive stripe in the villus. Scale bar: 50 μm

9. Wash slides two times in PBS, 5 min each on shaker.
10. Wipe of excess PBS and incubate sections with secondary antibody (1:200 dilution) at RT for 1 h.
11. Wash two times with PBS, 5 min each.
12. Add 3–4 drops of prepared ABC solution (Avidin-Biotin complex solution, from ABC KIT) per section and incubate for 1 h at RT.
13. Wash two times with PBS, 5 min each.
14. Add DAB substrate solution to the sections (*see* **Note 2**), keep the developing time constant for all the slides and wash the slides immediately in water.
15. Counterstain slides with hematoxylin for 5–7 min, rinse thoroughly, followed by acid/alcohol (1% HCl in 70% ethanol), wash with two quick dips, wash again with water and incubate in bluing solution (0.1% sodium bicarbonate) for 10 s, wash with water and then dehydrate back in ethanol series 70%. 90%, 100% to Xylene incubation each for 5–10 min.
16. Mount the sections with Cytoseal mounting media and go for microscopy (a bright field microscope, at 10× or 20× magnification) (Fig. 3) [6].

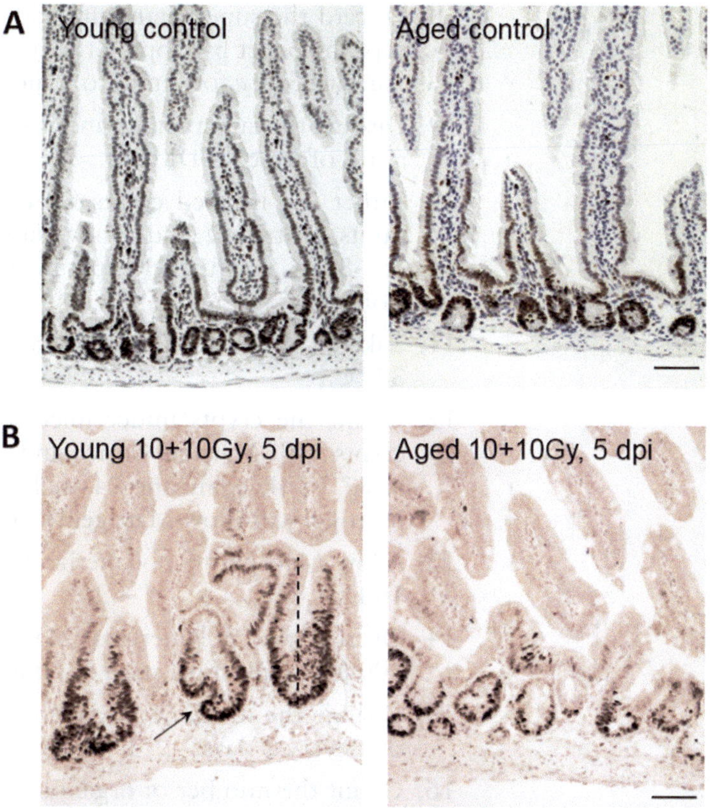

Fig. 3 Representative pictures of Ki67 staining in young and aged proximal intestine before (**a**) and 5 days after 10 + 10 GY IR (**b**). The dotted line in (**b**) determines the crypt height from the crypt base, arrow indicates crypt fission. Scale bar: 100 μm

3.4 Organoids

1. Dissect out mouse small intestine (*see* protocol to harvest, Subheading 3.1).
2. Take proximal 6–8 cm of intestinal piece and flush with ice-cold PBS.
3. Cut open longitudinally with villi facing upward and wash with ice-cold PBS.
4. Remove villi by gently scraping with glass slide and cut the 6–8 cm intestinal piece into small 1-in. pieces.
5. Transfer the intestinal pieces into 30 mL of ice-cold PBS, wash them by shaking them couple of times by hand within the tube.
6. Transfer the pieces with clean forceps from PBS to 30 mL of 5 mM EDTA in PBS (pH 8), followed by three 1-min shakings by hand, followed by a 10-min incubation at 4 °C.
7. Remove the intestinal pieces out with forceps and centrifuge the PBS/EDTA with crypts at 150 × *g* for 5 min.

8. Discard the supernatant and add 15 mL of ice-cold PBS, disrupt the pellet by tapping the tube couple of times and centrifuge at 85 × *g* for 2 min to remove single cells.
9. Discard the supernatant and resuspend the pellet (Crypts) in 5 mL of ice-cold DMEM.
10. Either use isolated crypts for organoid culture or make aliquots, centrifuge, and remove supernatant and freeze the pellet at −80 °C for further experiments (for example gene expression analyses).
11. Take 24-well tissue culture plate and keep it in incubator at 37 °C.
12. Count the crypts under microscope and mix 500 or 1000 crypts with Matrigel on ice and plate in a 24-well plate kept at 37 °C (keeping the 24 well plate at 37 °C before plating Matrigel will allow Matrigel to polymerize faster).
13. Allow Matrigel to polymerize in the incubator at 37 °C for 15 min.
14. Overlay the polymerized Matrigel containing crypts with 500 μL of intestinal stem cell medium.
15. Change the medium (completely) the next day and then further every 2 days.
16. Count the number of organoids and the number of lobes per organoid on day 6 after the initial plating.
17. Passage the organoids by taking out the medium and dissolve the Matrigel in 500 μL of ice-cold DMEM-F12.
18. Pipet the medium with the dissolved Matrigel for 25 times with a 200 μL pipette tip.
19. Disrupt the organoids by passing through a 26 G needle 5–6 times.
20. Centrifuge the Matrigel in DMEM-F12 with disrupted organoid mix in a sterile 15 mL tube at 4 °C for 5 min at 240 × *g*.
21. Discard the supernatant, resuspend the pellet in fresh Matrigel and replate the disrupted organoids (should be individual crypts). Allow Matrigel to solidify for 10–15 min at 37 °C and add 500 μL fresh intestinal stem cell medium (for each well in 48-well plate).
22. Count the number of organoids and lobes per organoids on day 5 or 6 (*see* **Note 3**).
23. Organoids are passaged the same way every time and change the medium every 2 days (Fig. 4) [6].

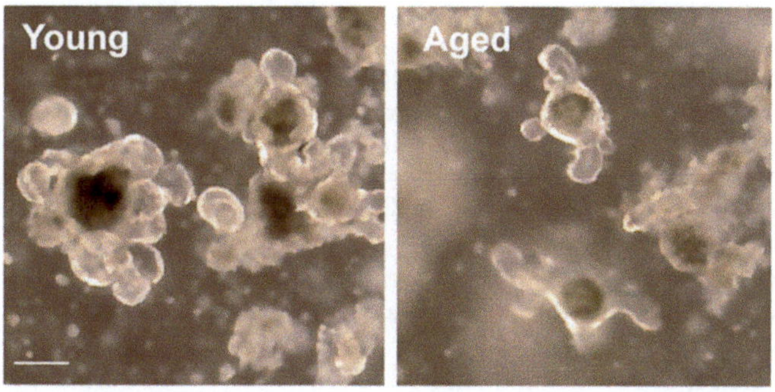

Fig. 4 Representative pictures of organoids derived from young and aged intestine (proximal part). Pictures are taken on the fourth day after the fifth passage. Scale bar 100 μm

4 Notes

1. In BrdU tracing experiments sometimes BrdU is not detectable anymore in the villus even on day 3. This might be due to a very fast cycling of ISCs at crypt base. Fast cycling ISCs might push cells up into the villus and then into the lumen of the intestine even within 3 days. In such a case, it would be good idea to rely on two-time points, like day 1.5 and day 2.5 after injection, to analyze the distance that BrdU cells are found from the crypt base to allow for further conclusions on the dynamics of regeneration [6].
2. For Ki67 staining, the time of incubation with the DAB substrate should be monitored carefully. Overincubation with the DAB substrate leads to false positive staining which usually starts from the crypt and then extends into villus zone. It is thus important to check already the staining directly under a microscope to determine the optimal time required for developing the DAB substrate. The time then needs to stay constant for all samples with regard to length for the DAB substrate incubation and might need to be rechecked every now and then again directly under a microscope.
3. For organoid cultures, avoid overseeding the well with crypts in Matrigel. Overseeding the well with crypts renders the precise quantification of organoids very difficult. If you are not sure, start with serial dilutions. Upon serial replating, around the second and/or third split, young organoids fill wells already within 3–4 days rather than usual 6 days. But also make sure to then split both young and aged organoids at the same time.

References

1. Nalapareddy K, Jiang H, Guachalla Gutierrez LM, Rudolph KL (2008) Determining the influence of telomere dysfunction and DNA damage on stem and progenitor cell aging: what markers can we use? Exp Gerontol 43 (11):998–1004
2. Rando TA (2006) Stem cells, ageing and the quest for immortality. Nature 441 (7097):1080–1086
3. Geiger H, de Haan G, Florian MC (2013) The ageing haematopoietic stem cell compartment. Nat Rev Immunol 13(5):376–389
4. Martin K, Kirkwood TB, Potten CS (1998) Age changes in stem cells of murine small intestinal crypts. Exp Cell Res 241(2):316–323
5. Martin K, Potten CS, Roberts SA, Kirkwood TB (1998) Altered stem cell regeneration in irradiated intestinal crypts of senescent mice. J Cell Sci 111(Pt 16):2297–2303
6. Nalapareddy K, Nattamai KJ, Kumar RS, Karns R, Wikenheiser-Brokamp KA, Sampson LL, Mahe MM, Sundaram N, Yacyshyn MB, Yacyshyn B, Helmrath MA, Zheng Y, Geiger H (2017) Canonical Wnt signaling ameliorates aging of intestinal stem cells. Cell Rep 18 (11):2608–2621
7. Potten CS, Kovacs L, Hamilton E (1974) Continuous labelling studies on mouse skin and intestine. Cell Tissue Kinet 7(3):271–283
8. Barker N, van Es JH, Kuipers J, Kujala P, van den Born M, Cozijnsen M, Haegebarth A, Korving J, Begthel H, Peters PJ, Clevers H (2007) Identification of stem cells in small intestine and colon by marker gene Lgr5. Nature 449(7165):1003–1007
9. Barker N, van de Wetering M, Clevers H (2008) The intestinal stem cell. Genes Dev 22 (14):1856–1864
10. Brack AS, Conboy MJ, Roy S, Lee M, Kuo CJ, Keller C, Rando TA (2007) Increased Wnt signaling during aging alters muscle stem cell fate and increases fibrosis. Science 317 (5839):807–810
11. Metcalfe C, Kljavin NM, Ybarra R, de Sauvage FJ (2014) Lgr5 stem cells are indispensable for radiation-induced intestinal regeneration. Cell Stem Cell 14(2):149–159
12. Sato T, Vries RG, Snippert HJ, van de Wetering M, Barker N, Stange DE, van Es JH, Abo A, Kujala P, Peters PJ, Clevers H (2009) Single Lgr5 stem cells build crypt-villus structures in vitro without a mesenchymal niche. Nature 459(7244):262–265

Chapter 4

Strategies for Measuring Induction of Fatty Acid Oxidation in Intestinal Stem and Progenitor Cells

Chia-Wei Cheng, Omer H. Yilmaz, and Maria M. Mihaylova

Abstract

This protocol describes a multipronged approach that we have created to determine the transcriptional induction of fatty acid oxidation (FAO) genes in Lgr5^{high} intestinal stem cells and a subsequent metabolomics-based approach for assessing fatty acid utilization in the mammalian intestinal crypt. More specifically, we describe methods for crypt isolation followed by a FACS-based purification of stem and progenitor populations and RNA-sequencing analysis. Using this workflow, we can determine both basal gene expression profiles of key metabolic genes as well as corresponding changes in response to altered metabolic states, such as fasting. Subsequently, we describe a complementary metabolomics-based approach that we have developed to assess fatty acid uptake and utilization in the crypt using ^{13}C stable isotope tracing. Combining these approaches, one can gain a better understanding of substrate utilization and the preceding transcriptional changes that accommodate these reactions in physiologic states of low carbohydrate utilization or during overabundance of dietary lipids.

Key words Fatty acid oxidation, RNA-sequencing, Stem cell metabolism, Metabolomics, Stable isotope tracing

1 Introduction

Understanding metabolic demands and fuel preferences in stem, progenitor and differentiated cells can help elucidate the role of metabolism in stem cell renewal and differentiation. From numerous previous studies, we now appreciate that the metabolic state of adult stem cells can be quite different than that of adjacent progenitor or terminally differentiated cells [1, 2]. One possibility to account for these differences is that adult stem cells may have unique metabolic gene expression signatures or metabolic flexibility, resulting in different metabolic flux compared to differentiated cells [3]. These dissimilarities could also be the consequence of higher proliferation in comparison to more terminally differentiated and post-mitotic cells. Additionally, environmental factors and paracrine signaling from adjacent niche cells can also play a

Paloma Ordóñez-Morán (ed.), *Intestinal Stem Cells: Methods and Protocols*, Methods in Molecular Biology, vol. 2171,
https://doi.org/10.1007/978-1-0716-0747-3_4,

significant role in somatic stem cell nutrient sensing and metabolism [4].

Measuring metabolic activities of adult stem cells in vivo remains a significant challenge due to (1) the low analyte abundances and limited sample volumes of the rare cell populations and (2) the small observation window of rapid metabolic changes. As cells regulate gene expression in response to environmental changes, gene-expression profiles of metabolic pathways have been used as the surrogate measurements of metabolic rearrangements in response to physiological status. Population RNA-Sequencing (RNA-Seq) can amplify the transcript sequences of rare cells and therefore becomes a popular method to study metabolic features of adult stem cells. However, despite a strong association between metabolic gene expression and metabolic phenotypes, experimental evidence shows that gene expression levels alone cannot accurately predict metabolic activities [5]. Therefore, subsequent validation by stable isotope labeling using stem cell–enriched tissue is crucial for providing a complementary metabolomic insight.

In this chapter, we describe the workflow (Fig. 1) that we have developed to identify and validate upregulation of fatty acid oxidation genes (FAO) in intestinal stem cells following a 24-h fasting protocol [1]. Understanding how these metabolic states in adult stem cells become altered in response to environmental signals, or in diseased states can potentially lead to therapeutic opportunities in regenerative medicine, as well as diseases such as cancer [3, 6–8].

2 Materials

2.1 Crypt Isolation Materials

1. 1× PBS (without calcium and magnesium).
2. 15 cm cell culture plates.
3. Glass microscope slides.
4. 20 mL Luer lock syringe
5. Gavage needle (18 G).
6. Sterile surgical tools.
7. 70-μm mesh.
8. 0.5 mM EDTA pH 8.5.
9. Inverted microscope.
10. 48-well cell culture plates.

2.2 Dissociation Methods

1. TrypLE™ Express Enzyme (1×).
2. S-MEM.
3. 40 μm mesh.

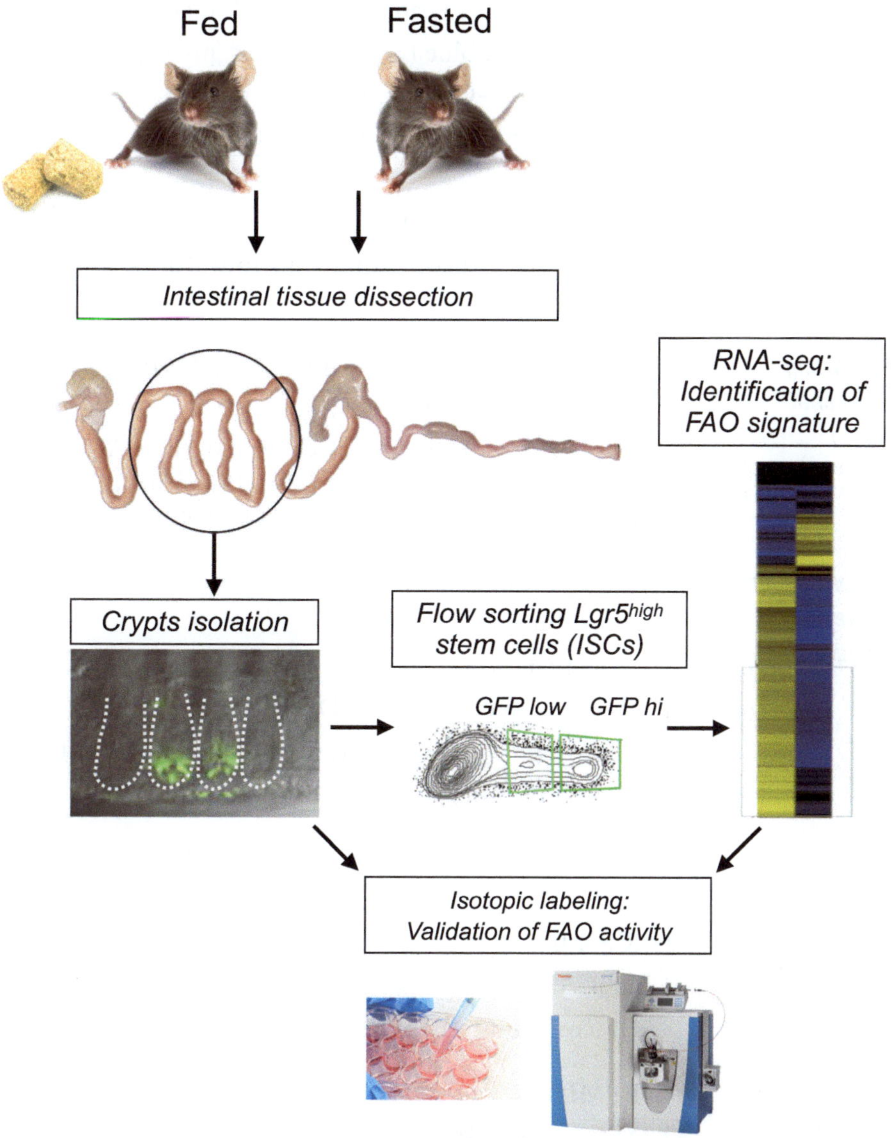

Fig. 1 Flowchart for isolation and dissociation of intestinal crypts followed by transcriptional and mass spec–based analysis of FAO induction. Ad libitum control mice or fasted mice are sacrificed and small intestine is dissected. Crypts are isolated following chemical and mechanical dissociation and further dissociated into a single cell suspension. Flow-based cell sorting enriches for intestinal stem and progenitor cells which are collected for RNA-Seq analysis. In parallel, purified crypts, which are enriched for intestinal stem and progenitor cells, are used for stable isotope tracing experiments to assess fatty acid oxidation and contribution to TCA cycle intermediates. Mouse images were adapted from [9]. Intestine image was adapted from [10]

2.3 Antibody and Viability Staining

1. Antibody cocktail:

   CD31-PE (2 μL) (0.2 mg/mL).

   CD45-PE (2 μL) (0.2 mg/mL).

   TER119-PE (2 μL) (0.2 mg/mL).

CD326-APC (Epcam) (3 μL) (0.2 mg/mL).

CD24-pacific blue (3 μL) (0.2 mg/mL).

CD117-APC-Cy7 (3 μL) (0.2 mg/mL)

2. 7-Aminoactinomycin D (7-AAD)
3. S-MEM.
4. 1000 μL filter pipette tip.
5. FACS tube: Falcon Round-Bottom Tubes with Cell Strainer Cap, 5 mL.

2.4 Flow-Sorting $Lgr5^{high}$ Intestinal Stem Cells

1. Low-binding 1.5 mL microcentrifuge tubes.
2. TRIzol™ Reagent.
3. Aria3: 4 laser, 11 color sorter running BDFACS Diva software on XP. Can sort into 1.5 mL, 5 mL, 15 mL, or 96-well plates. Can sort up to four populations at one time.

2.5 RNA Purification of Flow-Sorted Cells for RNA Sequencing

1. Chloroform.
2. GlycoBlue.
3. 75% ethanol.
4. Nuclease-free water (not DEPC-treated).
5. Centrifuge with temperature control (Eppendorf Microcentrifuge 5430 R).
6. Chemical fume hood.

2.6 Clustering Analysis and Heatmap Generation for FAO Gene Expressions

Software:

1. R 3.4.0.
2. javaGSEA.
3. Qlucore Omics Explorer 3.2.

2.7 Isotope Labeling of Purified Crypts

1. RPMI.
2. Murine EGF 40 ng/mL.
3. Recombinant Murine Noggin 200 ng/mL.
4. R-spondin 500 ng/mL.
5. *N*-acetyl-L-cysteine 1 μM.
6. N2 100×.
7. B27 50×.
8. Chir99021 10 μM.
9. Y-27632 10 μM.
10. Sterile Saline prepared with LC-MS grade water.
11. 13-C palmitate.

12. 80% methanol (LC/MS grade) containing internal standards (909 nM each of 17 isotopically labeled amino acids).
13. 4 °C centrifuge with 15 mL tube canisters.
14. Tissue culture incubator.
15. Thermo Scientific Q Exactive hybrid quadrupole-Orbitrap mass spectrometer.
16. Nitrogen dryer.

3 Methods

3.1 Crypt Isolation Protocol

1. Chill 1× PBS to 4 °C.
2. Mice are euthanized in designated chambers using fixed rates of carbon dioxide flow.
3. Following euthanasia, abdominal side is sprayed with 70% ethanol.

   Surgical tools are prepared, including 20 mL syringe and gavage needle (Fig. 2a).
4. After abdominal incision, small intestine is dissected and placed in ice cold sterile 1× PBS (Fig. 2b). Intestine is then flushed several times with 20 mL of ice cold 1× PBS and cut sagittally to remove any digested food debris (Fig. 2b–d).
5. Once the intestine is opened up, the mucus layer is gently removed using the index finger (Fig. 2e).
6. Next, the intestinal tissue is placed in 49 mL of ice cold PBS supplemented with 0.01 mM EDTA (49ml 1× PBS and 1ml 0.5 mM EDTA) and shaked on orbital shaker placed in 4C cold room.
7. Following the incubation, intestinal tissue and solution are returned to a 15 cm plastic dish on ice and are scraped using a sterile glass microscope slide (Fig. 2f). Tissue slurry is passed through 70-μm mesh continuously to separate villi from dissociated crypts. Repeat scraping and filtering steps and collect multiple 50ml fractions (*see* **Notes 1** and **2**).
8. Flow through and crypt fractions are assessed for purity and presence of villi contamination (*see* **Note 3**).
9. The filtrate is spun down at 4 °C at 300 × *g* in 50 mL conical tubes for 5 min.

3.2 Dissociation of Purified Crypts

1. To prepare samples for cell sorting by flow-cytometry, intestinal crypt pellets isolated from Subheading 3.1, **step 9** should be transferred to Falcon® 15 mL centrifuge tubes with 500 μL TrypLE per sample.

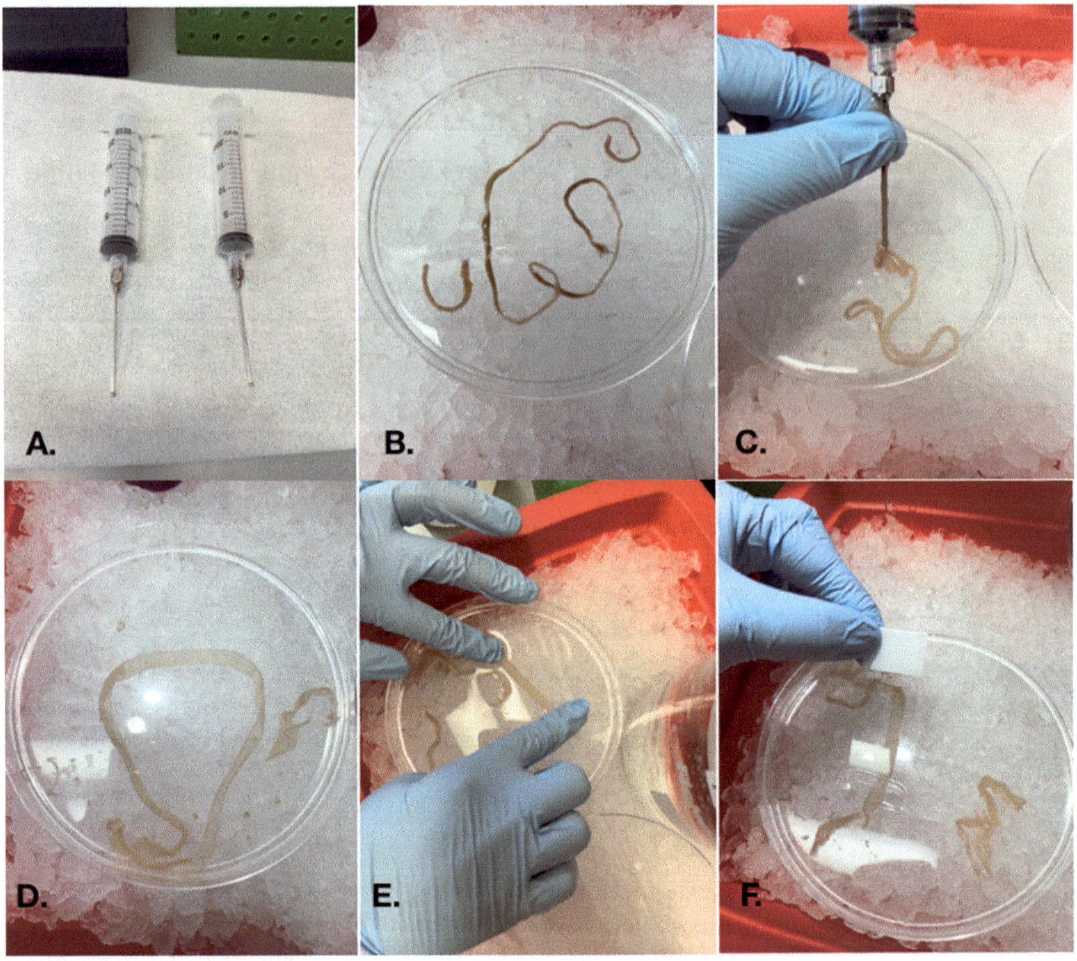

Fig. 2 Crypt isolation procedure. (**a**) 20 mL Luer lock syringes fitted for 18 G gavage needles are assembled and used in this procedure. (**b**) Following dissection of the small intestine, tissue is placed in 1× ice cold PBS. (**c**)The intestine is cleaned by applying pressure and passaging ice cold 1× PBS using the Luer lock syringe and gavage needle. The intestine is cut sagittally (**d**) and mucus and debris are further removed from the intestinal lining (**e**). Following an incubation in PBS containing EDTA on ice, a glass microscope slide is used to gently scrape the intestinal lining and remove villi and crypts (**f**). Tissue slurry is further filtered through 70 μm mesh to separate crypt fraction from villi. Purified crypts are then either used in metabolomic based analysis for fatty acid utilization using stable isotope tracing or further dissociated and used for gene expression analysis of intestinal stem and progenitor cells using RNA-seq

2. Crypts will become dissociated after 1 min incubation with TrypLE at 32 °C.
3. To remove TrypLE, wash cells with 10 mL ice cold S-MEM, invert twice and centrifuge at 300 × *g* at 4 °C for 5 min (*see* **Note 4**).

3.3 Antibody and Viability Staining

1. After removing the supernatant from **step 3**, resuspend each cell pellet in 250–400 μL of the following antibody cocktail:

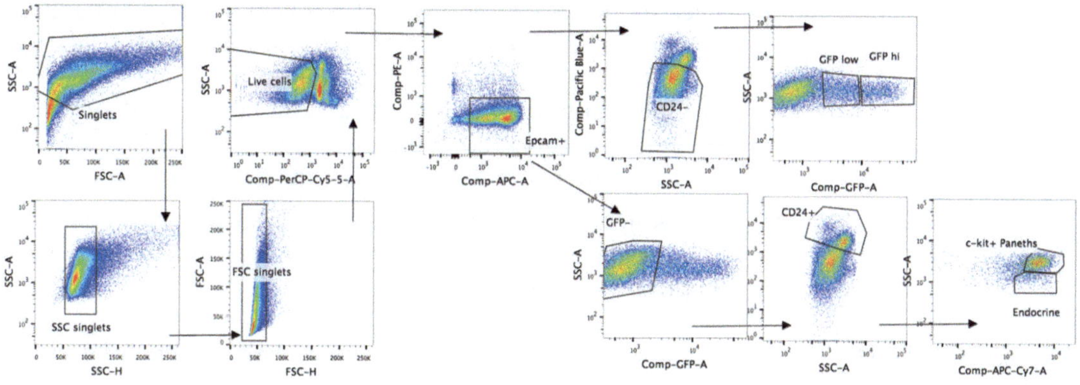

Fig. 3 Gating strategy and workflow described in Subheading 3.4

CD31-PE, CD45-PE, TER119-PE, CD326-APC (Epcam), CD24-pacific blue CD117-APC-Cy7 per 1 mL S-MEM.

2. Pipette vigorously up and down using a 1000 μL filtered pipette tip until most of the cells are dissociated and in suspension (*see* **Note 5**).
3. Incubate the cell mixture in the antibody cocktail for 15 min on ice.
4. Next, quench and wash the cell mixture with 10 mL ice cold S-MEM, invert twice and then centrifuge at 300 × *g*, 4 °C for 5 min.
5. Resuspend the pellet in 7AAD solution containing 10 μL of 7AAD reagent in 1 mL of S-MEM (amount may vary depending on 7AAD stock concentration). Cell suspension should be approximately 3× the volume of the cell pellet.
6. Next, filter the resuspended cell mixture through 40-μm mesh before transferring the cells through a cell strainer cap to a FACS tube (*see* **Note 6**).

3.4 Flow-Sorting *Lgr5*^{high} Intestinal Stem Cells

1. Gating strategy of Lgr5-eGFP^{high}, Lgr5-eGFP^{low}, and CD24^{+}C-kit^{+} Paneth cells is described below (Fig. 3).
2. First, gate cells for singlets (SSC-A vs. FSC-A, SSC-A vs. SSC-H, and FSC-A vs. FSC-H) and further gate for live cells by excluding 7AAD^{+} dead cells (PerCP-Cy5.5).
3. Next, use the following surface markers to further isolate different cell populations of 7AAD-live cells. Typically, Lgr5^{+} stem cells and progenitor cells are CD326/Epcam^{+} (APC)$^{+}$, CD24^{-} (Pacific Blue), and GFP^{high} (stem) and GFP^{low} (progenitor) cells. Paneth cells are CD24^{+} (Pacific blue), c-kit^{+} (APC-Cy7), and SSC^{high} cells.

4. Select the stem cell (Epcam^{+}/CD24^{-}/GFP^{high}), progenitor cell (Epcam^{+}/CD24^{-}/GFP^{low}), and Paneth cell (Epcam^{+}/GFP^{-}/CD24^{+}/c-Kit^{+}/SSC^{high}) populations and sort 25K cells per population into the assigned collection tube (low-binding tubes containing 400 μL TRIzol™ Reagent) with the setting of 4-way purity on the flow cytometry.
5. Following the completion of each 25K sort, vortex and place the collected sample on ice until all sample sorting has been completed (*see* **Note 7**).

3.5 RNA Purification of Flow Sorted Cells for RNA-Sequencing

1. Start this protocol by thawing cells at room temperature from Subheading 3.4, **step 5** (*see* **Note 8**).
2. To the 400 μL of TRIzol containing 25K stem cells, add 200 μL chloroform.
3. Vortex vigorously for 15 s and incubate at RT for 10 min.
4. Centrifuge at max speed (~12,000 × *g*) at 4 °C for 15 min.
5. Transfer aqueous phase to a clean, labeled tube.
6. Add 6 μL GlycoBlue to ~600 μL of the aqueous phase.
7. Add 600 μL Isopropanol to the GlycoBlue + aqueous phase samples, mix and store at −20 °C overnight or longer.
8. On the following day, remove the supernatant from the tube, leaving only the RNA pellet (blue).
9. Wash the pellet with 1 mL of 75% ethanol and centrifuge at maximum speed at 4 °C for 15 min.
10. Carefully remove the supernatant and air dry the pellet for 5 min in a chemical fume hood.
11. Resuspend and pool the samples (RNA from 200K cells) in 20 μL nuclease-free water.

3.6 Clustering Analysis and Heatmap Generation for FAO Gene Expression

1. Normalize aligned reads (against the mm10 murine genome assembly, with ENSEMBL 88 annotation) using the "geometric means" scaling method implemented in the DESeq R package [11].
2. Analyze potential enrichment of FAO-related gene sets using the command-line:
   version of the GSEA tool developed by Broad Institute.
3. Rank gene sets according to their log2 (FoldChange) values and analyzed using the "pre-ranked" mode of the GSEA software using the following parameters: -norm meandiv -nperm 5000 -scoring_scheme weighted -set_max 2000 -set_min 1 -rnd_seed timestamp.
4. Generate a merged gene expression matrix with the normalized reads and the metadata with sample annotations (column: sample ID/age/diet/cell population) and variable annotations

(row: ENSEMBL ID/gene symbol). Save the matrix as Tab Delimited Text file.

5. Prepare the FAO-related gene list by exporting FAO-related gene sets from the GSEA database.
6. Combine multiple gene lists and eliminate duplicated genes using R (unique()) or excel (=IFERROR(LOOKUP)).
7. Save the merged unique gene list as a variable list (VL_FAO genes.txt file).
8. In Qlucore, open the gene expression matrix (.txt file) with wizard.
9. Select samples "Young_AL_hi" for ISC from ad libitum fed mice (Ctrl), "Young_Fast_hi" for ISC from fasted mice.
10. Import the variable list VL_FAO genes.txt.
11. Click "VL_FAO genes" in the window of "Variable" to show only the normalized reads of genes listed in VL_FAO genes. Under Data, chose "Heat" as method.

3.7 Isotope Labeling of Purified Crypts (See Note 9)

1. Isolated crypts from Subheading 3.1 are divided into four equal fractions (50 mL fraction divided in 12.5 mL) and spun down at 200 × *g*, 4 °C for 5 min.
2. Next, three of the fractions per each biological replicate are resuspended in 1 mL volume of 100 μM ^{13}C-Palmitate in RPMI and placed in a 6 well plate.
3. 6-well plates containing crypts and ^{13}C labeled Palmitate are incubated at 37 °C in tissue culture incubators for 60 min or appropriate time point.
4. Following the incubation, crypts are spun down and washed once with saline.
5. Following the second spin and saline removal, the crypts are resuspended in LC/MS grade 80% methanol solution containing internal standards (909 nM each of 17 isotopically labeled amino acids and vortexed for 10 min) (*see* **Note 10**).
6. Samples are then spun down and dried in a vacuum dryer or dried under a stream of nitrogen and can be stored at −80 °C at this point (*see* **Note 11**).
7. Next, samples are resuspended in 100 μL LC/MS grade water and analyzed by LC/MS as described in [12].
8. 2 mL of each sample is injected onto a ZIC-pHILIC 2.1 × 150 mm (5 mm particle size) column.
9. Buffer A is 20 mM ammonium carbonate, 0.1% ammonium hydroxide; buffer B is acetonitrile.
10. The chromatographic gradient is run at a flow rate of 0.150 mL/min as follows: 0–20 min: linear gradient from

80% to 20% B; 20–20.5 min: linear gradient from 20% to 80% B; 20.5–28 min: hold at 80% B.

11. The Q Exactive Orbitrap mass spectrometer instrument is operated in full-scan, polarity switching mode with the spray voltage set to 3.0 kV, the heated capillary held at 275 °C, and the HESI probe held at 350 °C. The sheath gas flow is set to 40 units, the auxiliary gas flow is set to 15 units, and the sweep gas flow is set to 1 unit.
12. MS data acquisition is performed in a range of 70–1000 *m/z*, with the resolution set at 70,000, the AGC target at 10^6, and the maximum injection time at 80 ms.
13. Relative quantitation of polar metabolites is performed with XCalibur QuanBrowser 2.2 using a 5 ppm mass tolerance and referencing an in-house library of chemical standards.
14. The fraction m+2 acetylcarnitine is calculated as the raw peak area of m+2 acetylcarnitine, divided by the sum of raw peak areas of unlabeled acetylcarnitine, m+1 acetylcarnitine, and m +2 acetylcarnitine.
15. The same calculations are are used for M+2 citrate and other TCA metabolites.

4 Notes

1. During the intestinal cleaning procedure, it is important to remove the mucus layer and all connective adipose tissue to ensure proper dissociation and reduce contamination with other types of cells.
2. It is important to gently filter the slurry following the intestinal scrape in order to prevent the breakdown of villi and crypts. It is also important to maintain the slurry and intestinal tissue submerged in ice-cold PBS EDTA solution as much as possible to reduce the occurrence of cell death and enzymatic reactions due to temperature changes or tissue drying out.
3. Smaller clusters of villi cells can be a contaminant in the crypt filtrate and can potentially skew some of the downstream analysis, therefore it is integral to start with a highly pure crypt filtrate. Check the filtered crypt fraction under microscope (4× and 10×). If filtered crypt fraction is contaminated by large villi pieces, refiltering is recommended. If smaller, flat pieces of villi are present, new crypt isolation for that prep may be required.
4. Following the TrypLE dissociation, it is normal to observe the dissociated crypt cells aggregate into large clumps in S-MEM.
5. For Subheading 3.3, **step 2**, obtaining homogeneous cell suspension in the antibody cocktail is important for optimal staining results. Repeated pipetting up and down with a 1000 μL

filtered pipette tip is recommended until most of the cells are dissociated into the media. If the cell aggregates become difficult to resuspend, cut the end off of a pipette tip and use it to dissociate the cell clumps.

6. We recommend staining with ice-cold reagents/solutions and at 4 °C, as low temperature prevents the internalization of surface antigens. Internalization can cause a loss of fluorescence intensity. For Subheading 3.3, **step 6**, serial filtering of cell mixture in 7AAD/SMEM will eliminate any big cell aggregates and prevent clogs in the subsequent cell sorting step.
7. Vortexing collection tubes after each sort ensures that cells that have adhered to the sides of tubes are appropriately lysed.
8. It is important to select a clean, RNase-free space for all the RNA isolation steps. Pipettes, tube racks, and other surfaces should be thoroughly wiped down.
9. For the stable isotope tracing experiments, it is important to start with a healthy, viable prep for intestinal crypts. To ensure an equal amount of tissue distribution between all fractions for labeling, we divide the crypt filtrate prior to centrifugation into equal volumes and save two equal fractions for protein measurements and normalization. For the protein fractions, we lyse the crypts in 1× RIPA buffer and quantify the protein using BCA protein assay.
10. For the metabolite isolation step in methanol, we use a multitube vortex attachment and set the vortexer at the highest setting for 10 min.
11. Depending on the nitrogen drier flow, we adjust it such that the methanol–water–metabolite mixture does not "bubble over" and cause loss of material. Alternatively, a vacuum dryer can also be used.

Acknowledgments

We would like to thank Dr. David M. Sabatini for guidance and discussions. We would also like to thank Dr. Elizaveta Freinkman, Dr. Michael Pacold and Dr. Caroline A. Lewis for isotope tracing and metabolomics protocol assistance and discussions, as well as the Whitehead Institute Metabolite Profiling Core Facility. We thank Dr. Sue Jean Hong of the Bartel lab at the Whitehead Institute for suggestions on the RNA purification protocol, Glenn Paradis of the Koch Institute Flow Cytometry Core for suggestions on building multicolor panels, as well as the Whitehead Institute Flow Cytometry Core and the Koch Institute Flow Cytometry Core. We thank Dr. Stuart Levine of the MIT BioMicro Center for population RNA-sequencing. We would like to also thank the members of the Mihaylova lab for critical reading and comments. C.W.C. is supported by a Helen Hay Whitney postdoctoral fellowship and a NIH/NIDDK K99 fellowship. O¨.H.Y. is supported by NIH R00 AG045144, R01CA211184, R01CA034992, and U54CA224068; a V Foundation V scholar award; a Sidney Kimmel scholar award; a Pew-Stewart Trust scholar award; the Kathy and Curt Marble Cancer Research Fund; a Bridge grant; the American Federation of Aging Research (AFAR); and the MIT Stem Cell Initiative. M.M.M. is supported by NIH R00 AG054760 and the Glenn Foundation for Medical Research and AFAR Grants for Junior Faculty.

References

1. Mihaylova MM et al (2018) Fasting activates fatty acid oxidation to enhance intestinal stem cell function during homeostasis and aging. Cell Stem Cell 5:769–778.e4
2. Chandel NS et al (2016) Metabolic regulation of stem cell function in tissue homeostasis and organismal ageing. Nat Cell Biol 8:823–832
3. Rodriguez-Colman MJ et al (2017) Interplay between metabolic identities in the intestinal crypt supports stem cell function. Nature 7645:424–427
4. Mihaylova MM, Sabatini DM, Yilmaz OH (2014) Dietary and metabolic control of stem cell function in physiology and cancer. Cell Stem Cell 3:292–305
5. Vital-Lopez FG, Wallqvist A, Reifman J (2013) Bridging the gap between gene expression and metabolic phenotype via kinetic models. BMC Syst Biol 7(1):63
6. Wang B et al (2018) Phospholipid remodeling and cholesterol availability regulate intestinal stemness and tumorigenesis. Cell Stem Cell 2:206–220.e4
7. Schell JC et al (2017) Control of intestinal stem cell function and proliferation by mitochondrial pyruvate metabolism. Nat Cell Biol 9:1027–1036
8. Goncalves MD et al (2019) High-fructose corn syrup enhances intestinal tumor growth in mice. Science 6433:1345–1349
9. Stortz JA, Raymond SL, Mira JC, Moldawer LL, Mohr AM, Efron PA (2017) Murine models of sepsis and trauma: can we bridge the gap?. ILAR J 58(1):90–105
10. Treuting PM, Arends MJ, Dintzis SM (2018) Upper gastrointestinal tract. Comparative anatomy and histology, comparative anatomy and histology. https://doi.org/10.1016/B978-0-12-802900-8.00011-7
11. Anders S, Huber W (2010) Differential expression analysis for sequence count data. Genome Biol 11:R106
12. Birsoy K, Wang T, Chen WW, Freinkman E, Abu-Remaileh M, Sabatini DM (2015) An essential role of the mitochondrial electron transport chain in cell proliferation is to enable aspartate synthesis. Cell 162:540–551

Chapter 5

Visualization of Stem Cell Niche by Fluorescence Lifetime Imaging Microscopy

Irina A. Okkelman, Jens Puschhof, Dmitri B. Papkovsky, and Ruslan I. Dmitriev

Abstract

Fluorescence lifetime imaging microscopy (FLIM), enabling live quantitative multiparametric analyses, is an emerging bioimaging approach in tissue engineering and regenerative medicine. When combined with stem cell-derived intestinal organoid models, FLIM allows for tracing stem cells and monitoring of their proliferation, metabolic fluxes, and oxygenation. It is compatible with the use of live Matrigel-grown intestinal organoids produced from primary adult stem cells, crypts, and transgenic Lgr5-GFP mice. In this chapter we summarize available experimental protocols, imaging platforms (one- and two-photon excited FLIM, phosphorescence lifetime imaging microscopy (PLIM)) and provide the anticipated data for FLIM imaging of the live intestinal organoids, focusing on labeling of cell proliferation, its colocalization with the stem cell niche, measured local oxygenation, autofluorescence, and some other parameters. The protocol is illustrated with examples of multiparameter imaging, employing spectral and "time domain"--based separation of dyes, probes, and assays.

Key words Cell proliferation, FLIM, Intestinal organoids, Live cell imaging, PLIM, Real-time oxygenation, Stem cell niche

1 Introduction

The recent development and practical application of various stem cell-derived organoid cultures open new opportunities to study the tissue and organ development, stem cell niche organization and function, tumor induction and progression, and the aging. Organoid techniques were described for intestinal, gastric, esophagus, colon, brain, breast, lung, liver, and other tissues [1]. Organoids from these sources can be successfully grown, based on either adult stem cells or induced pluripotent stem cells, and genetically modified. These complex 3D structures comprising a stem cell niche, amplification zone of actively dividing daughter cells, and differentiated mature cells can be combined with different nonepithelial cell types or supplied with microbiota [2–4]. The architecture and

Paloma Ordóñez-Morán (ed.), *Intestinal Stem Cells: Methods and Protocols*, Methods in Molecular Biology, vol. 2171, https://doi.org/10.1007/978-1-0716-0747-3_5,

topological organization of organoids depend on the tissue they have been derived from: they can represent monolayers with/without protrusions and intrusions (e.g., gut organoids) or the multilayered complex structures such as taste buds, "mini-brains," and liver organoids [5]. Due to the complexity and heterogeneity of such cultures it is natural to use fluorescence microscopy as a principal method for analysis of cellular composition, cell proliferation, function, and metabolic parameters with precise single-cell and subcellular resolution [6]. Recent research in stem cell metabolism emphasizes that cellular interactions within the stem cell niche are much more complex than simply receptor recognition and triggering specific signaling molecules to activate signaling cascades: it also identifies metabolic signaling as one of the key parameters regulating cell fate and tissue self-organization [7–9]. Thus, the individual cell metabolism affects its microenvironment (e.g., by changing oxygenation, pH, concentration of ions, metabolic substrates and other factors) and protects genetic material from mutations and damage.

Fluorescence microscopy is applicable to end-point as well as continuous measurements ex vivo and in vivo and enables not only qualitative but also quantitative analyses of different physiological parameters, including localization of cell-specific biomarkers, labeling cell proliferation and cell cycle, dynamics of molecular oxygen (O_2), pH, redox ratio, metabolites, and others [10–14]. Live cell fluorescence microscopy enables close to real-time, multiparametric and direct measurements, avoiding production of artifacts normally encountered with analysis of fixed and "optically cleared" samples. The fluorescence lifetime imaging microscopy (FLIM) enables performing quantitative measurements of various biomarkers, protein interactions, allows separation of spectrally unresolved fluorophores under one- and multiphoton excitation modes and complements the super-resolution and light sheet microscopy modalities. Over the last decade, significant progress has been made in order to make the methodology more versatile and user-friendly and a number of "off-the-shelf" systems are now available commercially (Table 1). Importantly, FLIM becomes a widespread methodology, useful in autofluorescence imaging, FLIM-FRET applications, ion and other biomarker sensing, while the phosphorescence lifetime imaging microscopy (PLIM), highly useful for real-time analysis of cell and tissue oxygenation, still can be performed only on a limited range of available microscopes.

The successful use of the fluorescence microscopy and FLIM lies in selection of the most appropriate probes and sensors. These can represent endogenous fluorophores (NADH, FAD, luminal autofluorescence), endogenously expressed proteins (e.g., Lgr5-GFP, redox-sensitive proteins and others), and by a big variety of small molecule dyes and probes and the biosensing "hybrid" scaffold materials. In addition, live cell imaging can be further

Table 1
Available microscopy vendors providing FLIM platforms

Manufacturer	System	Comments	Examples of published studies with the microscope
Becker & Hickl GmbH	Broad selection of confocal scanners and modules, which can be attached to the existing confocal and two-photon microscopes (Zeiss, Olympus, Nikon etc.) (e.g., DCS-120). Complete FLIM systems.	Based on time-correlated single-photon counting (TCSPC), allows FLIM and PLIM measurements. No full integration between the microscope and FLIM scanner software (SPCImage software or open source, e.g., FLIMfit). https://www.becker-hickl.com/index.html	A number of works on autofluorescence FLIM [15–17], dyes, nanoparticles and probes [18–22]. Physiological studies of cell metabolism [23, 24], tumor spheroids [25–27], oxygenation of organoids [28]. Imaging scaffold materials for tissue engineering [29, 30].
PicoQuant	Selection of confocal scanners and modules for the FLIM system (detectors, lasers, etc.), compatible with existing confocal and two-photon microscopes (Zeiss, Olympus, Nikon, etc.). Complete FLIM systems.	Based on TCSPC. Allows only FLIM measurements. Good integration between microscope and FLIM scanner in software (Symphotime). https://www.tcspc.com/doku.php/start	Chromatin organization [31], sensing of phosphate [32], cAMP [33], protein interactions in plant roots [34].
Leica microsystems	SP8 Falcon module compatible with one-photon (white light laser and/or pulsed diode laser) and multiphoton excited systems (e.g., "Dive").	Based on TCSPC. Allows only FLIM measurements (up to ~400 ns). Fully integrated all-in-one system (e.g., confocal intensity + FLIM), user-friendly and broad choice of fitting algorithms (LAS X software). https://www.leica-microsystems.com/products/confocal-microscopes/p/falcon/	FLIM-FRET protein interaction studies, ion sensing, assessing mitochondrial polarization [35–37].
PCO.	FLIM camera compatible with existing widefield microscopes. Measurement range 100 ps–100 μs.	Frequency-modulation FLIM and PLIM [38]. https://www.pco.de/flim-camera/pcoflim/	pH sensing by FLIM [39, 40], fluorescence-guided neurosurgery using brain tumor imaging [41].

complemented by the analysis of fixed samples, for example, by using immunofluorescence or click-chemistry staining. Recently introduced gene manipulation methods such as CRISPR-Cas9 allow for cell-specific tracing with fluorescent proteins as well use of protein biosensors. However, the choice of high-performance fluorescent protein biosensors is still rather limited, compared to the dye-based FLIM and PLIM biosensors [18]. Altogether, available probes and sensors already provide the opportunity for live 3D imaging of organoids, labeling of the specific cell type and niche regions and analysis of dynamics of pH, proliferation, metabolism and mitochondrial function and overall oxygenation/hypoxia. Sensing can be done at high (subcellular) and low (extracellular sensing) resolution and in label-free modalities. It is highly anticipated that combined multiparametric FLIM and PLIM will allow understanding of mechanistic aspects of the stem cell niche function with attention to oxygenation, lipid content and activity of mitochondria in the near future.

In the present protocol, we summarize the application of FLIM and PLIM for imaging of stem cell niche, using genetically encoded proteins (green fluorescent protein, GFP), dyes which trace the microregions in the organoids, label mitochondria, and help visualize cell proliferation and oxygenation and other parameters (Table 2).

The protocol is based on the use of *Lgr5-EGFP-ires-CreERT2* intestinal organoids [47], in which the diffuse whole cell staining of GFP-positive cells is proportional to the levels of stem cell marker Lgr5 [48]. In dependence on the culturing conditions (e.g., presence of Wnt agonists, R-spondin, Noggin, histone acetylase inhibitors) organoids can become enriched or deprived in the number of stem cells (reflected in decrease of GFP-positive cells), with respective changes in proportion of differentiated enterocytes, goblet or dying cells. As we have shown before, Matrigel-grown organoids display significant heterogeneity in proliferating regions and crypts, which results in metabolic and proliferative variations [25, 26, 35 and 49].

Live organoids are relatively easy to stain with fluorescent and phosphorescent probes and dyes, within 5–120 min incubation and measure on FLIM (and or PLIM) microscope, in a way very similar to the conventional live cell confocal or two-photon excited microscopies. In addition to presented in the Table 2, label-free autofluorescence imaging can be performed in two variants: NADH/FAD (in recent literature often referred as "metabolic" imaging) [49] and measurement of the luminal autofluorescence (long-living broad range emission) [42]. Some other fluorescent probes can also be used for multiplexing with the presented dyes: for example, cell death can be assessed by staining with SYTOX Green/CellTox Green dyes (necrosis) and by various caspase-activated fluorescent substrates [18, 42, 50], antibodies and fluorescent protein

Table 2
Live imaging probes for multiparametric FLIM of the intestinal organoids used in this chapter

Measured parameter/probe	Exc./Em. (nm)	Staining concentration, time	Fluorescence lifetime range, application for FLIM or PLIM	Comments, localization
Lipid rafts/Cholera toxin, subunit B (CTX)-Alexa Fluor 488 conjugate	488/510 nm *Two-photon* exc. 985 nm	44 nM, 1–2 h	0.7–2.7 ns (FLIM)	Binds to ganglioside GM1 located in lipid rafts and endosomes (according to the manufacturer, Molecular Probes). In intestinal organoids stains regions very similar to the stem cell niche and amplification zone. CTX-Alexa Fluor 555 conjugate can be also used although this tracer shows less profound changes in membrane localization and fluorescence lifetimes.
Cell proliferation (S phase cells)/ Hoechst 33342 (+ BrdU)	405–440 nm/ 430–450 nm *Two-photon* exc. 660–705 nm	1–2 µM, 2 h	0.8–2.6 ns (FLIM)	Allows live imaging tracing of proliferating cells. The BrdU pulsing time and loading concentration may vary, depending on the model and used for visualizing different populations of proliferating cells [25, 42].
Nuclei and visualization of intracellular compartments/ SYTO 24 and others	488/ 500–530 nm *Two-photon* exc. 985 nm	0.5 µM, 1 h	3.4–4.5 ns (FLIM)	Nucleic acid stain originally described by Molecular Probes for live staining of cell nuclei. Stains the whole cell, including mitochondria, in cell cycle-dependent manner [35]. In dying and stressed cells SYTO 24 relocalizes to nucleus. This and related dyes (e.g., SYTO 13, SYTO 16) can be used to visualize different cell compartments in diverse types of cells in organoids. Caution: prolonged incubation with these dyes can result in mitochondrial toxicity,

(continued)

Table 2
(continued)

Measured parameter/probe	Exc./Em. (nm)	Staining concentration, time	Fluorescence lifetime range, application for FLIM or PLIM	Comments, localization
				affecting the mitochondrial membrane potential.
Polarized mitochondria/tetrame-thylrhodamine methyl ester (TMRM)	546/565 nm *Two-photon* exc. 1040 nm	2–20 nM, 15 min (has to be present in bathing solution continuously)	1.5–2.5 ns (FLIM)	Well-known dye for staining of mitochondrial membrane potential, one of the principal markers for assessing mitochondrial function [43]. Also works in FLIM mode [35]. Semiquantitative, distribution of fluorescence lifetimes (τ) can depend on staining concentration and "quenched/unquenched" modes of use.
Lipids and lipid droplets/Nile Red	488/550–600 nm	0.5–2 μg/mL, 1 h	2–4 ns (FLIM)	Lipophilic fluorescent stain, with emission intensity (ratiometric) and fluorescence lifetime affected by the polarity of its environment; stains membranes and intracellular neutral lipid droplets [44]. In FLIM displays strong concentration-dependent staining and can be used only as semiquantitative probe.
Cell oxygenation/Pt-Glc	405/650 nm	2 μM, 1 h	20–57 μs (PLIM)	Quantitative imaging of O_2 in 2D and 3D cell cultures. Can be multiplexed with any other FLIM probe. Pt-Glc [45] is not commercially available but can be easily synthesized [25, 26, 42]. Other O_2 probes can be also used instead as reviewed in [46].

biosensors [7]. It is also important to remember that not every chosen fluorescent probe or nanoparticle will provide efficient staining: for example, viscosity sensor BODIPY-C12 [51] and many tested O_2-sensitive and other fluorescent nanoparticles [26, 52] do not stain the intestinal organoids. The main "bottleneck" of the current FLIM measurements is the lack of well-established, standardized and widely accepted data analysis routine: here we explain the methodology accepted in our group for multi-parametric FLIM and discuss the future and desirable directions (*see* Subheadings 4 and 5) of its development. In Subheading 4 we describe staining and measured outcomes, which are expected with the mouse intestinal organoids grown under described conditions. Without establishing and optimizing organoid cultures, the protocol takes 1–2 days to be completed.

2 Materials

Prepare all solutions using ultrapure water (Milli-Q grade, sterile-filtered 0.22 μm, 18 MΩ cm). Store all reagents at 4 °C for no longer than 4 weeks, unless specified otherwise. To prevent contamination, perform all cell culture under laminar flow (class II Microbiological Safety Cabinet with HEPA filter) and aseptically. Wear gloves at all times and spray all used materials and surfaces with 70% ethanol. Unless provided sterile, autoclave all glass and plasticware (121 °C, 20 min) and filter-sterilize media and solutions prior to use.

2.1 Intestinal Organoid Culture

Mouse intestinal organoid culture containing Lgr5-GFP (Laboratory of Prof. H. Clevers, Hubrecht Institute, the Netherlands [47]) or equivalent.

2.2 Microscope Setup

Any of the systems described in Table 1, preferentially from Becker & Hickl, PicoQuant or Leica, having appropriate excitation sources (e.g., pulsed laser diodes, *see* Table 2 for the choice of the probes), acousto-optical modulators, detectors and software. The motorized control of the stage in *XYZ* directions is highly desirable. There is no difference in performance between the upright or inverted microscopes but the working distance of the objective must be considered. Water-dipping objectives must be used for upright setup (e.g., Zeiss 63×/1.0 W-Plan Apochromat) and loss of sterility during the imaging can occur. For inverted microscopes high NA objectives are preferred (e.g., Leica HC PL APO CS2 63×/1.3 Glyc).

The system must be also equipped with humidified temperature-controlled (37 °C) incubator and optionally with CO_2 and O_2 control. In the absence of CO_2 control, the 10–20 mM HEPES-buffered medium can be used for imaging.

Examples of the successfully used system for organoid imaging can be found here [25, 28, 42]. Below, we will focus on an upright one photon-excited (405 and 488 nm pulsed ps diode lasers) TCSPC FLIM/PLIM system (Becker & Hickl), inverted confocal (white light laser, providing excitation in range ~470–670 nm) and upright two-photon-excited ("Dive") TCSPC FLIM Falcon SP8 (Leica) systems. Depending on the vendor and the available software, the speed and ease of use of the FLIM microscopes will differ significantly: the most appropriate system must allow for completely motorized *XYZ* control, mosaic imaging and high speed of 3D reconstruction of FLIM images, ideally for preview purposes (1–5 min for FLIM *Z* stack from the acquisition to reconstruction).

2.3 Microscopy Imaging Supplies

For upright microscopy: tissue culture minidish (35/10 mm) cell$^+$. The growth area can be decreased (e.g., to minimize the use of expensive growth medium) by insertion of autoclave-sterilized silicon microchamber.

Optionally (also compatible with inverted microscope):

1. Glass bottom minidishes, 35 mm, No. 1.5 cover glass.
2. Glass bottom μ-dishes, 35 mm, low, Grid-500.
3. μ-chambers, 12 well. The silicon part is autoclavable and can be reused with any plastic or glass surface, when adhered to dish. Check the compatibility with the objective (e.g., that it can move freely within the sample) before use.
4. Immersion liquid (oil, glycerol) with appropriate refractive index.

See **Note 1**.

2.4 Chemicals and Plasticware

Reagents and plasticware for growth of intestinal organoid culture

1. 1 M HEPES solution pH 7.2, sterile.
2. Penicillin–streptomycin solution (P/S) 100× concentrate, sterile.
3. GlutaMax solution 100× concentrate.
4. Matrigel: thaw the original bottle at 4 °C overnight on ice. Pipette well on ice and divide the solution into 500 μL aliquots in 1 mL vials. The aliquots can be stored at −20 °C until the expiration date. One day before use thaw an aliquot of Matrigel on ice overnight (e.g., in the fridge), continue to keep on ice during the seeding procedure.
5. AdDF+++ medium (500 mL): advanced DMEM F12 Ham (Sigma) supplemented with 10 mM HEPES (from commercial stock), 100 U/mL streptomycin/penicillin (from 100× concentrate) and 2 mM GlutaMax (from 100× concentrate).

6. B-27 media Supplement, serum-free (50× concentrate). Aliquot aseptically and store at −20 °C prior use. Use as 1:50 dilution.
7. N2 Supplement, (100× concentrate) (Invitrogen). Aliquot aseptically and store at −20 °C before use. Use as 1:100 dilution.
8. 500 mM *N*-acetyl-L-cysteine (NAC) 400× concentrate solution in water: dilute 81.5 mg of NAC in 1 mL of water. Aliquots can be stored at −20 °C for 1 month.
9. Basal culture medium (BCM): supplement 50 mL of AdDF+++ media with 1 mL of 50× B-27 concentrate, 500 μL of 100× N2 concentrate, 125 μL of 400× NAC concentrate. Use for preparation of ENR medium. Store for 1 month at +4 °C.
10. Phosphate buffered saline (PBS), Ca^{2+} and Mg^{2+}-free.
11. Dimethyl sulfoxide (DMSO), "Hybri-Max" grade.
12. Albumin, from bovine serum (BSA).
13. Murine recombinant EGF (epidermal growth factor). Prepare 10,000× concentrate solution (500 μg/mL) in 0.1% (w/v) BSA in PBS. Aliquots can be stored at −20 °C for 1 month.
14. Murine recombinant Noggin. Prepare 1000× concentrate solution (100 μg/mL) in 0.1% (w/v) BSA in PBS. Aliquots can be stored at −20 °C for 1 month.
15. Human recombinant R-spondin. Prepare 1000× concentrate solution (1 μg/μL) in 0.1% (w/v) BSA in PBS. Aliquots can be stored at −20 °C for 1 month.
16. 3 mM CHIR99021 stock solution in DMSO (1000× concentrate). Aliquots can be stored at −20 °C for 1 month.
17. 500 mM sodium valproate (VA) solution in water (100× concentrate).
18. Intestinal organoids growth media (ENR): supplement 10 mL of BCM media with 100 ng/mL Noggin, 1 μg/mL R-spondin, and 50 ng/mL EGF. In order to increase the number of Lgr5-GFP^{+} stem cells, ENR media should be supplemented with 1 mM VA and 3 μM CHIR99021 (ENRVC media).
19. 12-well and 24-well flat bottom TC plates, sterile, tissue culture grade.
20. Sterile plastic tubes for 1.5 and 15 mL.
21. 10 mM ROCK inhibitor Y-27632 stock solution in PBS (1000× concentrate).
22. Cryovials.
23. Fetal bovine serum (FBS), sterile-filtered.

24. For cryopreservation of intestinal organoid culture prepare 2× concentrate of Freezing Media containing 20% FBS, 20% DMSO, 60% AdDF+++ medium and 10 μM ROCK inhibitor (to increase viability of organoids after defrosting). During the procedure of cryopreservation this medium will be diluted to final concentration.
25. Aphidicolin from *Nigrospora sphaerica*, 1 mg/mL 0.2 μm-filtered stock solution in DMSO.
26. Dulbecco's modified Eagle's medium, phenol red-, glucose-, pyruvate-, and glutamine-free.
27. D(+)-Glucose, powder. Prepare 1 M stock solution in sterile MQ-water and store at 4 °C.
28. 200 mM L-glutamine, sterile.
29. 100 mM sodium pyruvate solution, sterile.
30. "Imaging medium": DMEM supplemented with sodium bicarbonate (1.2 g/L), 10 mM HEPES-Na, pH 7.2, 1 mM sodium pyruvate, 2 mM L-glutamine, and 10 mM or 0.5 mM glucose, without phenol red.
31. Humidified CO_2/37 °C incubator, optionally with O_2 control.
32. Swinging bucket rotor–equipped centrifuge with cooling option (4 °C).

2.5 Live Imaging Probes

1. O_2-sensitive phosphorescent probe Pt-Glc synthesized as described [42], 1 mM solution in DMSO. Store in small aliquots at −20 °C or 4 °C for short-term.
2. 1 mM bis-benzimide Hoechst 33342, stock solution in water. Store in small aliquots at −20 °C or 4 °C for short-term.
3. 100 mM 5-bromo-2′-deoxyuridine stock solution in DMSO. Store at −20 °C.
4. 0.5 mg/mL Cholera toxin, subunit B (CTX), recombinant, Alexa Fluor 488 conjugate stock in PBS. Store in small aliquots at −20 °C or 4 °C for short-term. Do not vortex.
5. 20 μM Tetramethylrhodamine methyl ester (TMRM) stock solution in DMSO. Store in small aliquots at −20 °C or 4 °C for short-term.
6. 1 mg/mL Nile Red stock in DMSO. Store at −20 °C.
7. 5 mM SYTO 24 green fluorescent nucleic acid stain solution in DMSO. Store at −20 °C.

See Table 2 for description of properties and intended use.

Avoid unnecessary exposure of the probes to the ambient light and store in dark or wrapped in tinfoil. Mix well after defrosting and prior the use. If precipitates form during preparation of the working dilutions, check the calculations and remake the solutions.

2.6 Data Acquisition and Analysis Software

For collection and processing of FLIM data use the vendor-provided software (e.g., SPCImage for DCS-120 (Becker & Hickl), Symphotime for PicoQuant, and LAS X for Leica systems). Open source options are also available [12, 53]. The use of additional software for image organization, 3D reconstruction, colocalizations, segmentation, deconvolution and statistical evaluation is also recommended (e.g., Microsoft Excel, Origin (https://www.originlab.com), MatLab (https://www.mathworks.com/products/matlab.html), Fiji (ImageJ, http://fiji.sc/), SVI Huygens (https://svi.nl), and CellProfiler (https://cellprofiler.org)). Below we use the SPCImage (Becker & Hickl) and Microsoft Excel or LAS X FLIM/FCS 3.5.5 (as indicated). The final images are assembled using Adobe Photoshop and Illustrator CS2.

3 Methods

3.1 Defrosting the Intestinal Organoid Culture

1. For recovery of frozen organoids, remove the vial from the liquid nitrogen storage tank and thaw it quickly on a 37 °C water bath.
2. Pipet and collect the organoids with a 1000 μL pipette into a 15-mL centrifuge tube. Immediately add 10 mL of AdDF+++ medium dropwise and under constant shaking and spin the organoids down at 4 °C (500 × *g*, 5 min) in a swinging bucket rotor.
3. Using a pipette, gently remove and discard the supernatant and suspend the organoids in ice-cold Matrigel. For 1 vial use 50–200 μL (total volume) of the Matrigel.
4. Dispense droplets of 10 μL of the suspension to the 1–3 wells of a 37 °C prewarmed 12-well plate. Place the plate in a 37 °C incubator for 5–30 min to solidify the Matrigel.
5. Add 1 mL of ENR or ENRVC medium (depending on the experimental task) to each well and incubate at 37 °C.

   Optionally, to minimize anoikis, supplement culture medium with 10 μM Y-27632 (ROCK inhibitor) during the first 2 days of culture.
6. Medium should be refreshed 2–3 times per week. Passage organoids when they fill most of the Matrigel blob, stop growing, or start to appear dark due to cells being shed to the organoid lumen (routinely split in 1:5 ratio every 5–7 days) (*see* **Note 2**).

3.2 Passaging of Intestinal Organoids and Seeding for Microscopy Imaging

The night before passaging of organoids thaw aliquot of the Matrigel on ice. Prechill the centrifuge to 4 °C. Prewarm (20 min, 37 °C) new culture 24-well plates and/or microscopy dishes before the seeding.

1. Perform mechanical disruption of organoids in Matrigel with growth medium by pipetting them for 20 times with the 10 μL

tip on top of a 1 mL tip. Collect organoids in 15 mL tube (*see* **Note 3**).

2. Rinse wells once with additional volume of AdDF+++ media and collect it in the same tube. Add AdDF+++ media to 10 mL volume.
3. Centrifuge at 500 × *g* for 5 min at 4 °C. Carefully remove supernatant (*see* **Note 4**).
4. While keeping the tube on ice suspend the dissociated organoids in a fresh portion of cold Matrigel in a proportion 1:5 from the old Matrigel.
5. Dispense 50 μL of organoid–Matrigel mixture per well of pre-warmed 24-well plate or 20 μL of organoid–Matrigel mixture per microscopy dish. Leave to solidify at 37 °C for 5 min. Add 500 μL of growth media per well of 24-well plate or 200 μL to the imaging dish. Grow for 5–7 days before the next passage. Regularly monitor the dishes for excessive evaporation of the medium and add more growth medium if needed.

   Optional procedure, if the intact organoids have to be removed from the Matrigel:

1. Rinse Matrigel with organoid culture two times with ice-cold PBS without disturbing the Matrigel.
2. Add cold depolymerization solution (Cultrex "Organoid Harvesting Solution") using 500 μL per 50 μL of Matrigel. Incubate with gentle shaking for 15 min at 4 °C until complete dissolving of Matrigel. Collect organoids using 1 mL pipette tip in the 15 mL tube covered with BSA. If big organoids are to be collected, cut the tip of the 1 mL pipette tip to avoid breaking of organoids. Centrifuge at 500 × *g* for 5 min at 4 °C. Remove supernatant.
3. Wash with 10 mL of cold AdDF+++. Centrifuge at 500 × *g* for 5 min at 4 °C. Remove supernatant. Proceed with **steps 4** and **5**.

3.3 Sample Staining and Image Acquisition

The microscope and the associated equipment (lasers, camera, incubator system, computer, and other operating electronic blocks) have to be turned on 30 min before imaging to warm up and become equilibrated to measurement conditions (e.g., 37 °C, 5% CO_2, 20% O_2 or optionally to different O_2 values, temperature, humidity, CO_2). Prepare the microscope control software (e.g., SPCM or LAS X). Select the appropriate filter cubes or spectral settings for used fluorescent probes. If several samples have to be imaged on the same day we recommend having an appropriate gap time intervals (~30 min per sample, depending on the probe fluorescence collection time, number of probes used for multiplexed staining and the number of imaging replicates).

1. Dilute the required amount of fluorescent/phosphorescent probe in the intestinal organoid growth medium (minimum volume of media for microscopy dishes is 150 μL per dish) according to recommended staining concentrations (Table 2) (*see* **Note 5**). For HXT-BrdU method (*see* **Note 6**).
2. (Optional) In case of multiparametric imaging several probes can be mixed together in appropriate staining concentrations (*see* **Note 7**).
3. Aseptically replace the growth medium in the sample with the one containing premixed probes ("loading medium") and incubate in the incubator as required (Table 2).
4. Gently remove loading medium, rinse once with the imaging medium and add 2.5–3 mL of imaging media. If required, remove the microwell insert before the imaging (*see* **Note 8**).
5. Set up the microscopy dish with stained organoids on the preheated microscopy table (37 °C). Bring the objective to the working position.
6. Preview the sample in transmission light mode. Select the organoids and regions of interest (ROI) for imaging, adjust/set focus and *XYZ*(*t*) coordinates.
7. Do quick preview in fluorescence intensity mode (use appropriate excitation and emission parameters to collect signal, Table 2) and estimate the intensity and distribution of the probe staining. Adjust settings for the collection of fluorescence/phosphorescence signal (acquisition time sufficient for collection of fluorescence/phosphorescence signal, spatial resolution/pinhole, the desired optical section for 2D scanning or the top, the bottom and the desired range and increment for the *Z*-stack for 3D scanning, time for time-lapse measurements) taking into consideration the intensity of the stained organoid and the desired quality of the image (*see* **Note 9**).
8. Collect the image or the set of images (for the 3D scanning). Name the files or the work folder for appropriate saving. Proceed with the imaging of the next probe for the same optical section or *Z*-stack or the next organoid.
9. Process imaging data according to Subheading 3.4.

3.4 Processing of Imaging Data

Figure 1 illustrates the data processing routine for SPCM and SPCImage software (Becker & Hickl GmbH). For other FLIM and PLIM platforms (Table 1) data processing protocol can be modified accordingly: thus, different opportunities can be available for data analysis (e.g., the broader choice of fitting parameters, fluorescence lifetime separation based on χ^2 coefficient, improved batch processing and 3D reconstruction) and their export. The present protocol describes export and Excel-based analysis of the data obtained for the intensity image of Lgr5-GFP and PLIM data obtained with Pt-Glc-stained organoids.

1. Open several SPCImage software windows and import selected data files for different fluorescence or phosphorescence images, for example by viewing the fluorescence intensity data for Lgr5-GFP and Pt-Glc staining (Fig. 1a, c, e), for the same optical section of organoid.
2. Apply the appropriate fitting settings (e.g., two-component exponential fitting model by adjusting *t*1, *t*2, pixel binning, and other parameters) to the phosphorescence or fluorescence decay data. The quality of fitting can be estimated by the χ^2 coefficient, which ideally should be equal to or less than 1 for all analysed pixels of the image (*see* **Note 10**).
3. (Optional) Using ROI selection accurately choose the ROI on the fluorescence/phosphorescence intensity image. For example, select one bright area on Lgr5-GFP fluorescence intensity image, corresponding to localization of the GFP-expressing cells (as shown on Fig. 1a). Save the ROI mask and apply it to the image of Pt-Glc phosphorescence intensity made for the same optical section (*see* Fig. 1c). In order to calculate Pt-Glc phosphorescence lifetime for organoid epithelia avoid the inclusion of (auto)fluorescence signals from extracellular matrix and the lumen by choosing the appropriate ROI mask (*see* Fig. 1e) (*see* **Note 11**).
4. Calculate emission lifetime values ("Decay Matrix") for the chosen the ROI or the whole frame (*see* **Note 12**).
5. Export the phosphorescence/fluorescence intensity (photon counting), lifetime (color-coded values), and the distribution histogram (ASCII format) as well as the lifetime images as TIFF for all probes used in analysis. Note the range of used color scale.
6. Using Microsoft Excel or other relevant software open the obtained numerical data and using "conditional formatting" function produce color-coded images based on numerical values for intensity (Fig. 1b, f) and color-coded values (Fig. 1d) (*see* **Note 13**).
7. Choose the desired area in the intensity map for first probe (e.g., GFP^+ regions), note its coordinates (the starting cell number using for selection in the table in both directions) and apply them on "lifetime image" for the second probe (e.g., Pt-Glc staining). Copy the chosen values to the new file and use them for the following statistical analysis.
8. Repeat **step 7** to obtain the sufficient number of areas for statistical analysis (e.g., two groups of Pt-Glc phosphorescence lifetime data for GFP^+ and GFP^- areas) (*see* **Note 14**).

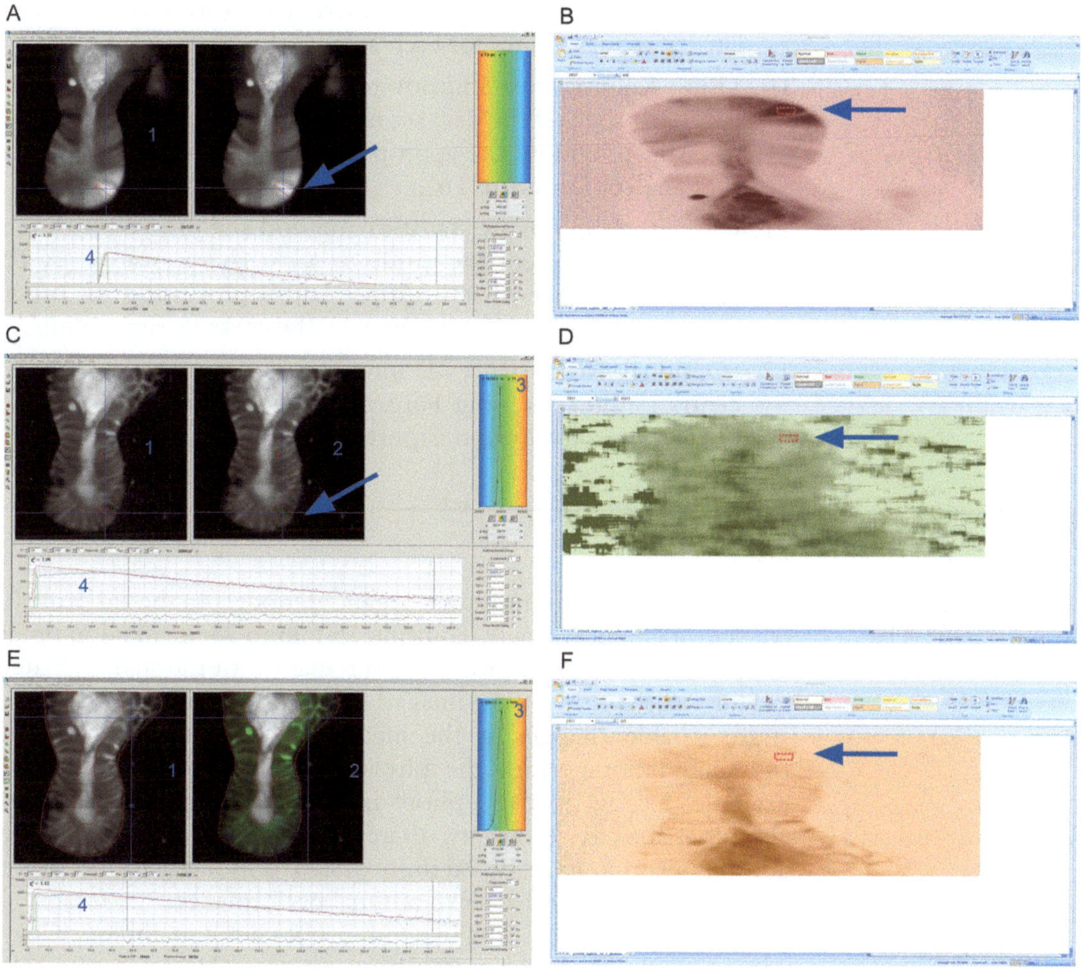

Fig. 1 Example of the data processing routine with Lgr5-GFP organoids, stained with O_2 probe Pt-Glc and measured by PLIM method, using SPCImage software (Becker & Hickl). (**a**) Lgr5-GFP data are imported in SPCImage and opened as FLIM data (calculation of fluorescence lifetimes for GFP are not necessary) and used for identifying of regions of interest, for example, the GFP^+ cell indicated by arrow. (**b**) Intensity data (photon counts) for GFP image has to be opened in Microsoft Excel and further color-coded with numbers. Note that the photon count data exported from SPCImage have "upside-down" orientation in Excel. (**c**) PLIM data for the same organoid has to be opened in a separate window and ROI selection drawn from GFP file can be copied here (arrow). After optimizing the fit, phosphorescence lifetimes are calculated for selected region. (**d**) Calculated PLIM data after color-coding procedure in Excel. The same ROI chosen in **a**, **b** and **c** can be copied for extracting the data. (**e**) More complex shape of ROI mask including only the epithelium (O_2 PLIM image). Note the difference in histograms between **c** and **e**. (**f**) Pt-Glc intensity data opened in Excel after color-coding procedure can help improving the selection of areas poorly expressing GFP. The arrow indicates the same ROI chosen in **a–d**. (1) Intensity image; (2) FLIM/PLIM images; (3) lifetime distribution histogram windows, (4) fluorescence decays and their fitting for selected pixels in 1 and 2

9. Calculate the average values of the measured parameter for each area in data set. Optionally, applying the calibration function recalculate the numerical intensity or lifetime data to the

values of real physiological parameter (*see* **Note 15**) and calculate its average values for each area.

10. Using Kolmogorov–Smirnov test or other relevant statistical approach check if the data set matches the normal distribution in order to choose the appropriate statistical method for further analysis (*see* **Note 16**).
11. Chose the appropriate statistical method and proceed with the calculations. Present the results as a bar charts, scatter (*X*:*Y*) charts or bubble charts. Apply batch processing to other images from the same experiment, 2D stacks or time points.
12. (Optional) Reconstruct 3D intensity, lifetime and/or O_2 concentration images using Fiji software (VolumeViewer Plugin) or relevant software and export/save them.

4 Anticipated Results

4.1 Selection of the "Right" FLIM Probe

Ideal FLIM probe for intestinal organoid model (1) should provide efficient staining, typically requiring 15 min to 2 h incubation time; (2) should distribute uniformly and remain with the stained material for sufficient time to be imaged (i.e., >2 h); (3) should demonstrate reliable changes in the lifetime and the calibration; and (4) should not perturb the physiological function of cells, for example, having no dark, photo-induced, genotoxicity or mitochondrial toxicity. In reality, there is no ideal fluorescent probe and the research must be carefully designed in order to introduce sufficient controls and achieve optimal performance.

Figure 2 illustrates staining with probes listed in Table 2. Conventional fluorescence microscopy image (HXT, Lgr5-GFP, TMRM, top left) shows typical appearance of the organoid having good expression levels of Lgr5-GFP. TMRM is a well-known marker in the intensity mode, is very bright, and provides immediate staining of mitochondria. In FLIM mode TMRM can show striking difference in mitochondrial membrane potential between different cell types or even intracellular compartments. Lipid droplet-specific probe Nile Red displays very high diversity of fluorescence lifetimes in cell cytoplasm, informing on the differences in lipid composition of different cell types. Longer lifetimes can demonstrate the presence of stem cell niche, shorter—differentiated cells. However, this probe shows very strong dependence of observed fluorescence lifetimes (Fig. 2) on the staining concentration, which makes it difficult to use for quantification. SYTO 24 is another interesting probe for studying polarized mitochondria and labeling of the organoids, showing different lifetimes between the nuclear and cytoplasmic fractions of stained cell (Fig. 2). For some unidentified rare cell types, SYTO 24 displays lack of staining of cytoplasm with weak nuclear staining. Potentially it is an interesting probe for studying organoids, but its possible inhibitory effect on the mitochondria has to be considered [35].

Cholera toxin subunit B (nontoxic) conjugated to Alexa Fluor 488 (CTX) or Alexa Fluor 555 display plasma membrane-confined accumulation in organoids, with shorter lifetimes at the basal membranes of some cell types (Fig. 2). CTX staining correlated with Lgr5-GFP and proliferating cells is a highly promising cell tracer, well accepted as nontoxic and widely commercially available.

The more "advanced" probes for multiparameter organoid imaging are Hoechst 33342 (allows tracing proliferating cells) and Pt-Glc (O_2 probe), described below (Figs. 3 and 4). The use of fluorescent probes can also be combined with the measurements of autofluorescence of cells (NADH and FAD) and lumen; however, currently there is very little information on the specific labeling of stem cell niche or other regions in organoids by the two-photon excited autofluorescence FLIM and the potential cross-talk (spectral and in the lifetime domain) between the NADH, FAD, and fluorescent exogenous or genetically encoded tracers.

4.2 Labeling Cell Proliferation by Hoechst-BrdU FLIM Method

This method [25] enables easy and widely compatible tracing of S phase cells (Fig. 3a). Briefly, cells are incubated with the fluorescent dye (Hoechst 33342, HXT), with or without 5-bromo-2-′-deoxyuridine (BrdU), which accumulates in cell nuclei proportionally to duration of cell cycle S phase progression. This is seen as quenching of the HXT blue fluorescence or decrease of fluorescence lifetime on a FLIM microscope.

Brief (1–4 h) loading with BrdU allows detection and tracing cells in S and following cell cycle phases, including mitosis (Fig. 3b). Combining these FLIM data with other markers provides additional information on the cell status: for example, combining of this method with imaging of Lgr5-GFP helps identifying populations of nondividing non-stem cells (1—$BrdU^-/GFP^-$ cells), dividing stem cells ($BrdU^+/GFP^+$ cells) and a rare group of dividing cells lacking Lgr5-GFP fluorescence (2—$BrdU^+/GFP^-$ cells) (Fig. 3b).

In principle, this method also helps measuring duration of S phase using calibration function [25]. However, with the complex heterogeneous organoid culture it is not straightforward due to the presence of mature nondividing cells as well as two or more populations of proliferating cells with distinct short and long cell cycles. The use of different synchronization methods (e.g., aphidicolin treatment, Fig. 3c) allows for "semi-calibration" of time-dependent BrdU uptake, which can be applied for studying of S phase duration in different cell types. Fluorescence lifetime distribution histograms show the proportion of S phase cells in regions of interest and

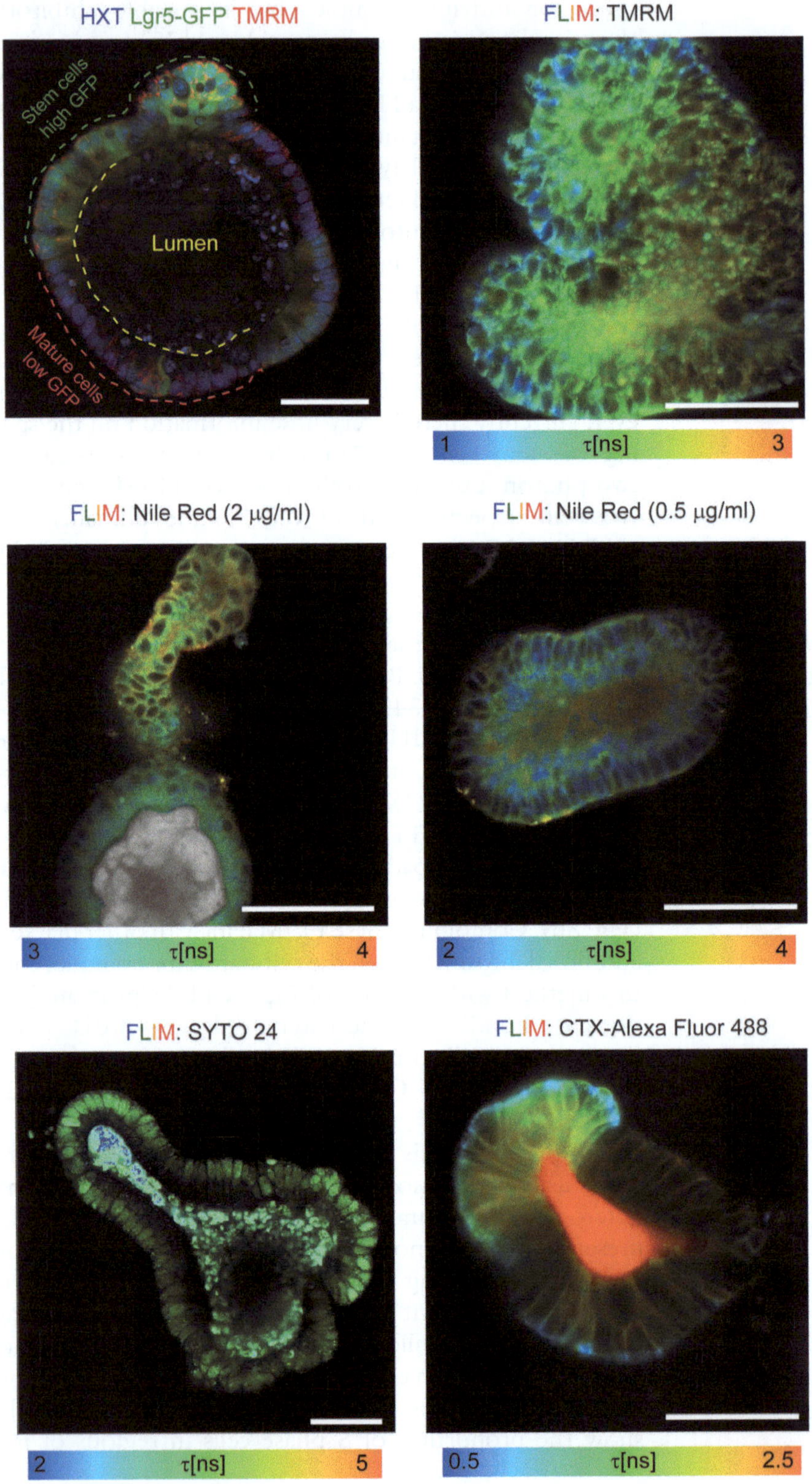

Fig. 2 Examples of live organoid staining with different FLIM probes. Top left: two-photon excited fluorescence microscopy of typical Lgr5-GFP organoid counter-stained with TMRM and Hoechst 33342 (HXT) dyes.

provide fluorescence lifetime values for BrdU^{+}/BrdU^{-} cell populations (Fig. 3d). More accurate fluorescence lifetime values can be calculated from numerical data in Excel or other software (the arrow on Fig. 3d indicates fluorescence lifetime of ~1.6 ns for BrdU-labeled nuclei, calculated from six individual BrdU^{+} nuclei taken from the image c). However, the aphidicolin treatment (even if used at <1 μg/mL concentration) and the other cell synchronization methods will inevitably affect the viability and proliferation of cells in intestinal organoids. Thus, the results of such synchronization experiments must be treated with caution.

Long-term (>12 h) loading with BrdU (Fig. 3e,f) allows for distinguishing several populations of cells with different cell cycle duration: (1) cells with high BrdU (average nuclear lifetimes of 0.963 ± 0.087 ns, shown in blue on Fig. 3e) and (2) cells with low BrdU uptake (average lifetime of 1.7 ± 0.125 ns, shown in green on Fig. 3e). Fluorescence lifetime distribution histogram (Fig. 3f) allowed for calculating the proportion of cells for these subpopulations: 87% of already completed S phase during 16 h and 13% of cell just entered S phase. We consider 1.575 ns (indicated with arrow) as a threshold value between these populations. The percentage of "high BrdU" nuclei was calculated as an area under the histogram between 0.5 and 1.575 ns using Origin 6.0 software.

BrdU accumulation itself can indeed affect the cell cycle but in practice it is used at minimal doses (e.g., 5–10 μM concentration for overnight or 50–100 μM for 1–2 h incubation, depending on the sensitivity of the used FLIM system). The use of blue-laser light excited (e.g., 405 nm) dyes usually leading to the photodamage of cells, rarely happens with pulsed diode lasers and FLIM [25] or the use of two-photon excitation. Thus, with minimal preliminary optimization experiments, HXT-BrdU FLIM method can be a very useful tool for characterization of stem cell niche and proliferation in intestinal organoids.

4.3 Combined Use of FLIM Probes and Multiparameter Assays

The spectral properties and fluorescence lifetimes of the listed dyes (Table 2and Fig. 2) enable various combinations of the multiparametric assays. Considering that all dyes and assays have been evaluated with the organoid model in preliminary experiments, their combined use can be easily realized. Figure 4 shows two types of the suggested assays on a one-photon excited FLIM-PLIM microscope (Becker & Hickl GmbH), with the Lgr5-GFP organoids

Fig. 2 (continued) Distribution of GFP-positive regions marks stem cells and stem cell niche. Lumen region (indicated by yellow line) displays the autofluorescence with broad spectrum and variable intensity. All the other sections represent FLIM images of organoids stained either with TMRM (mitochondrial membrane potential), Nile Red (lipid droplets), SYTO 24 (originally described as live nucleus-specific dye, two-photon FLIM image is shown) and Cholera toxin (CTX). Scale bar is 50 μm

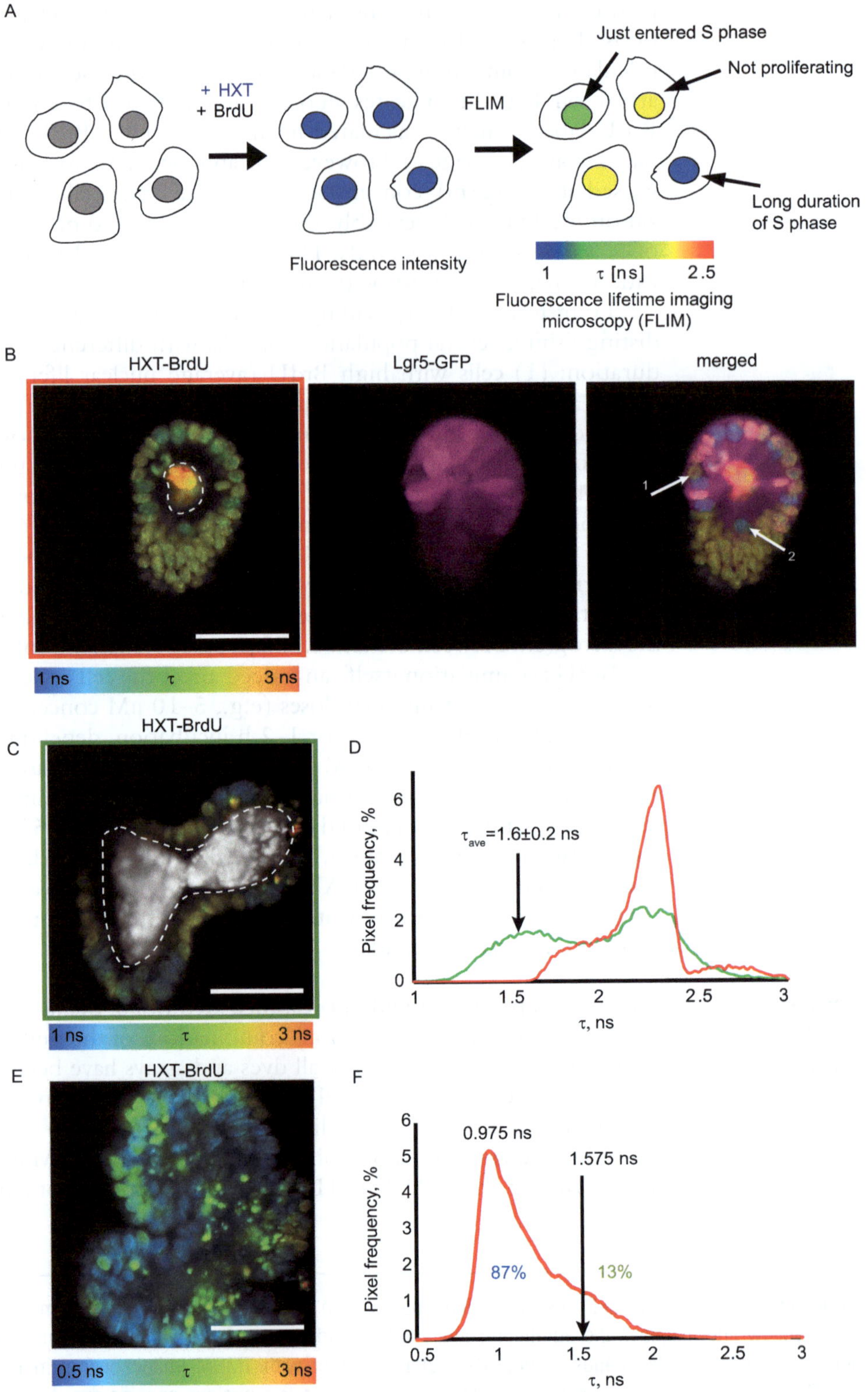

Fig. 3 Application of HXT-BrdU FLIM method to the intestinal organoids. (**a**) Summary of the method. (**b**) FLIM images of Lgr5-GFP organoids preincubated with 100 μM BrdU and 1.5 μM HXT (3 h). Lgr5-GFP is shown as

differing in use of the green fluorescence (exc. 488 nm, em. 500–530 nm) spectral channel. Note that crypt regions and proliferating stem cell niches can be labeled by HXT-BrdU (FLIM) method requiring the use of 405 nm laser; at the same time, live oxygenation can be assessed by using the same excitation source with the red (650 nm) emitting O_2 probe Pt-Glc (PLIM). Depending on cell growth conditions (e.g., ENRVC or ENR) medium, the number and expression of GFP-positive cells would differ; normally, exogenous fluorescent probes such as CTX-Alexa Fluor 488 will overlap and provide much brighter staining than GFP. Additionally, other orange and red FLIM dyes can be combined with the use of Pt-Glc: TMRM or Alexa Fluor 555-labeled CTX. However, depending on the available spectral resolving options and FLIM scanner, the spectral cross-talk can be still observed, for example, as with TMRM and Pt-Glc on a DCS-120 system (Fig. 4a).

Multiparametric and 3D scanning with FLIM and PLIM probes or the other dyes is the area where the key differences between the manufacturer/vendor can become dramatic: thus, annotating and exporting multiple ROIs, FLIM data and histograms, assembly of *XYZ*-stacks (3D reconstruction) of intensity or FLIM data can be very difficult or not even possible. In addition, PLIM platforms enabling high-resolution O_2 analysis are available currently only from few manufacturers.

Figure 4 shows data produced on Becker & Hickl DCS-120 (SPCImage software). Green and red selected ROIs show two chosen epithelial regions and corresponding fluorescence lifetime distribution histograms. Thus, HXT-BrdU, CTX and TMRM clearly display difference in fluorescence lifetimes between differentiated and proliferating niche regions, confirmed by Lgr5-GFP data (Fig. 4b). At the same time, cells are deoxygenated to some degree, displaying levels much lower than atmospheric 21% (200 μM) but the differences in oxygenation between stem and differentiated cells-enriched regions are not that profound and would need further analysis.

Multiparameter FLIM of organoids is also possible using two-photon excited microscopy. Figure 5 shows examples of data obtained on multiphoton FLIM microscope (Leica SP8 "Dive"

Fig. 3 (continued) intensity image in purple color. Numbers (1, 2) indicate different cell types revealed with this method. (**c**) FLIM analysis of organoids synchronized in S phase with aphidicolin (0.125 μg/mL, 18 h), subsequently pulsed with 100 μM BrdU (3 h after release of aphidicolin block) and 1.5 μM HXT. (**d**) Fluorescence lifetime distribution histograms obtained for asynchronous (section B, red color) and S phase-synchronized (section C, green color) organoid cultures. (**e**) FLIM image of organoid loaded with 5 μM BrdU (16 h) reveals different amplification zones. (**f**) Fluorescence lifetime distribution histogram for organoid E helps calculating the percentage of cells for two populations of cells with different duration of cell cycle. Scale bar is 50 μm

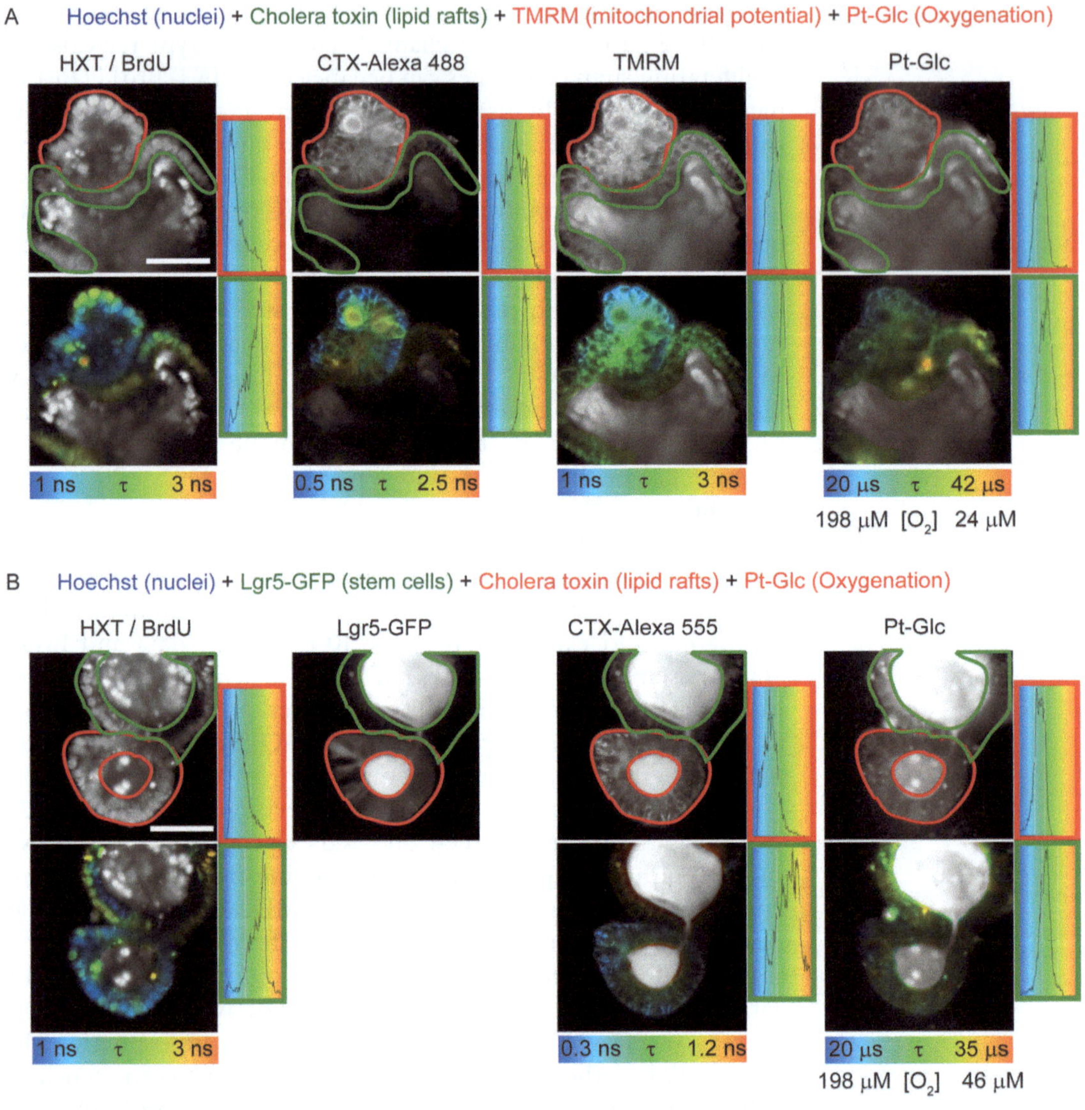

Fig. 4 Examples of multiparametric imaging of the organoids using one-photon excited FLIM-PLIM microscope. (**a**) Organoids were stained with HXT-BrdU (exc. 405 nm), CTX-Alexa 488 (exc. 488 nm), TMRM (exc. 488 nm) and Pt-Glc (exc. 405 nm), washed and subsequently imaged. Red and green lines indicate selection of distinct regions on the false-color FLIM and PLIM images, accompanied by the lifetime distribution histograms (for red and green selections) on the right. (**b**) Organoids stained with HXT-BrdU (exc. 405 nm), CTX-Alexa 555 (exc. 488 nm) and Pt-Glc (exc. 405 nm). Lgr5-GFP was excited by 488 nm (only intensity image is shown). The data were acquired using laser-scanning DCS-120 FLIM-PLIM microscope (Becker & Hickl GmbH). Scale bar is 50 μm

Falcon FLIM system, operated by the LAS X software). Two-photon excitation enables deeper light penetration, higher quality 3D reconstruction of the images, combined with much milder longwave illumination. Thus, it is possible to simultaneously image the NADH autofluorescence (this requires harmful ~360 nm

UV excitation on conventional widefield or confocal microscopes) and compare it with distribution of Lgr5-GFP cells. HXT-BrdU FLIM method can be also combined with Lgr5-GFP, TMRM, CTX, and SYTO 24 dyes in two-photon mode, thus enabling advanced 3D analysis of the stem cell niche regions in the live organoids.

Imaging of unstained live Lgr5-GFP organoids enables analysis of their redox status (NADH FLIM, Fig. 5a). In addition, the lumen autofluorescence provides strong signals and long emission lifetimes. Colocalization of GFP with NADH signals can help revealing heterogeneity of cell metabolism during proliferation and differentiation (Fig. 5a). Labeling of S phase cells by HXT-BrdU FLIM method is also compatible two-photon excitation: cell nuclei demonstrate uniform long lifetimes without BrdU and become shorter after BrdU accumulation (Fig. 5b–e). Three-dimensional reconstruction of HXT-BrdU-stained organoid helps visualizing stem cell-enriched regions and clearly distinguishes them from enterocytes and other differentiated cell types (Fig. 5d).

Time domain imaging also enables separation between luminescence lifetimes even if the spectral properties between the used probes/dyes/proteins are virtually the same. Figure 6 demonstrates the combined staining of live organoids with SYTO 24 and CTX-Alexa Fluor 488 dyes both excited and emission collected simultaneously. High-resolution imaging of SYTO 24-stained organoids allows visualizing nuclei and cytoplasmic foci, while staining with cholera toxin (CTX) helps labeling actively proliferating niche regions (Figs. 6a and 3d reconstruction). By choosing one of these regions in *XY* optical section the regions of interest can be rescanned with higher resolution (Fig. 6b–d) but the resulting intensity image will not allow for discrimination between the signals of SYTO 24 and CTX even though they have slightly different cellular localization. However, these two dyes have difference in fluorescence lifetimes (Table 2) and can be further resolved by using "Pattern fit" and decay diversity map (τ, τ distribution) features in LAS X FLIM/FCS software (Leica). Results of this separation are shown in Fig. 6e, f. Thus, two and potentially more fluorescent probes can be efficiently resolved in a single chosen spectral channel, enabling extended multiplexed capabilities in multiparametric assays (e.g., having more cell- and species-specific stains or combining it with biosensing) with simultaneous detection. Important to note that the brightness of the probes intended for such time domain-based resolving has to be very comparable and sometimes additional separate testing is required. For example, CTX staining can completely mask GFP signals in Lgr5-GFP organoids (Figs. 4 and 5).

Thus, the future development of two-photon excited FLIM systems such as Dive Falcon FLIM (Leica) and alike will significantly advance the area of 3D functional imaging of the engineered

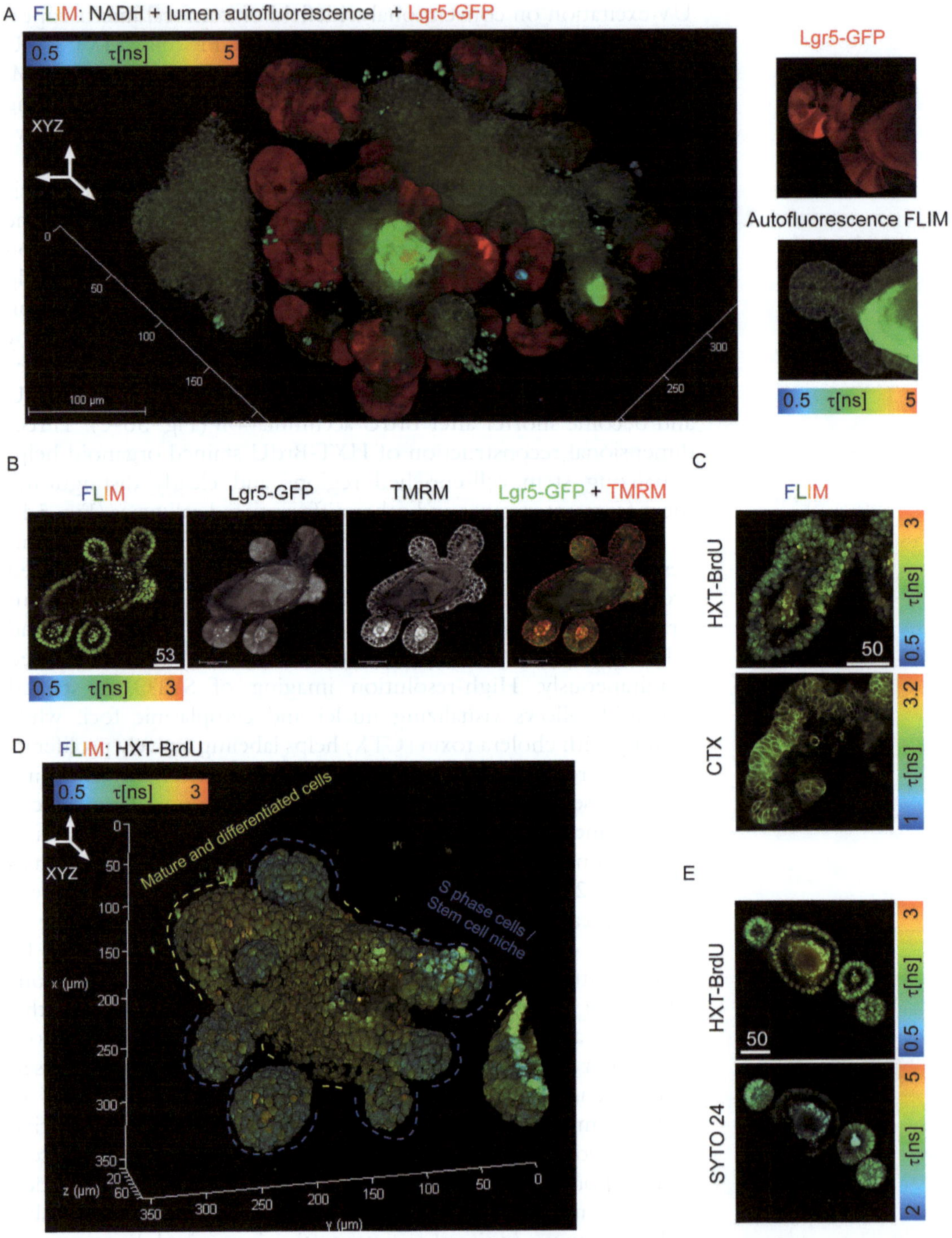

Fig. 5 Examples of multiparametric imaging using two-photon excited FLIM- microscope. (**a**) 3D reconstruction of the organoid autofluorescence (exc. 720 nm) FLIM images together with Lgr5-GFP (exc. 920 nm, shown in red). Right panel shows examples of optical sections. The lifetime scale of NADH FLIM is extended to include the luminal signals. (**b**) Two-photon imaging of Hoechst 33342-stained organoid (excited at 700 nm, FLIM image) costained with TMRM (5 nM, exc. 1040 nm). Lgr5-GFP signals are shown for reference. (**c**) Dual

live multicellular constructs. The overall goal of the imaging of 3D tissue models is the analysis and segmentation of 3D optical reconstructed images quickly and in live state—this can be achieved using two-photon FLIM.

4.4 Analysis of Intestinal Organoids Oxygenation by PLIM

While not frequently considered and experimentally assessed, the steady-state oxygenation and the oxygen consumption rate represent the direct indicators of the cell microenvironment and mitochondrial function, respectively [46]. We recently reported the heterogeneity of the organoid oxygenation using the primary organoids isolated from mouse intestinal crypts and grown in the absence of Wnt [26]. The observed heterogeneity is very similar to Lgr5-GFP organoids [47], grown either in ENRVC and ENR media. Growing in ENRVC can result in less profound heterogeneity, but the organoids are largely undeveloped in these conditions. Figure 7 shows examples of GFP-enriched and deprived regions and measured oxygenation. The presence of Lgr5-GFP-expressing cells facilitates more clear division and segmentation of the organoid regions and simplifies the analysis. Alternatively, stem cell niche can be visualized by the labeling with HXT-BrdU and/or CTX staining (Figs. 4, 5, 6, and 7).

Since the observed oxygenation is a result of metabolic function (balance between OxPhos, glycolysis, and other energy-production pathways), it can be affected by the addition of such drugs as mitochondrial activators, inhibitors or by changing medium composition. Thus, if the glucose content in the imaging medium is decreased from 10 to 0.5 mM (1 h preincubation), we can expect to see increased OxPhos and resulting decrease of organoid oxygenation [49]. However, the heterogeneity of individual organoids can mask these differences (Fig. 7c, d). The GFP-positive cells display slightly increased O_2 compared to differentiated cells (Fig. 7c) and on average the slightly lower O_2 in the cells in low glucose conditions (Fig. 7d). Indeed, use of different drugs, larger number of the experimental points, more specific labeling of cell types and more elaborated statistical analysis can reveal more striking differences in the oxygenation or oxygen consumption rate in the live organoids. The observed optical "sections" on Fig. 7 also do not exhibit high degree of "confocality" (typical size of pinhole at DCS-120 PLIM system is in range of

Fig. 5 (continued) FLIM imaging of HXT-BrdU and CTX-Alexa 488 (exc. at 985 nm) in organoids. (**d**) 3D reconstruction of two-photon FLIM-analyzed organoid, stained with HXT-BrdU FLIM method. The regions having shorter lifetimes indicate the presence of stem cells, while longer lifetime regions correspond to quiescent or mature cells (indicated). (**e**) Example of combined use of HXT-BrdU and SYTO 24 under two-photon excitation. The data were acquired using Dive Falcon SP8 FLIM system (Leica microsystems). Scale bar is 50 μm

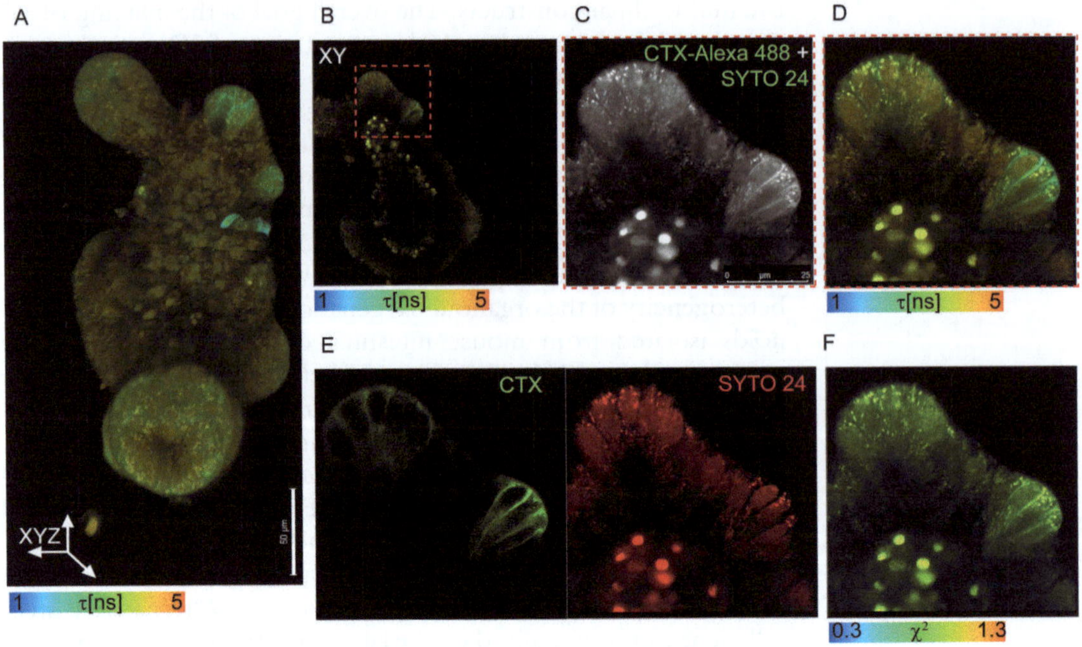

Fig. 6 FLIM-based separation of the SYTO 24 and CTX-Alexa Fluor 488 probes in live intestinal organoids. Organoids were stained with SYTO 24 (0.5 μM) and CTX (5 μg/mL) and measured using 488 nm exc. (510–550 nm em.). (**a**) 3D reconstruction of the organoid in FLIM. (**b**) FLIM image of the single optical section in organoid, with indicated selection for further analysis. (**c**) Intensity image for the selected section. (**d**) FLIM image for the selected section. (**e**) Intensity images obtained after lifetime-based separation ("pattern fit"), corresponding to distribution of probes. (**f**) Precision of calculations used for lifetime-based separation, expressed as colormap of $\chi 2$ (chi-square test). The data were acquired using confocal white light laser-based Falcon SP8 FLIM system (Leica microsystems)

0.25–5 mm) and can represent very thick optical sections (>5–10 μm): in this case, the measured oxygenation in the selected *XY* region of interest can be a result of underlying cell layers, lumen or other optical artifacts. Indeed, development of efficient 3D reconstruction of PLIM and FLIM data and the ability to perform annotation and segmentation of the images in 3D will be the next milestone in the live cell multiparametric analysis of organoids and related tissue-engineered constructs.

5 Notes

1. Upright design of the microscope is convenient for organoid imaging, especially with the use of long-distance focus objectives (working distance must be at least 0.3 mm). However, upright microscopes rarely have successfully implemented climate (and especially O_2) control. On the other hand, inverted microscope would need high-spec 40–63× immersion

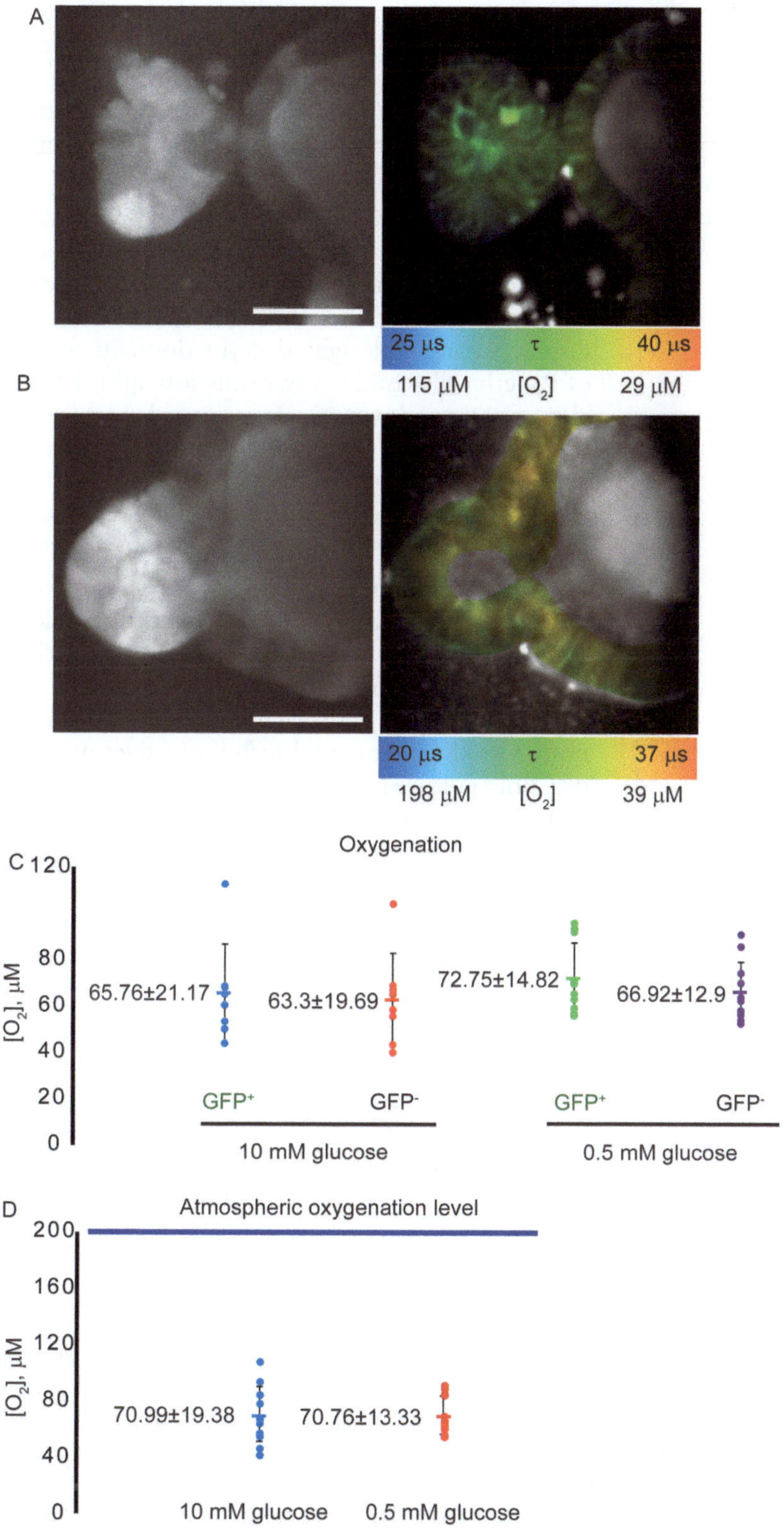

Fig. 7 Oxygenation of Lgr5-GFP organoids grown in ENRVC medium. (**a**) Representative Lgr5-GFP gray scale intensity and color O_2 PLIM images for the optical section of organoids preincubated (1 h) in 10 mM glucose-

objective with long working distance. We recommend 63× (NA 1 or higher) objective to achieve good spatial resolution and the coverage of imaged area, which can be significantly extended by the mosaic imaging. This high-magnification objective also helps avoiding too bright regions with luminal autofluorescence.

2. In our experience, ENRVC- and ENR-grown organoids display distinct phenotypes: 5 days after passaging ENRVC organoids consist of mostly small round organoids with minimal number of visible Paneth cells. In contrast, ENR organoids display strongly elongated crypt domains with high number of Paneth cells visible in transmission light. ENRVC organoids also can slow down if not passaged every 4–5 days. These findings are in an agreement with the data reported by Han and coworkers [54]. Simple changing to cultivation in ENR media leads to restoration of the phenotype and growth rate, giving flexibility to manipulate organoid growth conditions.
3. Pipetting technique and number of passes through the tip can drastically affect the size of disrupted organoid units. Ideally, mechanical disruption (at least 15–20 passes through the tip) of intestinal organoids allows separation of organoids into individual crypts, with each potentially forming a new organoid. Good experimental technique gives more homogeneous organoid culture in terms of size, which organoids would reach 3–4 days after passaging.
4. After centrifugation, the Matrigel with organoid agglomerates will be visible on the bottom of the tube. The Matrigel layer has to be removed as completely as possible without disturbing the organoids. If the organoid agglomerates do not separate clearly from the Matrigel layer, washing for another time with 10 mL ice cold AdDF+++ can resolve the issue.
5. Frequently, the staining concentration needs optimization due to different detector sensitivity, laser power, microscopy settings or the "new" experimental parameters, such as changing organoid metabolism and physiology or need in different staining medium. Please ensure that if the growth/culture medium

Fig. 7 (continued) containing medium. (**b**) Lgr5-GFP and O_2-PLIM images for the organoid preincubated (1 h) in 0.5 mM glucose-containing medium. Scale bar is 50 μm. (**c**) Calculated average oxygenation for GFP^+ and GFP^- regions in intestinal organoids treated different glucose concentrations. Nine separate ROIs were chosen for each group based on the GFP intensity. No statistical difference was observed between groups using *t*-test ($p = 0.05$). (**d**) Calculated oxygenation for two groups of organoids irrespectively to the GFP expression. $n = 8$ (10 mM glucose group), $n = 11$ (0.5 mM glucose group). No statistical difference was observed between groups using *t*-test ($p = 0.05$)

is changed prior the imaging and that organoids have sufficient time to adapt to it, typically 2 or more hours in the incubator.

6. In case of HXT-BrdU method, staining with Hoechst 33342 (HXT) can be done during BrdU loading (up to 3 h) or after BrdU uptake (e.g., with long-term BrdU loading). Since the HXT fluorescence lifetime can depend on staining concentration and loading time it is always a good idea to have $BrdU^-$ (negative control) to know the typical HXT fluorescence lifetime on the used system. However, due to the organoids heterogeneity the nondividing differentiated cells will be always present in the culture and can be used as internal "control" for unquenched lifetime. The synchronization of dividing cells in organoids is also possible but such calibration will not be as precise as in 2D culture and will surely affect the organoid viability. For our S phase synchronization experiments we recommend using 0.125 μg/mL of aphidicolin (*see* Subheading 4).
7. For multiparametric imaging the probes can have different excitation or/and emission spectra and can be separated spectrally. In some situations, time domain-based separation is also possible.
8. In some situations, it is important to keep the same media in which loading procedure was done. In this case instead of imaging medium the medium of choice should be used during the imaging. For our studies of glucose effect on intestinal organoids oxygenation we used standard imaging medium supplied with different amounts of D-glucose for loading as well as imaging procedure.
9. For multiplexing and further imaging data processing (application of ROI mask or data export with following analysis in Excel or other programs) all fluorescence/phosphorescence intensity images of probes (or fluorescent protein labeling) used for costaining should be collected with the same spatial resolution (e.g., 512×512) and pixel size.
10. As soon as fit settings for the defined probe fluorescence or phosphorescence are optimized keep them for the analysis of all microscopy data.
11. Several ROIs can be applied/measured simultaneously. Using of appropriate ROI mask allows easy initial sorting of the imaging data, for example, the analysis of oxygenation of stem cell (Lgr5-GFP^+ cells) in comparison to non-stem cells (Lgr5-GFP^- cells) (*see* Subheading 4) or comparison the oxygenation for dividing and nondividing cells (HXT-BrdU FLIM + Pt-Glc-based O_2 PLIM).
12. After calculation of emission lifetime values, SPCImage software shows the false-color lifetime distribution map and the distribution histogram for the whole image frame or the

chosen ROI (Fig. 1). It also produces numerical datasets for lifetime (color-coded values), distribution histogram and intensities (photons counting) on pixel-by-pixel basis, which can be exported and used for analysis in Excel software. However, exporting data for selected ROI is not always straightforward and has to be carefully controlled. The Subheading 3.4, **steps 5–10** of the suggested data processing protocol allow avoiding this limitation by using only numerical data sets.

13. To do this in Excel software, choose the whole *XY* dataset and apply specific rule ("Manage Rule" function) in "Conditional Formatting" function. This allows giving each numerical value (this can be a specified range of values or simple sorting from the lowest to the highest values of the data set) the false color to reconstruct the "image" in Excel table format.
14. In order to study the correlation between the intensity and the fluorescence lifetime of the first probe chosen for the selection of the ROI and the second probe intensity or lifetime it is necessary to produce an intensity/lifetime data set for this probe. For example, you can correlate the intensity of GFP^+ areas with the oxygenation values (based on O_2 PLIM measurements), fluorescence lifetime of HXT-BrdU-loaded organoids with oxygenation or fluorescence lifetime of TMRM (mitochondrial membrane potential).
15. Cell oxygenation data are produced by applying the Stern–Volmer calibration equation for this probe and intestinal organoids ("two-site" model, *see* refs. 26, 42) to the Pt-Glc phosphorescence lifetime data:

$$[O_2, \mu M] = (0.82587/(-0.17413 + \tau/54.8711) - 1)/0.01683),$$

where τ (phosphorescence lifetime) is in μs.

16. If the data distribution is normal then the parametric statistical test (e.g., Student's *t*-test) can be used for further analysis. Otherwise, nonparametric statistical tests (e.g., Mann–Whitney U-test or Fisher's exact test, depending on the data set size) have to be applied. It is also very important to have sufficient number of replicates in groups.

Acknowledgments

This work was supported by the Science Foundation Ireland grant 12/RC/2276 (D.B.P., I.A.O.) and by the Agilent University Research Program (ACT-UR) No. 4225 (R.I.D.). We thank Dr. H. Glauner, Dr. L. Alvarez and team at the Leica training

Centre (Mannheim, Germany) for support with demonstration of Leica SP8 Falcon systems.

References

1. Kretzschmar K, Clevers H (2016) Organoids: modeling development and the stem cell niche in a dish. Dev Cell 38(6):590–600
2. Dutta D, Clevers H (2017) Organoid culture systems to study host–pathogen interactions. Curr Opin Immunol 48:15–22. https://doi.org/10.1016/j.coi.2017.07.012
3. Fatehullah A, Tan SH, Barker N (2016) Organoids as an in vitro model of human development and disease. Nat Cell Biol 18:246. https://doi.org/10.1038/ncb3312
4. Shamir ER, Ewald AJ (2014) Three-dimensional organotypic culture: experimental models of mammalian biology and disease. Nat Rev Mol Cell Biol 15:647. https://doi.org/10.1038/nrm3873
5. Clevers H (2016) Modeling development and disease with organoids. Cell 165 (7):1586–1597. https://doi.org/10.1016/j.cell.2016.05.082
6. Fujii M, Matano M, Nanki K, Sato T (2015) Efficient genetic engineering of human intestinal organoids using electroporation. Nat Protoc 10(10):1474
7. Rodríguez-Colman MJ, Schewe M, Meerlo M, Stigter E, Gerrits J, Pras-Raves M, Sacchetti A, Hornsveld M, Oost KC, Snippert HJ (2017) Interplay between metabolic identities in the intestinal crypt supports stem cell function. Nature 543(7645):424
8. Schell JC, Wisidagama DR, Bensard C, Zhao H, Wei P, Tanner J, Flores A, Mohlman J, Sorensen LK, Earl CS, Olson KA, Miao R, Waller TC, Delker D, Kanth P, Jiang L, DeBerardinis RJ, Bronner MP, Li DY, Cox JE, Christofk HR, Lowry WE, Thummel CS, Rutter J (2017) Control of intestinal stem cell function and proliferation by mitochondrial pyruvate metabolism. Nat Cell Biol 19:1027. https://doi.org/10.1038/ncb3593. https://www.nature.com/articles/ncb3593 - supplementary-information
9. Yilmaz ÖH, Katajisto P, Lamming DW, Gültekin Y, Bauer-Rowe KE, Sengupta S, Birsoy K, Dursun A, Yilmaz VO, Selig M, Nielsen GP, Mino-Kenudson M, Zukerberg LR, Bhan AK, Deshpande V, Sabatini DM (2012) mTORC1 in the Paneth cell niche couples intestinal stem-cell function to calorie intake. Nature 486:490. https://doi.org/10.1038/nature11163. https://www.nature.com/articles/nature11163 - supplementary-information
10. Nguyen PD, Currie PD (2018) In vivo imaging: shining a light on stem cells in the living animal. Development 145(7):dev150441
11. Teodori L, Crupi A, Costa A, Diaspro A, Melzer S, Tarnok A (2017) Three-dimensional imaging technologies: a priority for the advancement of tissue engineering and a challenge for the imaging community. J Biophotonics 10(1):24–45
12. Conway JR, Warren SC, Timpson P (2017) Context-dependent intravital imaging of therapeutic response using intramolecular FRET biosensors. Methods 128:78–94
13. Sarder P, Maji D, Achilefu S (2015) Molecular probes for fluorescence lifetime imaging. Bioconjug Chem 26(6):963–974
14. Dmitriev RI (2017) Multi-parametric live cell microscopy of 3D tissue models, vol 1035. Springer, Cham
15. Kalinina S, Breymayer J, Schäfer P, Calzia E, Shcheslavskiy V, Becker W, Rück A (2016) Correlative NAD (P) H-FLIM and oxygen sensing-PLIM for metabolic mapping. J Biophotonics 9(8):800–811
16. Lukina M, Orlova A, Shirmanova M, Shirokov D, Pavlikov A, Neubauer A, Studier H, Becker W, Zagaynova E, Yoshihara T (2017) Interrogation of metabolic and oxygen states of tumors with fiber-based luminescence lifetime spectroscopy. Opt Lett 42 (4):731–734
17. Evers M, Salma N, Osseiran S, Casper M, Birngruber R, Evans CL, Manstein D (2018) Enhanced quantification of metabolic activity for individual adipocytes by label-free FLIM. Sci Rep 8(1):8757. https://doi.org/10.1038/s41598-018-27093-x
18. O'Donnell N, Dmitriev RI (2017) Three-dimensional tissue models and available probes for multi-parametric live cell microscopy: a brief overview. In: Multi-parametric live cell microscopy of 3D tissue models. Springer, Cham, pp 49–67
19. Baggaley E, Botchway SW, Haycock JW, Morris H, Sazanovich IV, Williams JG, Weinstein JA (2014) Long-lived metal complexes open up microsecond lifetime imaging microscopy under multiphoton excitation: from

FLIM to PLIM and beyond. Chem Sci 5 (3):879–886

20. Jenkins J, Dmitriev RI, Papkovsky DB (2015) Imaging cell and tissue O2 by TCSPC-PLIM. In: Advanced time-correlated single photon counting applications. Springer, Cham, pp 225–247
21. Aigner D, Dmitriev RI, Borisov S, Papkovsky DB, Klimant I (2014) pH-sensitive perylene bisimide probes for live cell fluorescence lifetime imaging. J Mater Chem B 2 (39):6792–6801
22. Dmitriev RI, Borisov SM, Düssmann H, Sun S, Müller BJ, Prehn J, Baklaushev VP, Klimant I, Papkovsky DB (2015) Versatile conjugated polymer nanoparticles for high-resolution O2 imaging in cells and 3D tissue models. ACS Nano 9(5):5275–5288
23. Zhdanov AV, Okkelman IA, Collins FW, Melgar S, Papkovsky DB (2015) A novel effect of DMOG on cell metabolism: direct inhibition of mitochondrial function precedes HIF target gene expression. Biochim Biophys Acta Bioenergetics 1847(10):1254–1266
24. Meleshina AV, Dudenkova VV, Shirmanova MV, Shcheslavskiy VI, Becker W, Bystrova AS, Cherkasova EI, Zagaynova EV (2016) Probing metabolic states of differentiating stem cells using two-photon FLIM. Sci Rep 6:21853
25. Okkelman IA, Dmitriev RI, Foley T, Papkovsky DB (2016) Use of fluorescence lifetime imaging microscopy (FLIM) as a timer of cell cycle S phase. PLoS One 11(12):e0167385
26. Okkelman IA, Foley T, Papkovsky DB, Dmitriev RI (2017) Live cell imaging of mouse intestinal organoids reveals heterogeneity in their oxygenation. Biomaterials 146:86–96
27. Jenkins J, Borisov SM, Papkovsky DB, Dmitriev RI (2016) Sulforhodamine nanothermometer for multiparametric fluorescence lifetime imaging microscopy. Anal Chem 88 (21):10566–10572
28. Dmitriev RI, Zhdanov AV, Nolan YM, Papkovsky DB (2013) Imaging of neurosphere oxygenation with phosphorescent probes. Biomaterials 34(37):9307–9317
29. Jenkins J, Dmitriev RI, Morten K, McDermott KW, Papkovsky DB (2015) Oxygen-sensing scaffolds for 3-dimensional cell and tissue culture. Acta Biomater 16:126–135. https://doi.org/10.1016/j.actbio.2015.01.032
30. O'Donnell N, Okkelman IA, Timashev P, Gromovykh TI, Papkovsky DB, Dmitriev RI (2018) Cellulose-based scaffolds for fluorescence lifetime imaging-assisted tissue engineering. Acta Biomater 80:85–96
31. Sherrard A, Bishop P, Panagi M, Villagomez MB, Alibhai D, Kaidi A (2018) Streamlined histone-based fluorescence lifetime imaging microscopy (FLIM) for studying chromatin organisation. Biology Open: bio. 031476
32. Paredes JM, Giron MD, Ruedas-Rama MJ, Orte A, Crovetto L, Talavera EM, Salto R, Alvarez-Pez JM (2013) Real-time phosphate sensing in living cells using fluorescence lifetime imaging microscopy (FLIM). J Phys Chem B 117(27):8143–8149
33. Klarenbeek JB, Goedhart J, Hink MA, Gadella TW, Jalink K (2011) A mTurquoise-based cAMP sensor for both FLIM and ratiometric read-out has improved dynamic range. PLoS One 6(4):e19170
34. Long Y, Stahl Y, Weidtkamp-Peters S, Postma M, Zhou W, Goedhart J, Sánchez-Pérez M-I, Gadella TW, Simon R, Scheres B (2017) In vivo FRET–FLIM reveals cell-type-specific protein interactions in Arabidopsis roots. Nature 548(7665):97
35. Okkelman IA, Papkovsky DB, Dmitriev RI (2019) Estimation of the Mitochondrial Membrane Potential Using Fluorescence Lifetime Imaging Microscopy. Cytometry Part A 0 (0). doi:10.1002/cyto.a.23886
36. Schützhold V, Fandrey J, Prost-Fingerle K (2018) Fluorescence lifetime imaging microscopy (FLIM) as a tool to investigate hypoxia-induced protein-protein interaction in living cells. In: Hypoxia. Springer, Cham, pp 45–53
37. Anzilotti C, Swan DJ, Boisson B, Deobagkar-Lele M, Oliveira C, Chabosseau P, Engelhardt KR, Xu X, Chen R, Alvarez L, Berlinguer-Palmini R, Bull KR, Cawthorne E, Cribbs AP, Crockford TL, Dang TS, Fearn A, Fenech EJ, de Jong SJ, Lagerholm BC, Ma CS, Sims D, van den Berg B, Xu Y, Cant AJ, Kleiner G, Leahy TR, de la Morena MT, Puck JM, Shapiro RS, van der Burg M, Chapman JR, Christianson JC, Davies B, McGrath JA, Przyborski S, Santibanez Koref M, Tangye SG, Werner A, Rutter GA, Padilla-Parra S, Casanova J-L, Cornall RJ, Conley ME, Hambleton S (2019) An essential role for the Zn^{2+} transporter ZIP7 in B cell development. Nat Immunol. https://doi.org/10.1038/s41590-018-0295-8
38. Koren K, Mosshammer M, Scholz VV, Borisov SM, Holst G, Kühl M (2019) Luminescence lifetime imaging of chemical sensors – a comparison between time-domain and frequency-domain based camera systems. Anal Chem. https://doi.org/10.1021/acs.analchem.8b05869
39. Dalfen I, Dmitriev RI, Holst G, Klimant I, Borisov SM (2019) Background-free fluorescence decay time sensing and imaging of pH with highly photostable diazaoxotriangulenium dyes. Anal Chem 91(1):808–816

40. Chen H, Holst G, Gratton E (2015) Modulated CMOS camera for fluorescence lifetime microscopy. Microsc Res Tech 78 (12):1075–1081
41. Erkkilä MT, Bauer B, Hecker-Denschlag N, Madera Medina MJ, Leitgeb RA, Unterhuber A, Gesperger J, Roetzer T, Hauger C, Drexler W (2019) Widefield fluorescence lifetime imaging of protoporphyrin IX for fluorescence-guided neurosurgery: an ex vivo feasibility study. J Biophotonics 12(6): e201800378
42. Okkelman IA, Foley T, Papkovsky DB, Dmitriev RI (2017) Multi-parametric imaging of hypoxia and cell cycle in intestinal organoid culture. In: Dmitriev R (ed) Multi-parametric live cell microscopy of 3D tissue models, Advances in experimental medicine and biology, vol 1035. Springer, Cham, pp 85–103. https://doi.org/10.1007/978-3-319-67358-5_6
43. Brand MD, Nicholls DG (2011) Assessing mitochondrial dysfunction in cells. Biochem J 435(2):297–312
44. Diaz G, Melis M, Batetta B, Angius F, Falchi AM (2008) Hydrophobic characterization of intracellular lipids in situ by Nile Red red/yellow emission ratio. Micron 39(7):819–824
45. Dmitriev RI, Kondrashina AV, Koren K, Klimant I, Zhdanov AV, Pakan JM, McDermott KW, Papkovsky DB (2014) Small molecule phosphorescent probes for O2 imaging in 3D tissue models. Biomater Sci 2(6):853–866
46. Papkovsky DB, Dmitriev RI (2018) Imaging of oxygen and hypoxia in cell and tissue samples. Cell Mol Life Sci 75(16):2963–2980
47. Barker N, Van Es JH, Kuipers J, Kujala P, Van Den Born M, Cozijnsen M, Haegebarth A, Korving J, Begthel H, Peters PJ (2007) Identification of stem cells in small intestine and colon by marker gene Lgr5. Nature 449 (7165):1003
48. Sato T, Vries RG, Snippert HJ, Van De Wetering M, Barker N, Stange DE, Van Es JH, Abo A, Kujala P, Peters PJ (2009) Single Lgr5 stem cells build crypt–villus structures in vitro without a mesenchymal niche. Nature 459(7244):262
49. Okkelman IA, Neto N, Papkovsky DB, Monaghan MG, Dmitriev RI (2020) A deeper understanding of intestinal organoid metabolism revealed by combining fluorescence lifetime imaging microscopy (FLIM) and extracellular flux analyses. Redox Biology 30:101420. https://doi.org/10.1016/j.redox.2019.101420
50. Dmitriev RI, Borisov SM, Jenkins J, Papkovsky DB (2015) Multi-parametric imaging of tumor spheroids with ultra-bright and tunable nanoparticle O2 probes. In: Imaging, manipulation, and analysis of biomolecules, cells, and tissues XIII, 2015. International Society for Optics and Photonics, p 932806
51. Kuimova MK (2012) Mapping viscosity in cells using molecular rotors. Phys Chem Chem Phys 14(37):12671–12686
52. Müller BJ, Zhdanov AV, Borisov SM, Foley T, Okkelman IA, Tsytsarev V, Tang Q, Erzurumlu RS, Chen Y, Zhang H (2018) Nanoparticle-based fluoroionophore for analysis of potassium ion dynamics in 3D tissue models and in vivo. Adv Funct Mater 28(9):1704598
53. Arena ET, Rueden CT, Hiner MC, Wang S, Yuan M, Eliceiri KW (2017) Quantitating the cell: turning images into numbers with ImageJ. Wiley Interdiscip Rev Dev Biol 6(2):e260
54. Han S-H, Shim S, Kim M-J, Shin H-Y, Jang W-S, Lee S-J, Jin Y-W, Lee S-S, Lee SB, Park S (2017) Long-term culture-induced phenotypic difference and efficient cryopreservation of small intestinal organoids by treatment timing of Rho kinase inhibitor. World J Gastroenterol 23(6):964

Chapter 6

Generation and Quantitative Imaging of Enteroid Monolayers

Laura E. Sanman, Ina W. Chen, Jake M. Bieber, Curtis A. Thorne, Lani F. Wu, and Steven J. Altschuler

Abstract

The intestinal epithelium is a single layer of cells that plays a critical role in digestion, absorbs nutrients from food, and coordinates the delicate interplay between microbes in the gut lumen and the immune system. Epithelial homeostasis is crucial for maintaining health; disruption of homeostasis results in disorders including inflammatory bowel disease and cancer. The advent of 3D intestinal epithelial organoids has greatly advanced our understanding of the molecular underpinnings of epithelial homeostasis and disease. Recently, we developed an enteroid monolayer (2D) culture system that recapitulates important features of 3D organoids and the in vivo intestinal epithelium such as tissue renewal, representation of diverse epithelial cell types, self-organization, and apical–basolateral polarization. Enteroid monolayers are cultured in microtiter plates, enabling high-throughput experiments. Furthermore, their 2D nature makes it easier to distinguish individual cells by fluorescent microscopy, enabling quantitative analysis of single cell behaviors within the epithelial tissue.

Here we describe experimental methods for generating enteroid monolayers and computational methods for analyzing immunofluorescence images of enteroid monolayers. We outline experimental methods for generating enteroid monolayers from freshly isolated intestinal crypts, frozen intestinal crypts, and 3D organoids. Fresh crypts are easily obtained from murine or human intestinal samples, and the ability to derive enteroid monolayers from both frozen crypts and 3D organoids enables genetic modification and/or biobanking of patient samples for future studies. We outline computational methods for identifying distinct epithelial cell types (goblet, stem, EdU+) in immunofluorescence images of enteroid monolayers and, importantly, individual nuclei, enabling truly single cell measurements of epithelial cell behaviors to be made. Taken together, these methods will enable detailed studies of epithelial homeostasis and intestinal disease.

Key words Intestinal organoids, Enteroid monolayer, Intestinal stem cells, Quantitative image analysis, Confocal microscopy

Laura E. Sanman and Ina W. Chen contributed equally.

Paloma Ordóñez-Morán (ed.), *Intestinal Stem Cells: Methods and Protocols*, Methods in Molecular Biology, vol. 2171,
https://doi.org/10.1007/978-1-0716-0747-3_6, © Springer Science+Business Media, LLC, part of Springer Nature 2020

1 Introduction

The development of three-dimensional (3D) organoids has greatly advanced our ability to study the intestinal epithelium in a controlled manner in vitro [1]. Indeed, studies in 3D organoids over the past 10 years have revealed a multitude of insights into mechanisms of homeostatic maintenance and intestinal epithelial dysfunction in disease [2, 3]. In order to study the intestinal epithelium at the single-cell level in high-throughput, we recently developed an enteroid monolayer culture system which recapitulates key features of 3D organoids and the in vivo intestinal epithelium. Specifically, enteroid monolayers are composed of the major intestinal epithelial cell types (stem, transit-amplifying, goblet, Paneth, tuft, and enteroendocrine) and they also renew, self-organize, and polarize with apical face exposed. Enteroid monolayers are readily cultured for up to 2 weeks and maintain both distinct crypt-like regions composed of stem cells and villus-like regions composed of differentiated cells throughout the course of treatment. Enteroid monolayers are two-dimensional (2D) and can be cultured in 96-well imaging plates, facilitating high-throughput investigation of tissue-level and single-cell behaviors [4].

Here, we first outline three methods (Fig. 1) for deriving enteroid monolayer cultures: from freshly isolated and frozen murine small intestinal crypts (Subheadings 3.1 and 3.2) and from 3D organoids (Subheading 3.3). Fresh intestinal crypts are plentiful and highly reproducible when derived from laboratory mouse strains, enabling the quantitative comparison of hundreds of different experimental conditions with crypts from a single mouse. In contrast, freezing crypts or propagating crypts as 3D organoids prior to generating enteroid monolayers enables banking and/or genetically modifying precious samples from patients or genetically engineered mice. We also describe an immunofluorescence protocol optimized for enteroid monolayers (Subheading 3.4).

Second, we outline computational methods for segmenting (identifying) individual nuclei, EdU+ nuclei, Lgr5+ stem cells, and Muc2+ goblet cells in immunofluorescence images of enteroid monolayers (Subheading 3.5) (*see* **Note 1**). The synopsis of each method is accompanied by step-by-step instructions and links to a GitHub repository containing example code and sample images that should enable others to implement segmentation methods in their own research.

Taken together, the methods discussed below provide a detailed experimental and computational framework with which to generate and analyze enteroid monolayer cultures. These methods provide the opportunity to (1) disentangle the contributions of morphogens vs. 3D tissue architecture to tissue homeostasis,

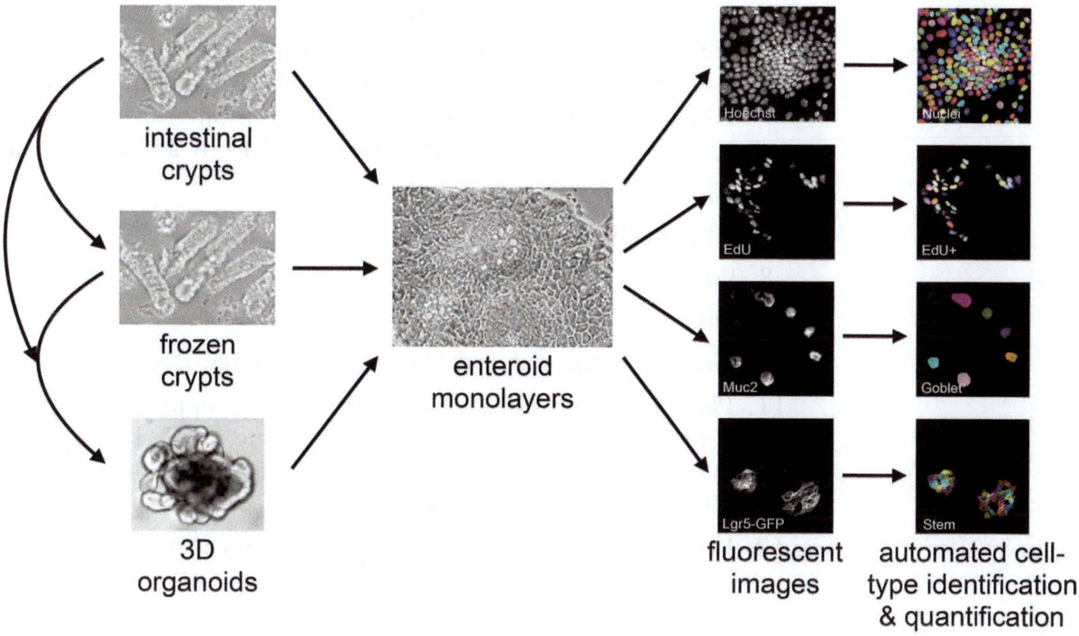

Fig. 1 Workflow for generating and analyzing enteroid monolayers. Enteroid monolayers can be derived from fresh or frozen intestinal crypts and from 3D organoids (Subheadings 3.1–3.3). Immunofluorescence assays (Subheading 3.4) are used to identify the identity of individual cell types in enteroid monolayers. Numbers of each cell type in the resulting fluorescent images are quantified using the methods described in Subheading 3.5

(2) easily access the luminal face of the epithelium, enabling studies of topics including host–microbiome interactions and drug transporters, and (3) study single-cell identity and signaling in the tissue context in high throughput. With these advantages over more traditional 3D organoid models, enteroid monolayer cultures enable investigations to further our understanding of epithelial homeostasis and dysregulation in disease.

2 Materials

2.1 Culturing Enteroid Monolayers from Freshly Isolated Murine Intestinal Crypts

1. Phosphate buffered saline (PBS): 137 mM NaCl, 2.7 mM KCl, 10 mM Na_2HPO_4, 1.8 mM KH_2PO_4 (no Ca^{2+}/Mg^{2+}).
2. Intestine washing buffer: PBS supplemented with 100 U/mL penicillin/100 μg/mL streptomycin (*see* **Note 2**).
3. Intestine harvest buffer: PBS supplemented with 1 mM EDTA, 2 mM dithiothreitol (DTT), 100 U/mL penicillin/100 μg/mL streptomycin, and 10 μM Y-27632.
4. Crypt dissociation buffer: PBS supplemented with 3 mM EDTA, 2 mM DTT, 100 U/mL penicillin/100 μg/mL streptomycin, and 10 μM Y-27632 (*see* **Note 3**).

5. Organoid basal medium: Advanced DMEM/F12 with nonessential amino acids and sodium pyruvate and without L-glutamine, supplemented with 2 mM GlutaMAX, 10 mM HEPES, 100 U/mL penicillin/100 μg/mL streptomycin, 1 mM *N*-acetylcysteine, 1× N-2 supplement, and 1× B-27 supplement.
6. 100 μm cell strainer.
7. 70 μm cell strainer.
8. Forceps.
9. Dissecting scissors.
10. Growth factor–reduced Matrigel, phenol red-free.
11. Plating medium: organoid basal medium supplemented with 3 μM CHIR-99021, 50 ng/mL murine EGF, 100 ng/mL murine Noggin, 500 ng/mL murine R-spondin-1, and 10 μM Y-27632 (*see* **Note 4**).
12. Long-term culture medium: organoid basal medium supplemented with 50 ng/mL EGF, 100 ng/mL Noggin, and 500 ng/mL R-spondin-1 (*see* **Note 5**).
13. Glass microscope slides (e.g., 75 mm × 25 mm).
14. Brightfield or phase contrast inverted microscope.
15. 96-well clear polystyrene bottom imaging plates (*see* **Note 6**).
16. Wildtype or Lgr5-eGFP-DTR mice (*see* **Note 7**).

2.2 Culturing Enteroid Monolayers from Frozen Crypts

1. Matrigel, organoid basal medium, plating medium, and long-term culture medium (*see* Subheading 2.1).
2. Freezing medium: DMEM supplemented with 10% FBS and 10% dimethyl sulfoxide (DMSO).

2.3 Culturing Enteroid Monolayers from 3D Organoids

1. Matrigel, organoid basal medium, plating medium, and long-term culture medium (*see* Subheading 2.1).
2. TrypLE Express.
3. Fire-polished Pasteur pipettes.
4. Hemocytometer or automated cell counter.

2.4 Performing Immunofluorescence on Enteroid Monolayers

1. Fixation buffer: 4% PFA in PBS (*see* **Note 8**).
2. Permeabilization buffer: 0.3% Triton-X-100 in PBS.
3. Blocking buffer: 3% bovine serum albumin (BSA) in PBS.
4. Antibody buffer: 1% BSA, 0.3% Triton-X-100 in PBS.
5. Washing buffer: 0.1% Tween-20 in PBS (PBS-T).
6. Click reaction buffer: 1 mM $CuSO_4$, 5 μM fluorophore-azide (e.g., sulfo-Cy5-azide; LumiProbe), and 100 mM sodium ascorbate (*see* **Note 9**) in PBS.

7. Nuclear staining buffer: 5 μg/mL Hoechst 33342 in antibody buffer or PBS.
8. Primary antibody of interest.
9. Dye-conjugated species-specific secondary antibody (e.g., Alexa-conjugated antibodies).

2.5 Equipment

1. Automated point-scanning confocal or epifluorescent microscope.

2.6 Image Analysis Software

2.6.1 Software List

1. Miniconda3 (https://docs.conda.io/en/latest/miniconda.html).
2. Python 3.7.2 (https://www.python.org/downloads/, see below for installation).
3. Github repository of custom Python code (https://github.com/AltschulerWu-Lab/EnteroidSeg).

2.6.2 Installation

1. Install Miniconda3 for Python 3.7, which can be downloaded at the link above.
2. Download the Github repository linked above. The repository can be downloaded in following ways: clone the repository using the Github desktop app (under **Clone Repository**, enter the link to the repository) or clone the repository using the command **git clone** followed by link to the repository.
3. Navigate to the **EnteroidSeg** directory in the downloaded repository. Install Python 3.7.2 and associated Python packages required for running the code using the following command.

   conda env create --file=environment.yaml
4. Activate the conda environment with the following command.

   source activate enteroidseg

3 Methods

3.1 Culturing Enteroid Monolayers from Freshly Isolated Intestinal Crypts

1. Prepare intestine washing buffer (>30 mL per intestine), intestine harvest buffer (10 mL per intestine), and crypt dissociation buffer (10 mL per intestine) and keep on ice.
2. Prepare organoid basal medium (500 mL). Store a 50 mL aliquot at 4 °C and warm the remainder to 37 °C.
3. Thaw 10 mL vial of Matrigel on ice and make 1 mL aliquots. Prior to each experiment, thaw a Matrigel aliquot on ice. Avoid freeze-thaw cycles.
4. Thaw EGF, Noggin, and R-spondin-1 aliquots on ice.
5. Bring CHIR-99021 and Y-27632 aliquots to room temperature.

6. After organoid basal medium is warmed and EGF, Noggin, R-spondin-1, CHIR-99021, and Y-27632 are thawed, make plating medium (20 mL per intestine) and long-term culture medium (20 mL per intestine) (*see* **Note 10**).
7. Isolate small intestine from a male or female mouse between 6 and 12 weeks of age. We typically harvest the jejunum because it is the largest section of the mouse small intestine. Filet open longitudinally and wash in intestine washing buffer until fecal matter and debris are cleared.
8. Transfer washed intestine to intestine harvest buffer in 50 mL conical tube and incubate for 15 min on ice to loosen mucus and debris.
9. Shake intestine in intestine harvest buffer in 50 mL conical tube for 1 min. Use forceps to transfer intestine to crypt dissociation buffer in 50 mL conical tube and incubate for 1 h on ice with gentle rocking (*see* **Note 11**).
10. During the 1 h incubation period, coat plates with Matrigel (*see* **Note 12** for alternative coating material). In short: pipet Matrigel up and down gently to homogenize, mix Matrigel with ice-cold organoid basal medium at ratio of 1:40, aliquot 100 μL of Matrigel–medium mixture into each well of 96-well imaging plate, and place plate in 37 °C tissue culture incubator for at least 30 min prior to using.
11. Shake intestine in crypt dissociation buffer for 1 min or until solution is cloudy. Shaking time can be extended if increased crypt yield is desired, though excessive shaking can cause deterioration of the crypt structures.
12. Remove intestine from crypt dissociation buffer and discard (*see* **Note 13**).
13. Centrifuge crypt dissociation buffer at 300 × *g* for 3 min at room temperature. An epithelial cell pellet will be observed at the bottom of the tube.
14. Aspirate buffer. Resuspend epithelial cell pellet gently with 10 mL warm organoid basal medium (*see* **Note 14** for alternatives for this step and washes in **steps 15** and **16**). Centrifuge at 300 × *g* for 3 min at room temperature.
15. Repeat **step 14**.
16. Aspirate medium. Resuspend in 10 mL warm organoid basal medium. Pass through 100 μm filter then 70 μm filter. Centrifuge at 300 × *g* for 3 min at room temperature.
17. Resuspend in 2–3 mL plating medium. Evaluate success of intestinal crypt harvest by observing a 10 μL aliquot on a glass slide under a brightfield or phase contrast microscope.
18. Determine crypt concentration by aliquoting 10 μL of crypt/medium solution (1:10 dilution or no dilution) onto a glass

slide and counting number of crypts under a brightfield or phase contrast microscope.

19. Dilute crypts to a final concentration of 3000 crypts per mL in plating medium.
20. Remove Matrigel-coated 96-well imaging plates from tissue culture incubator and flick out medium into waste container in a laminar flow biosafety cabinet. Aliquot 100 μL of crypt/medium solution into each well. Assess plating density and consistency under brightfield or phase contrast microscope.
21. Transfer plate to tissue culture incubator and incubate for 4 h (*see* **Note 15**).
22. After 4 h, flick medium out of plates and wash once with organoid basal medium. Add 100 μL of long-term culture medium to each well. Assess seeding efficiency under bright-field or phase contrast microscope. At this point, crypts should have flattened out into disk-shaped enteroid monolayers (*see* **Note 16**).
23. Add perturbations of interest. If perturbing cell-type composition, we typically treat with morphogens for 48 h prior to fixation and analysis. Change medium every 2 days to maintain optimal growth (*see* **Notes 17** and **18**).

3.2 Culturing Enteroid Monolayers from Frozen Crypts

3.2.1 Freezing Crypts

1. Prepare freezing medium.
2. Follow **steps 1–18** of Subheading 3.1 to harvest intestinal crypts.
3. Centrifuge crypts at 300 × *g* for 3 min at room temperature.
4. Resuspend crypts to a final concentration of 3000–5000 crypts per mL in freezing medium.
5. Transfer 1 mL of crypts suspended in freezing medium to a cryovial and freeze in freezing container in −80 °C overnight and then transfer to liquid nitrogen for long-term storage.

3.2.2 Deriving Enteroid Monolayers from Frozen Crypts

1. Coat 96-well imaging plates with Matrigel as described in **step 10** of Subheading 3.1.
2. Prepare organoid basal medium, plating medium, and long-term culture medium.
3. Revive crypt aliquots by thawing cryovials in 37 °C water bath.
4. Transfer crypts to 15 mL conical tube and add 9 mL warm organoid basal medium.
5. Centrifuge crypts at 300 × *g* for 3 min at room temperature.
6. Resuspend in warm plating medium to a final concentration of 3000 crypts per mL.
7. Follow **steps 20–23** of Subheading 3.1 to generate enteroid monolayers.

3.3 Culturing Enteroid Monolayers from 3D Organoids

1. Generate 3D organoid cultures: embed freshly isolated intestinal crypts in Matrigel (200 crypts/100 μL Matrigel). Pipet 100 μL of Matrigel slowly into each well of a 24-well plate to form a dome. Place in 37 °C incubator for 10 min to stiffen Matrigel. Once Matrigel has stiffened, add 500 μL long-term culture medium per well. Propagate 3D cultures as long as desired.
2. Coat 96-well imaging plates with Matrigel as described in **step 10** of Subheading 3.1.
3. Prepare organoid basal medium, plating medium, and long-term culture medium.
4. Warm TrypLE Express to 37 °C.
5. Aspirate medium from around Matrigel domes containing 3D organoids.
6. Dissolve Matrigel dome and 3D organoids in cold organoid basal medium by adding organoid basal medium to well and then pipetting up and down several times to dislodge Matrigel.
7. Transfer cold medium, Matrigel, and 3D organoid mixture to 15 mL conical tube.
8. Add approximately 10 mL cold organoid basal medium to tube to further dissolve remaining Matrigel.
9. Centrifuge at 300 × *g* for 5 min at 4 °C. A small organoid pellet should be visible at the bottom of the tube. If there appears to still be Matrigel in the pellet, aspirate medium and repeat **step 8**. The Matrigel will usually fully dissolve after a second wash with cold medium.
10. Aspirate medium and resuspend in a small volume (500–1000 μL) of cold organoid basal medium.
11. Shear organoids by running them through a fire-polished glass Pasteur pipet 8–10 times.
12. Centrifuge at 300 × *g* for 5 min at 4 °C.
13. If 3D organoids contained many dead luminal cells then aspirate medium, resuspend in cold organoid basal medium, and then centrifuge at 300 × *g* for 5 min at 4 °C again.
14. Aspirate medium from organoids. Add 500 μL TrypLE Express and resuspend thoroughly. Incubate at 37 °C for 5–10 min, shaking or triturating with a P1000 pipette a few times to break up cell clumps.
15. Add 5 mL warm organoid basal medium. Centrifuge at 300 × *g* for 5 min at room temperature.
16. Aspirate medium, resuspend in 5 mL warm organoid basal medium, and then centrifuge at 300 × *g* for 5 min at room temperature again.

17. Resuspend in small volume of plating medium. At this point, organoids should be single cells and some small clumps of cells.
18. Count cells using hemocytometer or automated cell counter.
19. Dilute cells to a final concentration of 50,000 cells per mL in plating medium.
20. Remove Matrigel-coated 96-well imaging plates from tissue culture incubator and flick out medium into waste container. Aliquot 100 μL of cell/medium solution into each well. Assess plating density and consistency under brightfield or phase contrast microscope.
21. After 18–24 h, flick medium out of plates. Add 100 μL of long-term culture medium to each well. There should be some cell aggregation at this point which will proceed over the course of the next 4–7 days to re-form crypt-villus–like patterning [4]. Change medium every 2 days to maintain optimal growth.

3.4 Performing Immunofluorescence on Enteroid Monolayers

1. Prepare PBS, washing buffer, fixation buffer, permeabilization buffer, blocking buffer, and antibody buffer.
2. If assaying proliferating cells, add 10 μM EdU in culture medium to enteroid monolayers for 1–2 h prior to fixation.
3. Flick medium out of plates. Wash once with 50 μL/well PBS.
4. Add 50 μL/well fixation buffer and incubate for 15 min at room temperature.
5. Wash three times with washing buffer. Add 50 μL/well permeabilization buffer and incubate for 10 min at room temperature (*see* **Note 19**).
6. Flick permeabilization buffer out of plates. Wash three times with washing buffer. Add blocking buffer and incubate for 30 min at room temperature.
7. Flick blocking buffer out of plates. Wash three times with washing buffer. Add 25 μL/well antibody buffer containing primary antibody of interest. Incubate overnight at 4 °C (*see* **Note 20**).
8. Wash three times with 50 μL/well washing buffer for 5+ min each (*see* **Note 21**).
9. Add 25 μL/well antibody buffer containing species-specific fluorescent secondary antibody of interest. Incubate for 2 h at room temperature in the dark.
10. Repeat **step 8**.
11. If assaying proliferating cells, prepare click reaction buffer. Wash three times with PBS and then add 50 μL click reaction buffer to each well. Incubate for 30 min at room temperature in the dark.

12. Wash three times with washing buffer and then add 50 μL nuclear staining solution to each well. Incubate for 30 min at room temperature in the dark.
13. Repeat **step 8**.
14. Store and image in washing buffer. Wrap in Parafilm for extended storage.

3.5 Quantitative Analysis of Immunofluorescence Images

3.5.1 Nuclear Segmentation

Nuclei in enteroid monolayers are highly heterogeneous in size and density with small, densely packed nuclei in crypt-like regions and large, sparsely distributed nuclei in villus-like regions [4]. To accommodate the range of nuclear characteristics, we employed a two-pass segmentation process. The first pass detects large, sparse nuclei and the second pass segments small dense nuclei. All parameters are set in **DNA Segmentation** section of the **config/seg_params.yaml** file. To skip parameter tuning and to run the script on the provided sample image, go to **step 9**.

Directions:

1. Smooth image with a bilateral filter. Set parameters for bilateral filter: BILATERAL_SIGMA_COLOR (standard deviation for pixel value range over which pixels are averaged), BILATERAL_SIGMA_SPATIAL (standard deviation for spatial distance range over which pixels are averaged).
2. Threshold image using a modified Otsu threshold method. Set parameter for thresholding: THRESHOLD_FACTOR (adjustment factor on Otsu threshold).
3. Detect location of nuclei using a multiscale Laplacian of Gaussian (LoG) method parameterized for large, sparse nuclei. Set parameters for sparse segmentation under LOG_SPARSE: MIN_SIG (lower bound for standard deviation of LoG filters), MAX_SIG (upper bound for standard deviation of LoG filters), NUM_SIG (number of standard deviations), THRESH (minimum intensity of peaks), OVERLAP (overlap allowance for neighboring objects).
4. Segment using watershed to separate connected nuclei. Set parameters for watershed: WATERSHED_CONN (neighborhood connectivity), WATERSHED_COMPACTNESS (compactness of segmented objects), WATERSHED_MIN_SZ (minimum size of segmented objects).
5. Detect and remove clumped nuclei from the sparse segmentation result. Clumped nuclei are detected based on both size and shape. Set parameters for clump detection: SEG_SINGLE_MIN_SZ (minimum size of single nuclei. Objects below this threshold are considered single nuclei), SEG_SINGLE_MAX_SZ (maximum size of single nuclei. Objects above this threshold are considered clumps of nuclei), SEG_CLUMP_SOLIDITY (threshold for object irregularity to classify objects

between the minimum and maximum nuclear sizes as clumps), SEG_CLOSE_HOLES (maximum size of holes to remove in clump objects).

6. Detect locations of nuclei in clumped regions using a multiscale LoG method parameterized for small, densely packed nuclei. Set parameters for dense segmentation under LOG_DENSE. These parameters are the same as LOG_SPARSE in **step 3** above.
7. Segment using watershed to separate connected nuclei. The same parameters as **step 4** above are used. No additional parameters are needed for this step.
8. Combine sparse and dense segmentation outputs for the final nuclear segmentation. No parameters are needed for this step.
9. Set path to image in the **enteroidseg/nuclear_segmentation.py** file. The path is already set for the provided sample image.
10. Navigate to the **enteroidseg** folder and make sure the **enteroidseg** conda environment is activated (*see* **item 4** of Subheading 2.6.2). Run nuclear segmentation using the following command: **python nuclear_segmentation.py**. The output results will be stored in the **output** folder.

3.5.2 EdU+ Nuclear Segmentation

We detect proliferating (S phase) cells by staining for EdU incorporation. EdU+ nuclei can be identified using a method similar to the nuclear segmentation described in Subheading 3.5.1 (*see* **Note 22**). The pipeline for segmenting EdU+ objects using the nuclei segmentation method is provided. To skip parameter tuning and to run the script on the provided sample image, go to **step 3**.

Directions:

1. Set parameters for EdU segmentation under **EdU Segmentation** in **seg_params.yaml**. The parameters are the same as nuclear segmentation.
2. Set path to image in the **edu_segmentation.py** file. The path is already set for the provided sample image.
3. Navigate to the **enteroidseg** folder and make sure the **enteroidseg** conda environment is activated (*see* **item 4** of Subheading 2.6.2). Run the EdU segmentation using the following command:

   python edu_segmentation.py

   The output results will be stored in the **output** folder.

3.5.3 Goblet Cell Segmentation

Cell types in enteroid monolayers can be detected by staining with cell type–specific antibodies or by deriving enteroids from transgenic mice expressing cell-type markers. We identify goblet cells using anti-Mucin-2 (Muc2) antibody. The staining pattern encompasses cytoplasmic regions above the nuclear plane that may overlap multiple nearby nuclei. Thus, goblet objects are identified solely based on the Muc2 staining pattern. All parameters are set in the **seg_params.yaml** file under **Goblet Segmentation**. To skip parameter tuning and to run the script on the provided sample image, go to **step 6**.

Directions:

1. Smooth image using a median filter. Set parameters for smoothing filter: MEDIAN_FILTER_SZ (radius of median filter).
2. Threshold image using a modified Otsu threshold method. Objects in thresholded image are expanded to the convex hull to create whole objects from partial membrane stains. Set parameters for thresholding: THRESHOLD_FACTOR (adjustment factor on Otsu threshold).
3. Detect goblet cell-object locations using a multiscale Laplacian of Gaussian method. Set parameters for segmentation under LOG_BLOB. These parameters are the same as LOG_SPARSE in **step 3** of Subheading 3.5.1.
4. Segment connected objects using watershed. Set parameters for watershed. The same parameters are used as in **step 4** of Subheading 3.5.1.
5. Set path to image in the **goblet_segmentation.py** file. The path is already set for the provided sample image.
6. Navigate to the **enteroidseg** folder and make sure the **enteroidseg** conda environment is activated (*see* **item 4** of Subheading 2.6.2). Run the goblet segmentation pipeline using the following command:

   python goblet_segmentation.py

   The output results will be stored in the **output** folder.

3.5.4 Stem Cell Segmentation

Stem cells are labeled by staining with anti-GFP antibodies in enteroid monolayers derived from Lgr5-eGFP-DTR mice [5] (*see* **Note 23**). In enteroid monolayers derived from Lgr5-eGFP-DTR mice, GFP signal localizes to cell membranes [5], requiring first segmentation of GFP+ "crypt-base" regions followed by identification of stem cells in crypt-base regions using nuclear segmentation information. Therefore, both Lgr5-GFP and Hoechst stain images are required for stem segmentation. Optionally, for more accurate stem segmentation, Paneth segmentation is used to remove Paneth nuclei in crypt-base regions from the final result. Sample images for Lgr5-GFP stain, Hoechst stain, and Paneth segmentation are provided in the **images** folder. All parameters are set in the **seg_params.yaml** file under **Stem Segmentation**. To skip parameter tuning and to run the script on provided sample images, go to **step 5**.

Directions:

1. Threshold image to detect crypts. Set parameters for thresholding: DNA_FACTOR (removes any bleedthrough into GFP channel from Hoechst channel using Hoechst stain image), THRESH (manual threshold for Lgr5 stain).
2. Thresholded image is further processed using morphological operations to connect holes in the crypt base-like regions. Set parameters for morphological operations: MORPH_CLOSING_SZ (radius of closing filter), MORPH_OPENING_SZ (radius of opening filter), MIN_SZ (minimum size of crypt-base regions in pixels).
3. Stem segmentation is finalized by identifying nuclei in the crypt regions and then (optionally) filtering out nuclei associated with Paneth cells. Set parameters for identification of nuclei: PARTIAL_RATIO (minimum ratio of nuclear area outside the crypt to the nuclear area inside the crypt to qualify as residing in the crypt).
4. Set path to image in the **stem_segmentation.py** file. The path is already set for the provided sample images.
5. Navigate to the **enteroidseg** folder and make sure the **enteroidseg** conda environment is activated (*see* **item 4** of Subheading 2.6.2). Run the stem segmentation pipeline by calling the **stem_segmentation.py** script using the following command: **python stem_segmentation.py**.

   The output results will be stored in the **output** folder.

4 Notes

1. It should be noted that other images of enteroid monolayers may require additional parameter optimization, additional processing steps, or different analysis setups depending on differences in staining and imaging properties.
2. If contamination is an issue, Primocin is another antimicrobial option that we have found to be effective and gentle on enteroid cultures; use at manufacturer's recommended concentration of 100 μg/mL.
3. Increase EDTA concentration to 5–10 mM and/or extend shaking and time in crypt dissociation buffer to harvest ileal crypts or colon crypts.
4. The BMP receptor inhibitor LDN-193189 (100 nM–1 μM) can be substituted for recombinant Noggin to save costs.
5. R-spondin-1 conditioned medium (15%) can be substituted for recombinant R-spondin-1. We often observe more robust growth in R-spondin-1 conditioned medium, though it can

suffer from batch-to-batch variability. Regardless of source, R-spondin-1 must be of high quality and concentration for optimal enteroid monolayer growth.

6. There is batch-to-batch variability in plate manufacture that can affect enteroid monolayer growth and imaging. Specifically, plates that are as flat as possible and have minimal blue channel autofluorescence are ideal.
7. Kind gift of Frederic de Sauvage via Ophir Klein under MTA #OM-216813. *See* also ref. 5.
8. We have also seen improvements in immunofluorescence signal-to-noise ratio using 4% PFA + 4% sucrose fixative.
9. We find that it is best to prepare sodium ascorbate stock solutions fresh in water prior to preparing click reaction buffer.
10. We store organoid basal medium for 2 months maximum. We store both plating and long-term culture medium for 1 week maximum.
11. If crypt yield is insufficient, one can also shake the conical tube containing intestine and crypt dissociation buffer every ~10 min during the incubation period.
12. We have also successfully cultured enteroid monolayers on Collagen I (Corning)-coated plates. Imaging is more difficult due to the thickness of the collagen coating. To perform Collagen I coating, add 50 μL/well of 1.6 mg/mL Collagen I diluted in long-term culture medium and NaOH (see also manufacturer's instructions).
13. Can keep intestine if one wants to have the option of shaking off more crypts. If so, place into a separate tube of crypt dissociation buffer rather than discarding.
14. Washes can be conducted in other buffers or media that contain Ca^{2+}/Mg^{2+}, we have personally tested Hank's Buffered Saline Solution (HBSS) and DMEM supplemented with 10% FBS.
15. Length of incubation in plating media can vary from 4 to 24 h depending on the needs of the experiment, but 4 h is used as a default. Enteroid monolayers should not be cultured in media containing Y-27632 longer than 24 h.
16. After changing into long-term medium, enteroid monolayers can be imaged on a brightfield point-scanning microscope to assess plating consistency. We find that between 10% and 30% confluent enteroid monolayers tend to grow to a consistent density and cell-type composition.
17. When changing medium, if there is excessive amounts of debris, wash once with organoid basal medium prior to putting fresh long-term culture medium on enteroid monolayers.

18. Generally, enteroid monolayers will increase in size for the first 3–4 days. A subset (~10%) of seeded crypts survives beyond 3–4 days but can be cultured for weeks. Enteroid monolayers derived from 3D cultures tend to not have this drop off in crypt survival.
19. Ice-cold methanol can be substituted for permeabilization buffer for specific antibodies; proceed according to manufacturer's instructions.
20. We find it helpful to wrap imaging plates in wet paper towels and plastic wrap for overnight antibody incubation to maintain moisture.
21. Washes can be extended for antibodies that exhibit nonspecific staining.
22. EdU+ nuclei can also be identified using the EdU staining intensity in previously identified nuclear objects.
23. Enteroids derived from Lgr5-CreERT2 mice have mosaic Lgr5-GFP expression, which makes it impossible to segment all stem cells. Therefore, we greatly prefer enteroids derived from Lgr5-eGFP-DTR mice for stem cell segmentation purposes.

Acknowledgments

This work was supported by NIH GM112690 (S.J.A.), NCINIH R01 CA184984 (L.F.W.), the UCSF Program for Breakthrough Biomedical Research which is partly funded by the Sandler Foundation (L.F.W.), NIH NRSA fellowship F32DK120102 (L.E.S.), NSF GRFP fellowship 1650113 (I.W.C.), and NIH R00 DK10312 (C.A.T.).

References

1. Sato T, Vries RG, Snippert HJ et al (2009) Single Lgr5 stem cells build crypt-villus structures in vitro without a mesenchymal niche. Nature 459:262–265
2. van de Wetering M, Francies HE, Francis JM et al (2015) Prospective derivation of a living organoid biobank of colorectal cancer patients. Cell 161:933–945
3. Farin HF, Jordens I, Mosa MH et al (2016) Visualization of a short-range Wnt gradient in the intestinal stem-cell niche. Nature 530:340–343
4. Thorne CA, Chen IW, Sanman LE et al (2018) Enteroid monolayers reveal an autonomous WNT and BMP circuit controlling intestinal epithelial growth and organization. Dev Cell 44:624–633.e4
5. Tian H, Biehs B, Warming S et al (2012) A reserve stem cell population in small intestine renders Lgr5-positive cells dispensable. Nature 482:120

Chapter 7

Autophagy Detection in Intestinal Stem Cells

Jumpei Asano, Taku Sato, and Toshiaki Ohteki

Abstract

Autophagy is a lysosomal degradation pathway with important roles in physiological homeostasis and disease. We previously showed that intrinsic autophagy in intestinal stem cells (ISCs) is important for ISC homeostasis. Here we describe the detailed methods for detecting autophagy in ISCs by observing autophagosomes in GFP-LC3 transgenic mice and quantifying the p62 protein levels. We also describe methods for detecting mitophagy in these cells, by analyzing the mitochondrial transmembrane potential and reactive oxygen species (ROS) level by MitoTracker and CellROX solution, respectively.

Key words Autophagy, Intestinal stem cells (ISCs), Lgr5, Autophagosome, GFP-LC3, p62, *Atg5*, Intestinal epithelial cells (IECs), Mitochondria, Reactive oxygen species (ROS)

1 Introduction

Autophagy is a process by which cellular components are transferred to lysosomes for degradation. During autophagy induction, parts of the cytoplasm are sequestered into double-membraned structures, called autophagosomes, which fuse with lysosomes to allow their contents to be degraded [1]. Autophagy protein 5 (ATG5), an E3 ubiquitin ligase, forms a complex with ATG12 and ATG16L1 [2], and this complex is necessary for autophagosome elongation. Concomitantly, the cytosolic microtubule associated protein 1A/1B-light chain 3 (LC3) is recruited to the autophagosomal membrane. In green fluorescent protein (GFP)-LC3 transgenic (tg) mice, autophagosomes, an indicator for autophagy, are identified as GFP-LC3 dots in various tissues [3]. In the intestine, ISCs express leucine-rich repeat–containing G protein–coupled receptor 5 ($Lgr5^+$ ISCs) and are located at the bottom of crypts [4, 5]. Therefore, using confocal microscopy, we observed autophagosomes as the GFP-LC3 dots in the $Lgr5^+$ ISCs of intestinal sections prepared from GFP-LC3 tg mice (Fig. 1). In

Jumpei Asano and Taku Sato contributed equally with all other contributors.

Paloma Ordóñez-Morán (ed.), *Intestinal Stem Cells: Methods and Protocols*, Methods in Molecular Biology, vol. 2171, https://doi.org/10.1007/978-1-0716-0747-3_7,

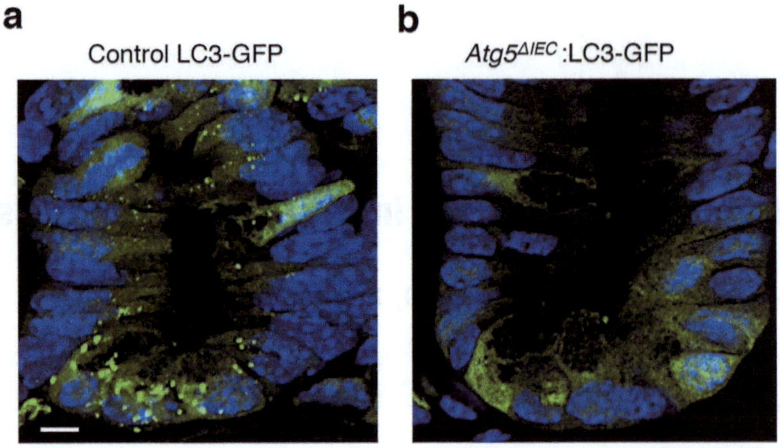

Fig. 1 Autophagosome detection at the bottom of intestinal crypts. Sections were prepared from the small intestine (Jejunum) of GFP-LC3 tg mice (**a**) and *Atg5*$^{\Delta IEC}$: GFP-LC3 tg mice (**b**). GFP-LC3 dots were observed by confocal laser microscopy. Nuclei in IECs stained with DAPI are in blue. GFP-LC3 dots, which represent autophagosome formation and are positively correlated with the amounts of LC3-II, are detected in the crypts under steady-state conditions. Scale bar: 10 μm. Data are representative of five independent experiments. (With permission from [11])

addition, since p62 is a well-known autophagy substrate that is degraded upon autophagy induction [6, 7], p62 is used as a biomarker to study autophagic flux [8]. Thus, we also quantified the p62 protein level in sort-purified Lgr5$^+$ ISCs (Fig. 2). Furthermore, autophagy is one of the major pathways for eliminating damaged or excessive mitochondria from cells, through a process called mitophagy [9, 10]. In Lgr5$^+$ ISCs, mitochondrial functions are maintained through the elimination of excessive reactive oxygen species (ROS) [11]. Therefore, we also describe methods for analyzing the mitochondrial membrane potential and ROS level in Lgr5$^+$ ISCs (Fig. 3).

2 Materials

Prepare all solutions using water purified by Milli-Q Integral 5 and analytical grade reagents. Prepare and store all reagents at room temperature unless otherwise stated. We do not add sodium azide to reagents. Strictly follow the institutional regulations for animal experiments.

2.1 Autophagosome Detection in the Crypt Bottom

1. *Atg5*$^{fl/fl}$ LC3-GFP tg mice: To obtain this strain, cross *Atg5*$^{fl/fl}$ mice with LC3-GFP tg mice (both provided by N. Mizushima). Maintain the obtained *Atg5*$^{fl/fl}$ LC3-GFP tg (hereafter called control LC3-GFP tg) mice in a specific pathogen free (SPF) facility.

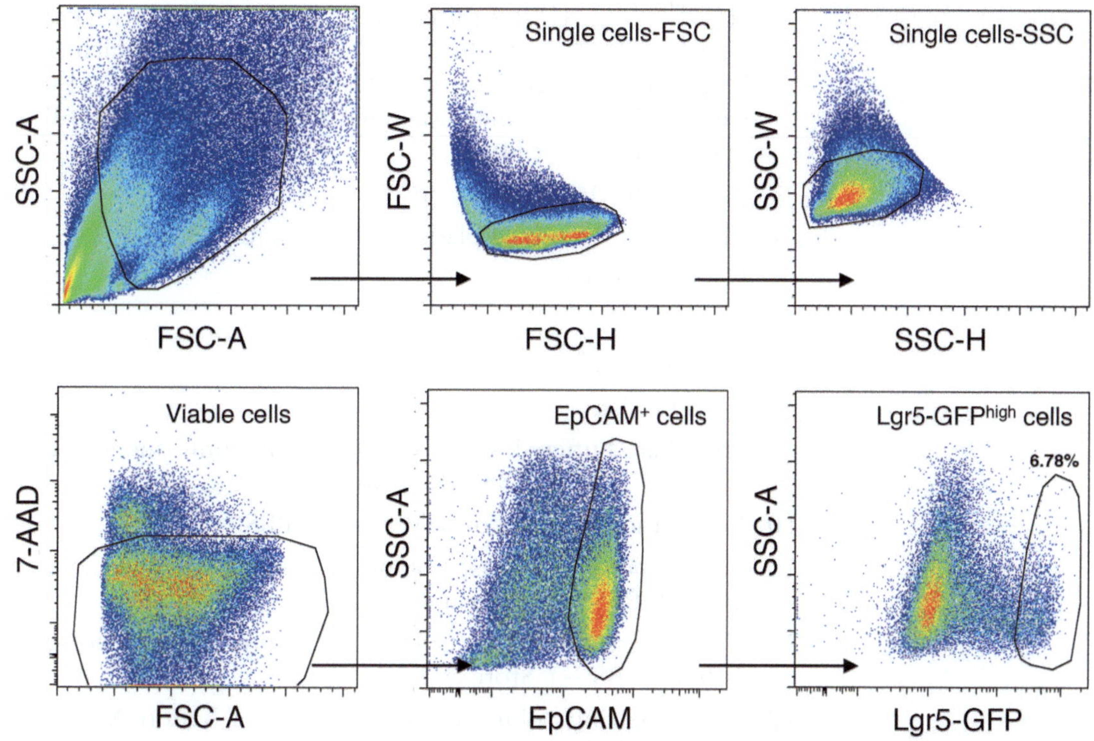

Fig. 2 Gating strategy to isolate Lgr5$^+$ ISCs (Lgr5-GFP^{high} cells). Debris is gated out based on FSC-A/SSC-A, then doublets are gated out based on FSC-W/FSC-H and SSC-W/SSC-H. Dead cells are eliminated by gating on 7-AAD$^-$ viable cells. Finally, EpCAM$^+$Lgr5-GFP^{high} cells are isolated on a FACS cell sorter. *FSC* forward scatter, *SSC* side scatter

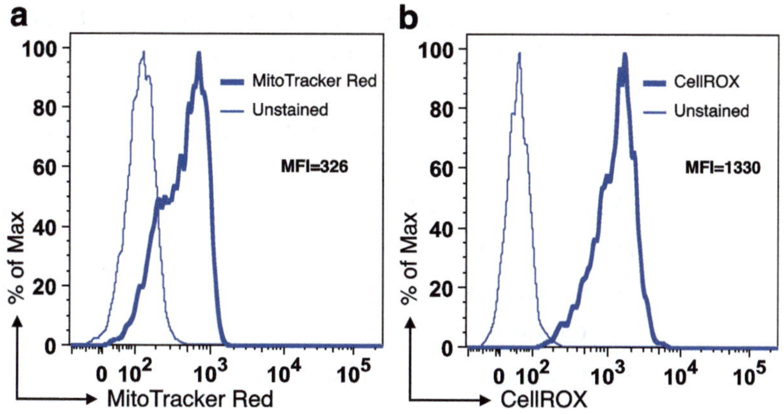

Fig. 3 Mitochondrial transmembrane potential and ROS level in Lgr5$^+$ ISCs. Bold solid lines represent Lgr5$^+$ ISCs (CD45.2$^-$Lgr5-GFP^{high} cells) stained with MitoTracker Red CMXRos (**a**) and CellROX (**b**), and thin solid lines show unstained controls (**a**, **b**). Mean fluorescence intensity (MFI) of CMXRos and CellROX were calculated as MFI of CMXRos or CellROX − MFI of an unstained control. Data are representative of three independent experiments

2. *Atg5*$^{\Delta IEC}$ *LC3-GFP tg* mice: To obtain this strain, cross *Villin-Cre:Atg5*$^{flox/flox}$, (hereafter *Atg5*$^{\Delta IEC}$) mice [12], in which *Atg5* is specifically deleted in the intestinal epithelial cells (IECs), with LC3-GFP tg mice. Maintain the *Atg5*$^{\Delta IEC}$:LC3-GFP tg mice in an SPF facility.
3. Tissue-Tek OCT compound.
4. Tissue-Tek Cryomold plastic embedding dishes.
5. Plastic containers.
6. Cork board.
7. Cell culture plates (24 well).
8. 27-G needles.
9. Phosphate-buffered saline (PBS), magnesium- and calcium-free (Mg^-/Ca^-) (PBS(−)). Store at 4 °C.
10. Fix solution: 4% paraformaldehyde in PBS.
11. Fifteen percent sucrose solution: Dissolve 15 g sucrose in 100 mL PBS(−). Store at 4 °C.
12. Thirty percent sucrose solution: Dissolve 30 g sucrose in 100 mL PBS(−). Store at 4 °C.
13. Permeabilizing solution: Add 500 μL of Triton X-100 to 500 mL of PBS(−). Store at 4 °C.
14. Blocking solution: Add 10 mL fetal calf serum (FCS) to 90 mL of PBS(−). Store at 4 °C.
15. Washing solution: Add 250 μL of Triton X-100 to 500 mL of PBS(−). Store at 4 °C.
16. Anti-GFP mouse antibody-Alexa Fluor 488: Dilute 1 mg/mL antibody stock solution to 5 μg/mL in washing solution (*see* **Note 1**).
17. DAPI dye solution: Dilute 1 mg/mL DAPI stock solution to 0.5 μg/mL in PBS.
18. Fluoromount-G.
19. Glass slides.
20. Cover glasses.
21. Cryostat.
22. Confocal laser microscope.

2.2 Isolation of Lgr5$^+$ ISCs, Measurement of p62 Protein Level in Lgr5$^+$ ISCs

1. *Atg5*$^{fl/fl}$ *Lgr5-EGFP-ires-creERT2* mice: To obtain this strain, cross *Atg5*$^{fl/fl}$ mice with *Lgr5-EGFP-ires-creERT2* mice [4] (Jackson Laboratory). Maintain the *Atg5*$^{fl/fl}$ *Lgr5-EGFP-ires-creERT2* (hereafter control *Lgr5*) mice in an SPF facility.
2. *Atg5*$^{\Delta IEC}$ *Lgr5* mice. To obtain this strain, cross *Atg5*$^{\Delta IEC}$ mice with control *Lgr5* mice. Maintain the *Atg5*$^{\Delta IEC}$ *Lgr5* mice in an SPF facility.

3. Plastic centrifuge tube (15 and 50 mL).
4. Eppendorf tube (1.5 mL).
5. Plastic petri dish (90 mm).
6. Nylon mesh (40- and 70-μm) (*see* **Note 2**).
7. PBS(−): Store at 4 °C. Keep on ice during experiments.
8. PBS(−)/10% FCS: Store at 4 °C. Keep on ice during experiments.
9. PBS(−)/10 mM EDTA: Add 10 mL of 500 mM EDTA/PBS stock solution to 490 mL of PBS(−). Store at 4 °C. Keep on ice during experiments.
10. Advanced DMEM/F12 (×1) Store at 4 °C.
11. TrypLE™ Express Enzyme (×1). Store at 4 °C.
12. Suspension buffer: Add 10 mL of FCS and 2 mL of 500 mM EDTA/PBS stock solution to 488 mL of PBS(−). Store at 4 °C.
13. APC-conjugated anti-CD326 (EpCAM) antibody: Dilute 0.2 mg/mL antibody stock solution to 0.2 μg/mL with suspension buffer.
14. 7-Amino-actinomycin D (7-AAD) (0.1 μg/μL): Store at 4 °C.
15. Pierce™ BCA Protein Assay Kit.
16. p62 ELISA kit.
17. Refrigerated centrifuge.
18. Micro refrigerated centrifuge.
19. Multicolor FACS sorter.
20. Plate reader.

2.3 Examination of Mitochondrial Transmembrane Potential and ROS Level in Lgr5⁺ ISCs

1. Suspension buffer: Add 10 mL of FCS and 2 mL of 500 mM EDTA/PBS stock solution to 488 mL of PBS(−).
2. PBS(−)/2% FCS: 2% FCS in PBS(−). Store at 4 °C.
3. Pacific Blue anti-mouse CD45.2 antibody (10^4): Dilute 0.5 mg/mL antibody stock solution to 5 μg/mL with suspension buffer.
4. MitoTracker Red CMXRos solution: Dilute 1 mM MitoTracker stock solution to 50 nM in PBS(−)/2% FCS (*see* **Note 3**).
5. CellROX Oxidative Stress Reagent (Deep Red Reagent) solution: Dilute 2.5 mM CellROX stock solution to 5 μM in PBS (−)/2% FCS (*see* **Note 4**).
6. Multicolor FACS.

3 Methods

Carry out all procedures at room temperature, unless otherwise specified.

3.1 Autophagosome Detection at the Bottom of the Intestinal Crypts

1. Harvest the small intestine from a control LC3-GFP tg mouse, open it longitudinally with scissors, and wash it thoroughly with PBS(−).
2. Cut the intestine into 5-cm-long pieces.
3. Prepare a corkboard and a plastic container to hold it. Extend and fasten the incised intestines on the corkboard with a 27 G needle, then place the corkboard in the plastic container (*see* **Note 5**).
4. Fix the intestine on the corkboard with fix solution at room temperature for 4 h.
5. Wash the intestine with PBS then cut it into 1-cm-long pieces.
6. Place each 1-cm-long piece in each well of a 24-well culture plate.
7. Treat each intestine piece with 15% sucrose solution at 4 °C for 4 h and then with 30% sucrose solution at 4 °C overnight, as reported previously [3].
8. Fill a Tissue-Tek Cryomold plastic embedding dish with Tissue-Tek OCT compound, and embed one piece of intestine in each mold (*see* **Note 6**).
9. Prepare 5-μm-thick sections of the intestine using a cryostat. Place each section on a glass slide and store it at −80 °C.
10. Dry the tissue sections with a dryer.
11. Treat the sections with permeabilizing solution for 5 min, then with PBS(−) twice for 5 min each.
12. Treat the sections with blocking solution at room temperature for 45 min.
13. Incubate the sections with an Alexa 488–conjugated anti-GFP antibody (1:200 dilution, diluted in washing solution) at 37 °C for 90 min.
14. Wash with washing solution for 5 min. Repeat this step three times.
15. Incubate with DAPI dye solution (1:2000 dilution, diluted in PBS) at room temperature for 2 min.
16. Wash with PBS for 5 min.
17. Mount a cover glass with Fluoromount-G on the sections, then observe autophagosomes at the crypt by confocal microscopy (×63) (Fig. 1) (*see* **Note 7**).

3.2 Isolation of $Lgr5^+$ ISCs, Measurement of the p62 Protein Level in $Lgr5^+$ ISCs

1. Harvest the small intestine from control *Lgr5* and $Atg5^{\Delta IEC}$ *Lgr5* mice, open the intestine longitudinally with scissors, and it wash with PBS(−). Remove the mucus by gently rubbing the intestine between the fingers in PBS(−) as described previously [13].
2. Cut the intestine into 5-mm-long pieces with a small scalpel blade. Place the pieces in cold PBS(−)/10 mM EDTA solution in a 50-mL tube, and incubate them for 40 min on ice (*see* **Note 8**).
3. Decant the supernatant and replace it with PBS(−)/10 mM EDTA, and incubate the pieces on ice for 30 min with intermediate vigorous shaking by hand every 5 min. Collect the supernatant, then resuspend the pieces again in PBS(−)/ 10 mM EDTA and incubate them on ice for 15 min with intermediate vigorous shaking by hand every 5 min, and pool the supernatants. Repeat this step twice.
4. Filter the pooled supernatants through a 70-μm nylon mesh into a 50-mL tube. The resulting cell suspension mainly consists of crypts.
5. Centrifuge the sample at 190 × *g*, 4 °C for 8 min, and remove the supernatant.
6. Resuspend the pellet in 10 mL of PBS(−)/10% FCS, and transfer it to a 15 mL tube.
7. Repeat **step 5**.
8. Resuspend the pellet in 1.5 mL of TrypLE Express and incubate the sample in a water bath at 37 °C for 30 min, with gentle pipetting (*see* **Note 9**).
9. Add 10 mL of PBS(−)/10% FCS to the sample, then filter the dissociated crypt epithelial cells through 70-μm nylon mesh into a 15 mL tube (*see* **Note 10**).
10. Centrifuge the sample at 630 × *g*, 4 °C for 5 min, and remove the supernatant.
11. Resuspend the pellet in 10 mL of PBS(−)/10% FCS, then filter the cells through 40-μm nylon mesh into a 15 mL tube.
12. Repeat **step 10**.
13. Resuspend the cells in an appropriate volume of PBS(−)/10% FCS and count the viable cells by trypan blue exclusion.
14. Add 1 μL of APC-anti-CD326 (EpCAM) antibody (1:1000 dilution, diluted in suspension buffer) to the cell suspension (1×10^6 to 1×10^7 cells/mL), and incubate the mixture at 4 °C for 30 min.
15. Wash the cells with 10 mL of suspension buffer, centrifuge the sample at 630 × *g*, 4 °C for 5 min, and remove the supernatant.

16. Resuspend the cells in suspension buffer at 1×10^7 cells/mL, and transfer them into a 1.5-mL Eppendorf tube.
17. Add 5 μL of 7-AAD to 1×10^6 cells, and incubate the sample at 4 °C for 5 min.
18. Wash the cells with 1 mL of suspension buffer, centrifuge them at $1500 \times g$, 4 °C for 5 min, and remove the supernatant.
19. Resuspend the cells in suspension buffer at 1×10^7 cells/mL, and isolate the $Lgr5^+$ ISCs ($EpCAM^+$ Lgr5-GFP^{high} cells) using a multicolor FACS sorter into a 1.5-mL Eppendorf tube containing 850 μL of Advanced DMEM/F12 (×1) (Fig. 2) (*see* **Note 11**).
20. Centrifuge the sample at $1500 \times g$, 4 °C for 5 min, and remove the supernatant.
21. Resuspend the cells in an appropriate volume of lysis buffer containing protease inhibitor and DNase supplied in the p62 ELISA kit, according to the manufacturer's instructions (*see* **Note 12**).
22. Prepare the plate for the assay according to the manufacturer's instructions.
23. Zero the plate reader against the blank wells and read the optical density at 450 nm on a plate reader.
24. Calculate the amount of p62 protein in the $Lgr5^+$ ISCs (*see* **Note 13**).

3.3 Estimation of Mitochondrial Transmembrane Potential and ROS Level in $Lgr5^+$ ISCs

1. Prepare the crypt cells from control *Lgr5* mice as described in Subheading 3.2 (**steps 1–15**).
2. Suspend the cells in suspension buffer at 5×10^6 to 1×10^7 cells/mL, then transfer them to a 1.5 mL Eppendorf tube.
3. Centrifuge the cells at $1500 \times g$, 4 °C for 5 min, and remove the supernatant.
4. Incubate the cells with a Pacific Blue–conjugated anti-mouse CD45.2 antibody (1:100 dilution, diluted in suspension solution) at 4 °C for 30 min.
5. Wash the cells with 1 mL of suspension buffer, then centrifuge them at $1500 \times g$, 4 °C for 5 min, and remove the supernatant.
6. To determine the mitochondrial transmembrane potential and ROS level, incubate the cells (1×10^6 to 1×10^7 cells/tube) with 500 μL of MitoTracker Red CMXRos or CellROX solution at 37 ° C for 15 min (*see* **Note 14**).
7. Wash the cells with PBS(−)/2% FCS, then centrifuge them at $1500 \times g$, 4 °C for 5 min, and remove the supernatant. Repeat this step twice.
8. Resuspend the cells in suspension buffer at 1×10^7 cells/mL.

9. Add 5 μL of 7-AAD to 1×10^6 cells, and incubate them at 4 °C for 5 min.
10. Wash the cells with 1 mL of suspension buffer at $1500 \times g$, 4 °C.
11. Resuspend the cells in suspension buffer at 1×10^7 cells/mL and analyze the mitochondrial transmembrane potential and ROS level in the $Lgr5^+$ ISCs on a multicolor FACS (Fig. 3) (*see* **Note 15**).

4 Notes

1. Although LC3-GFP dots can be observed under a microscope without using this antibody, better images will be obtained if this antibody is used to enhance the signal. To prevent nonspecific antibody binding, sufficient blocking should be performed. In addition, the conditions of the antibody reaction (e.g., the antibody concentration and reaction time) should be determined for each experiment.
2. If necessary, sterilize the nylon mesh in an autoclave before use.
3. Although the recommended final working concentration is 25–500 nM, we found 50 nM to be appropriate for staining crypt cells under our experimental conditions. Prepare it at the time of use.
4. CellROX Oxidative Stress Reagent is a fluorogenic probe designed to reliably measure ROS in live cells. This reagent is localized to the cytoplasm. As the reagent is sensitive to light and air, do not to keep the vial open for a long time. Once the vial is opened, divide the reagent into microtubes and store it at −20 °C.
5. Carefully extend the intestine with the villus side up on the corkboard. After setting the corkboard in a plastic container, pour fix solution into the container until the intestine is immersed.
6. Embed the piece of intestine with the villus side facing the outside, and freeze it in liquid nitrogen. Use the small Cryomold size (Cryomold <1>, 10 × 10 × 5 mm) which fits the size of the intestine pieces. Gently attach the frozen block to the Tissue-Tek object holder, to avoid crushing the tissue.
7. As a negative control for the LC3-GFP dots detection, observe the crypt bases of $Atg5^{\Delta IEC}$: LC3-GFP tg mice. GFP-LC3 dots are found at the crypt base of control LC3-GFP tg mice but not of $Atg5^{\Delta IEC}$: LC3-GFP tg mice.

8. Pour 40 mL of PBS(−)/10 mM EDTA solution into a 50 mL tube. Add the intestine pieces to the tube and incubate them on ice for 40 min with gentle shaking by hand every 10 min, which removes the intestinal villi. Do not shake vigorously.
9. After Subheading 3.2, **step 8**, add 3 mL of TrypLE Express to the pellet in a 15 mL tube, then divide the cell suspension into two 15 mL tubes (2.25 mL/tube). Since the viscosity of the suspension becomes high at this step, the pipetting should be done thoroughly for about 15 min after the TrypLE Express addition. This process is very important to keep the suspension smooth, to achieve a high recovery of the crypt cells.
10. During this step, use a new mesh every time the cells become clogged and do not pass through the mesh.
11. Sort the $Lgr5^+$ ISCs from control *Lgr5* and $Atg5^{\Delta IEC}$ *Lgr5* mice. Since there are fewer ISCs in the crypts of $Atg5^{\Delta IEC}$ *Lgr5* mice compared with control Lgr5 mice [11], five or six $Atg5^{\Delta IEC}$ *Lgr5* mice are needed for each experiment.
12. Before storing the lysates at −20 °C, measure the total amount of protein. We use the Pierce™ BCA Protein Assay Kit. The BCA assay is little affected by the types of proteins, surfactants, and chelating agents.
13. Measure the p62 protein amount in $Lgr5^+$ ISCs from the absorbance at 450 nm, and calculate it as the ng/mg protein based on the total protein amount.
14. The expression amount of EpCAM is somewhat attenuated after the incubation with CMXRos and CellROX at 37 °C. Thus, cells incubated at 37 °C for 15 min must be used as a control for the FACS analysis. When comparing the intensity of CMXRos and CellROX between different samples, the same cell number must be used for the staining and analysis. If the cell number is different, the intensity may be altered.
15. Calculate the mitochondrial transmembrane potential and ROS level in $Lgr5^+$ ISCs from the mean fluorescence intensity (MFI) of the CMXRos- or CellROX-treated cells – the MFI of unstained control cells.

Acknowledgments

This work was supported in part by a Grant-in-Aid for Young Scientists (B) from the Ministry of Education, Culture, Sports, Science, and Technology of Japan (J.A., 25860158), the Japan Science and Technology Agency, Precursory Research for

Embryonic Science and Technology (PRESTO) (T.S., JPMJPR13M4), a Grant-in-Aid for Scientific Research on Innovative Areas "Stem Cell Aging and Disease" (#25115002) from MEXT, Japan (T.O., 17H05635), Takeda Science Foundation (T.O.), and Nanken-Kyoten, TMDU. Jumpei Asano and Taku Sato contributed equally to this work.

References

1. Deretic V, Levine B (2009) Autophagy, immunity, and microbial adaptations. Cell Host Microbe 5:527–549
2. Mizushima N (2007) Autophagy: process and function. Genes Dev 21:2861–2873
3. Mizushima N, Yamamoto A, Matsui M et al (2004) In vivo analysis of autophagy in response to nutrient starvation using transgenic mice expressing a fluorescent autophagosome marker. Mol Biol Cell 15:1101–1111
4. Barker N, van Es JH, Kuipers J et al (2007) Identification of stem cells in small intestine and colon by marker gene Lgr5. Nature 449:1003–1007
5. Tian H, Biehs B, Warming S et al (2011) A reverse stem cell population in small intestine renders Lgr5-positive cells dispensable. Nature 478:244–259
6. Biørkøy G, Lamark T, Brech A et al (2005) p62/SQSTM1 forms protein aggregates degraded by autophagy and has a protective effect on huntingtin-induced cell death. J Cell Biol 171:603–614
7. Biørkøy G, Lamark T, Pankiv S et al (2009) Monitoring autophagic degradation of p62/SQSTM1. Methods Enzymol 452:181–197
8. Kirkin V, McEwan DG, Novak I et al (2009) A role for ubiquitin in selective autophagy. Mol Cell 34:259–269
9. Kim I, Rodriguez-Enriquez S, Lemasters JJ (2007) Selective degradation of mitochondria by mitophagy. Arch Biochem Biophys 462:245–253
10. Tal MC, Sasai M, Lee HK et al (2009) Absence of autophagy results in reactive oxygen species-dependent amplification of RLR signaling. Proc Natl Acad Sci U S A 106:2770–2775
11. Asano J, Sato T, Ichinose S et al (2017) Intrinsic autophagy is required for the maintenance of intestinal stem cells and for irradiation-induced intestinal regeneration. Cell Rep 20:1050–1060
12. Madison BB, Dunbar L, Qiao XT et al (2002) Cis elements of the villin-gene control expression in restricted domains of the vertical (crypt) and horizontal (duodenum, cecum) axes of the intestine. J Biol Chem 277:33275–33283
13. Yilmaz ÖH, Katajisto P, Lamming DW et al (2012) mTORC1 in the Paneth cell niche couples intestinal stem-cell function to calorie intake. Nature 486:490–495

Part II

Single-Cell Transcriptional Profiling of the Intestinal Epithelium

Chapter 8

Single-Cell Transcriptional Profiling of the Intestinal Epithelium

Claudia Capdevila, Ruben I. Calderon, Erin C. Bush, Kismet Sheldon-Collins, Peter A. Sims, and Kelley S. Yan

Abstract

Emerging single-cell technologies, like single-cell RNA sequencing (scRNA-seq), enable the study of heterogeneous biological systems at cellular resolution. By profiling the set of expressed transcripts in each cell, single-cell transcriptomics has allowed for the cataloging of the cellular constituents of multiple organs and tissues, both in health and disease. In addition, these technologies have provided mechanistic insights into cellular function, cell state transitions, developmental trajectories and lineage relationships, as well as helped to dissect complex, population-level responses to environmental perturbations. scRNA-seq is particularly useful for characterizing the intestinal epithelium because it is a dynamic, rapidly self-renewing tissue comprised of more than a dozen specialized cell types. Here we discuss the fundamentals of single-cell transcriptomics of the murine small intestinal epithelium. We review the principles of proper experimental design and provide methods for the dissociation of the small intestinal epithelium into single cells followed by fluorescence-activated cell sorting (FACS) and for scRNA-seq using the 10× Genomics Chromium platform.

Key words Single-cell, Transcriptomics, RNA-sequencing, Intestinal epithelium, Flow cytometry

1 Introduction

The intestinal epithelium is a single-layer columnar epithelium that lines the small intestine. It is a highly dynamic tissue that performs distinct functions, including nutrient absorption, chemosensation, mucin production, interaction with the outside environment and the immune system, and hormone production [1–4]. These functions are carried out by differentiated cells. Absorptive enterocytes, hormone-producing enteroendocrine cells, goblet cells, Paneth cells, and tuft cells are the major lineages within the intestinal epithelium. All five major lineages are produced from the differentiation of Lgr5+ intestinal stem cells (ISCs) [5] (Fig. 1). These Lgr5+ ISCs reside at the base of the crypt compartment and maintain the epithelium during homeostasis, rapidly generating replacement cells

Paloma Ordóñez-Morán (ed.), *Intestinal Stem Cells: Methods and Protocols*, Methods in Molecular Biology, vol. 2171, https://doi.org/10.1007/978-1-0716-0747-3_8, © Springer Science+Business Media, LLC, part of Springer Nature 2020

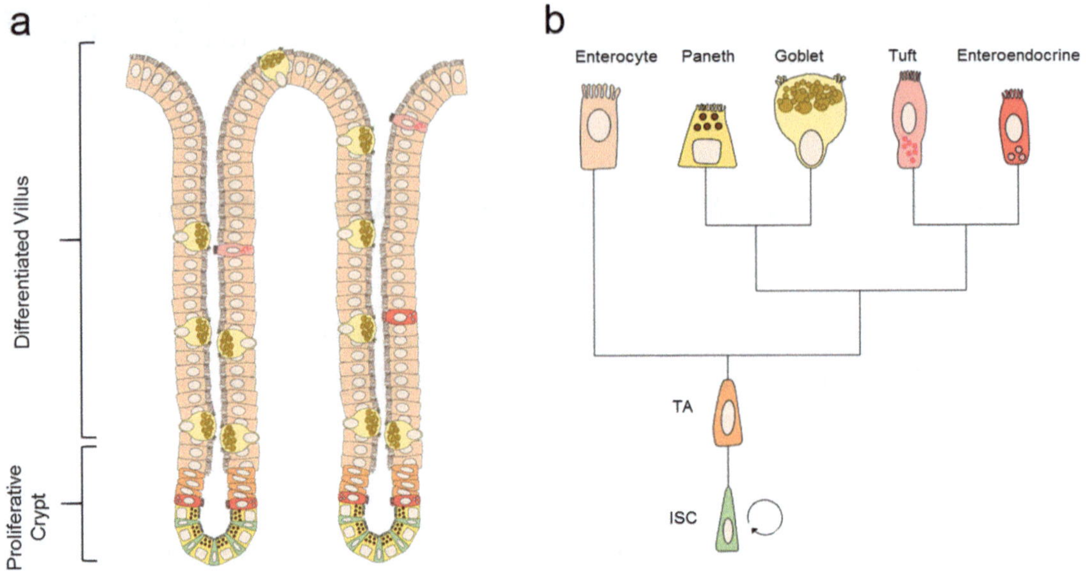

Fig. 1 Lineage hierarchy and cellular diversity of the intestinal epithelium (**a**). Organization of the intestinal epithelium (**b**). Lineage hierarchy of the intestinal epithelium. Lgr5+ intestinal stem cells (ISCs) at the crypt base regenerate the epithelium during homeostasis. ISCs give rise to transit-amplifying (TA) cells that differentiate into absorptive and secretory cell lineages. Absorptive enterocytes comprise the bulk of the intestinal epithelium. Secretory lineage cells include Paneth cells, interspersed between ISCs in the crypt, as well as mucin-secreting goblet cells, hormone-secreting enteroendocrine cells, and chemosensory tuft cells

that support the high rate of basal cellular turnover every 3–5 days [5]. Lgr5+ ISCs divide daily and give rise to transit-amplifying (TA) daughter cells that have limited self-renewal potential but are ultimately committed to becoming terminally differentiated. Thus, in addition to its remarkably high level of proliferation, the intestinal epithelium is extremely diverse in its cellular composition.

Historically, immunohistochemistry, quantitative PCR (qPCR), and flow cytometry have been used to study heterogeneous cell populations like those of the intestinal epithelium. One problem with these approaches is that they all rely on a limited number of predefined markers and thus prior knowledge. As such, these strategies are inherently biased and give rise to shallow classification schemes that are neither accurate nor quantitative, and that are limited in their ability to capture the true variability found in otherwise "defined" populations [6, 7].

Advances in sequencing technologies have started to erode this bias. Messenger RNA sequencing (mRNA-seq) is a technique that employs next generation sequencing (NGS) for the detection and quantification of expressed mRNA transcripts in a bulk biological sample [8]. This technology can be applied to study unfractionated cells within a tissue, like a jejunum-derived RNA extract, or particular cell populations enriched by other methods, like E-cadherin+

intestinal epithelium isolated by fluorescence-activated cell sorting (FACS). This can provide an unbiased assessment of the transcriptome of the cells under study with a greater dynamic range than that offered by comparable microarray-based technologies. Its applications include, but are not limited to, novel transcript discovery, analysis of alternative splicing, and analysis of differential gene expression.

However, cellular heterogeneity is a universal property of multicellular organisms, and also one that has confounded interpretation of bulk transcriptional profiling studies until the recent development of robust single-cell resolution methodologies. Indeed, the investigation of tissues or cell populations by traditional bulk approaches is inherently limited by the fact that any pooled assay that uses bulk tissue as an input will invariably represent a weighted, not necessarily real, average of that population's cellular constituents [7]. Thus, intrinsic cellular heterogeneity has been masked in the typical ensemble studies that have been reported during the last few decades—at least until the recent emergence of single-cell RNA sequencing (scRNA-seq) technology in 2009 [9].

A powerful approach to systematically dissect the heterogeneity found in a wide range of tissues, scRNA-seq aims to quantitatively profile the transcriptome of defined cell populations at single-cell resolution. In particular, scRNA-seq measures the distribution of the expression levels for each individual gene across a population of cells. This can overcome some of the previous limitations and is in remarkable contrast to bulk RNA-sequencing, which instead measures the average expression level for each gene across a mixture of cells. As such, scRNA-seq allows us to study complex heterogeneous systems and address new biological questions in which cell-specific changes are important, like the identification of new cell types and their associated transcriptomic signatures [10–13], the heterogeneity of cellular responses to treatment [14–16], the stochastic nature of gene expression in otherwise homogeneous populations [17, 18], or the inference of gene regulatory networks underlying cell function, lineage hierarchies, and fate decisions [19–23], among others [24]. Indeed, the characterization and quantification of the genes that are expressed at the single-cell level at a particular time and under specific conditions is thought to provide a highly accurate readout of cellular identity and function—rather than the study of cellular morphological characteristics or limited surface marker expression profiles. However, it is important to keep in mind that single-cell transcriptome analysis still provides an undersampled and skewed approximation of the true transcript distribution, and this is most notably because the mRNA capture efficiency in scRNA-seq is less than 100%.

Below is a brief description of the typical scRNA-seq experimental pipeline [25] (Fig. 2).

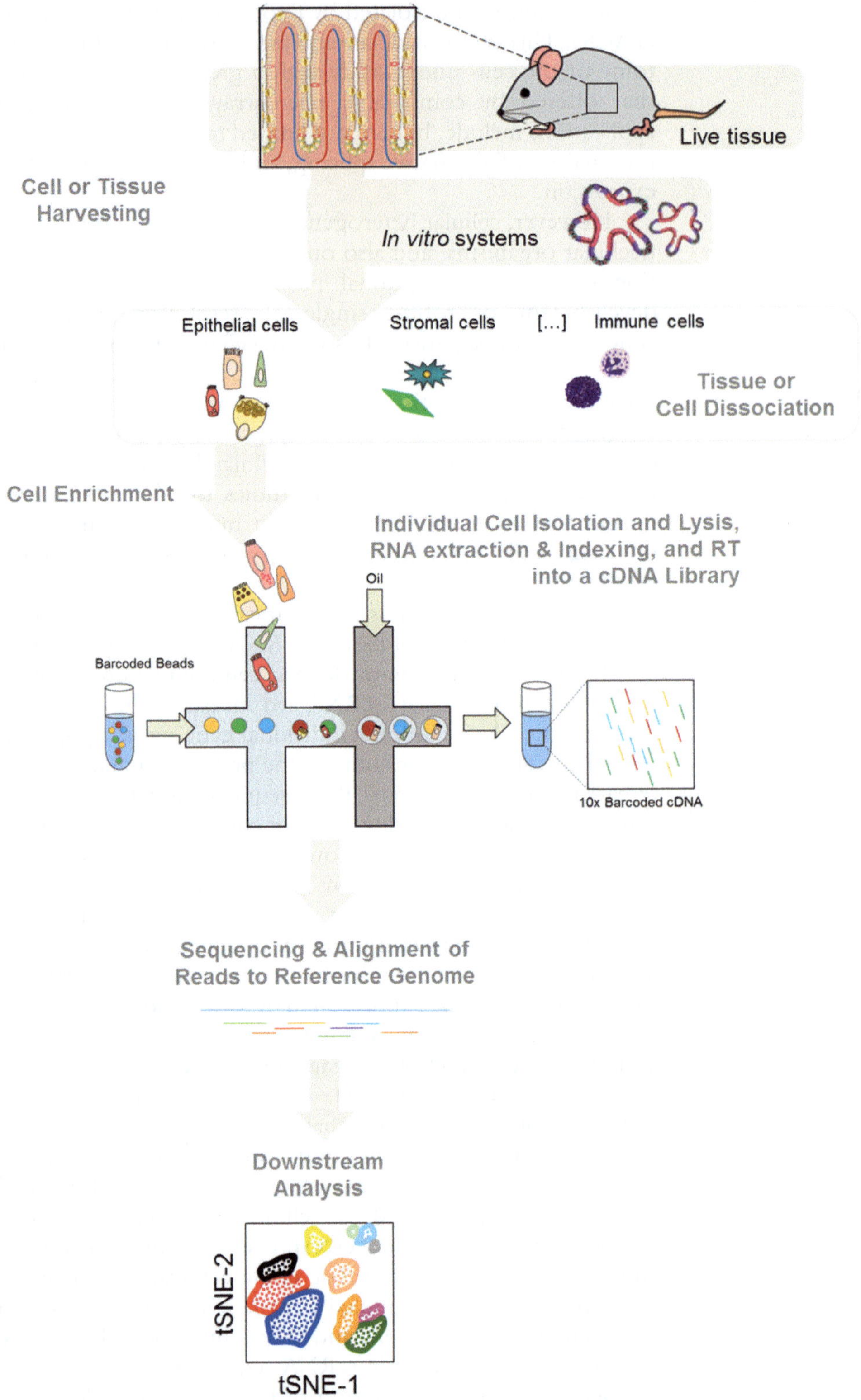

Fig. 2 Experimental pipeline for scRNA-seq studies. Individual cells are dissociated from intestinal tissue or 3D organoids, and further enriched based on viability and/or other characteristics, like surface epithelial marker

- Tens to thousands of single cells are isolated from a tissue or culture using different methods and physically separated during cell lysis. Some platforms use unique cellular barcodes to identify individual cells.
- The mRNA from each individual cell is extracted and carefully isolated, and reverse-transcribed into cDNA.
- cDNA is amplified and further processed for NGS.
- Sequenced fragments (reads) are obtained.
- Splice-aware alignment is used to map reads to a reference genome with a transcriptome annotation, producing estimates of normalized gene expression levels in an $N \times L$ gene expression matrix (N = number of cells, L = total number of genes detected).
- Statistical analysis is used to identify trends in gene expression across many individual cells, which in turn can allow for the assignment of individual cells to clusters [25, 26].
- Further bioinformatics analyses can be performed [24], such as differential gene expression analysis and/or novel biomarker identification, pseudotime analysis for the inference of developmental trajectories, regulatory network analysis, etc.

Here, we will focus on the use of single-cell transcriptomics for the study of the intestinal epithelium. We will review the basic principles of experimental design and single-cell dissociation for scRNA-seq. We will also present detailed methods for the characterization of the murine intestinal epithelium, including protocols for single-cell dissociation, epithelial cell enrichment by FACS, and scRNA-seq using the 10× Genomics Chromium platform.

1.1 Tissue Dissociation and Cell Isolation Methods for Single-Cell Sequencing

Getting a viable single-cell suspension out of a biological tissue or culture is essential. Indeed, our ability to generate highly informative, high-quality scRNA-seq data is ultimately dependent on our ability to release individual cells out of their supporting stroma, and to do that in a way that preserves both viability and endogenous cell physiology. These suspensions can be achieved through various cellular dissociation methods. Mechanical or enzymatic dissociation methods can be used to process samples into single-cell suspensions, and most protocols feature a combination of these two. This processing step is typically performed before cells are stained, if

Fig. 2 (continued) expression. Single cells are individually captured and lysed for cDNA library construction. After amplification, the library is sequenced and the obtained reads are aligned to the reference genome, providing estimates for gene expression in individual cells. Further analysis can be applied to each individual transcriptome to identify trends in gene expression and differentially expressed genes across cell types

needed, and the two methods each have benefits and caveats. Ultimately, one must empirically determine what works best for the specific tissue and the questions to be addressed.

- *Mechanical dissociation*: This method makes use of tools like a glass mortar and a pestle or similar tissue homogenizers to dissociate tissues. In general, mechanical dissociation entails mincing, cutting, and sieving the tissue into smaller fragments.
  - Benefit: Mechanical dissociation can be a rapid technique that works well for samples that are loosely associated with the underlying extracellular matrix, such as mouse spleens, bone marrow, and lymph nodes.
  - Caveat: It is also usually associated with inconsistent cell yields or poor cell viability, and as a consequence the derived results can vary widely between operators—multicellular aggregates of different sizes (incomplete dissociation) not being uncommon.
- *Enzymatic dissociation*: Different enzymes, such as collagenase, trypsin, Pronase, or hyaluronidase, are used to dissociate tissue samples. The methods and conditions of cellular dissociation can influence transcriptomics. It is important to consider that cells should be exposed to enzymes for a minimal amount of time to preserve maximum viability and avoid the potential digestion of plasma membrane proteins that may be further needed for population enrichment purposes. Temperature is an additional variable to consider. There are recent descriptions of cold active proteases that may reduce activation of heat shock and stress response genes during the dissociation process [27]. The optimal conditions and concentrations of enzymes employed will usually need to be worked out empirically.
  - Benefit: Specific enzyme cocktails are commercially available for certain types of tissue, and some of them work with great efficacy—even in denser tissues.
  - Caveat: Enzymatic dissociation can modify proteins on the cell surface, which can alter cell function or binding of antibodies.

Once we get our cellular suspensions, we will need to isolate the viable, individually dissociated single cells. This is of particular importance for scRNA-seq for a number of reasons. First, in order to be individually captured, cells will sometimes need to flow through microfluidic channels that can otherwise clog if loaded with multicellular aggregates. Second, if two or more cells are processed together for scRNA-seq, this confounds the analysis by presenting mixed cell types (two or more cell transcriptomes are combined, yielding an artifactual hybrid expression profile) that otherwise don't exist in the population. Finally, by selecting for viable cells, we will glean a greater number of informative reads from live cells and will avoid problems of contamination from dead/dying cells. The main ways in which one can do this are briefly presented below:

- *FACS*: Flow cytometry is usually the method of choice to purify thousands of single cells, exclude doublets, separate dead from alive by the addition of a viability dye like propidium iodide, and perhaps enrich for a particular cell population based on the expression of one or a combination of surface markers. However, FACS may not be suitable for some tissues with very complex morphologies such as the adult mammalian brain.
  - With nonrestrictive sorting gates, heterogeneous cell samples can be isolated.
  - With restrictive sorting gates, we can enrich for a certain population of interest, while depleting for others.
- *Magnetic bead-based cell enrichment*: Magnetic bead-based cell enrichment may be a good alternative if cell isolation is needed. This involves the use of magnetic beads coated with antibodies against a cell surface antigen that will allow for the capture of the cells of interest, which will be isolated by using a special type of column and exposing it to a magnetic field. Many different types of antibody-coated beads and both positive enrichment and exclusion columns are commercially available. Although positive enrichment can provide high yields of cell types-of-interest, it can also drive aggregation and must be used with caution. Some of the main advantages of magnetic bead enrichment is that the procedure is usually fast, involves minimal labelling, and does not make cells undergo the physical stress of flow cytometry—which can potentially damage or alter the physiology of sensitive cell types.
- *Micromanipulation*: When only few cells are available and visual inspection of a cell is desired prior to sequencing, micromanipulation can be used. Here, cells are aspirated with a glass micropipette under a microscope. However, this is highly laborious and not suited to high-throughput analyses.
- *Use of microfluidic devices*: These allow for the sorting of single cells into individual compartments, where they can be visually monitored prior to processing. An example of this is the Fluidigm C1 autoprep.

Either way, until further processing, it is extremely important to keep single-cell suspensions ice-cold and in a suitable medium to maintain viability—usually a source of glucose, amino acids, and other nutrients supplemented or not with fetal bovine serum (FBS) or a similar substitute. In addition, and in order to prevent the cells from entering into anchorage-dependent forms of cell death like *anoikis* (a common cause of sample disruption that occurs when cells detach from the surrounding extracellular matrix), some small-molecule compounds may be needed. This is the case of the Rho-associated protein kinase inhibitor Y27632, commonly used to preserve the viability of dissociated cells [28]. But in general, any protocol that is able to generate a single-cell suspension that shows high viability prior to sequencing (as evidenced by dead/live staining) and from which high-quality RNA can be extracted (discussed below) has a high chance of producing high-quality scRNA-seq data.

1.2 Choosing the Right Platform for scRNA-seq

Once individual cells have been isolated, different scRNA-seq technologies implement the previously described pipeline in slightly different ways. In order to select the appropriate technology, one has to consider several issues [18, 26].

- *The goal of the experimental study:*
  - If the goal is to study *gene expression changes* across cells (e.g., for cell type identification), then technical variability needs to be minimized and a technology that allows for the integration of unique molecular identifiers (UMIs) [29, 30] should be prioritized (*see* for example Zeisel et al. [13]). UMIs are random sequences of bases (usually 8–15 nt) commonly employed in the most recent scRNA-seq methods that can be used to tag each individual transcript prior to library amplification, thereby aiding in the identification of PCR duplicates after sequencing. UMIs are added in excess, so that the number of unique barcodes is much larger than the number of transcripts in each individual cell. Therefore, virtually every transcript within the cell will receive a different UMI. Thus, if two reads align to the same gene and contain the same UMI, it is highly likely that they are PCR duplicates originating from the same fragment prior to amplification—and as such, one of them should be discarded. By collapsing all the reads sharing the same genomic coordinates and UMI into a single representative read, we can obtain an accurate estimate of the true transcript levels in the original sample. Counting UMIs instead of reads can lead to a significant reduction in technical noise that arises from exponential amplification by PCR; however, UMIs can only be used for methods that sequence a single end from a given transcript molecule. If one is interested in applying the study of gene expression to characterize the cellular composition of a tissue, then a high-throughput droplet-based method that allows for a very large number of cells to be sequenced will be preferred. Sequencing a large number of cells is also particularly important if one is to confidently identity rare cell types in an agnostic manner [31, 32]. However, if one is interested in

characterizing a rare cell population for which there are known surface markers, then it will be probably best to enrich using FACS or other methods and then sequence a smaller number of cells [32].

- If *information along the entire transcript is required*, for instance, to study splice variants, then a technology that yields whole-transcript coverage should be chosen instead [33, 34].

• *Ease of the experimental procedure, size of the dataset, and sequencing cost per cell*: Each scRNA-seq technology has its own particularities in terms of user-friendliness, cell capture efficiency, and cost, among others, and it is important that any person interested in scRNA-seq is aware of them. Development of scRNA-seq technology is a very active area of research, with most commonly used methods having been deeply reviewed, compared, and contrasted elsewhere [35–37]. Essentially, these different methods can be categorized in different ways—the two most important aspects for classification are the type of quantification and capture.

- *Classification of single-cell RNA-seq methods according to quantification*: According to the way in which transcript quantification is performed, we will distinguish between full-length and tag-based methods.

  Full-length-based methods [34, 38]: These try to achieve a uniform read coverage of each transcript and are preferred if interested in studying different mRNA isoforms.

  Tag-based methods [39–42]: These methods only capture either the 5′ or the 3′ end of the transcript. Their main advantage is that they can be combined with UMIs to improve quantification. However, because they are restricted to one end of the transcript, there is a reduction in the overall mappability of the transcriptome and it is harder to distinguish different mRNA isoforms and allelic variants.

- *Classification of single-cell RNA-seq methods according to mechanism of capture*: The strategy used for the capture will impact, among others, the throughput of the experiment (that is, the number of cells that can be targeted or dataset size) and what other additional information can potentially be obtained. The main capture strategies are briefly explained below.

  Plate-based strategies [40]: Cells are isolated (either with a pipette, a FACS instrument, or by laser capture) and placed individually in wells. This potentially allows for the visual monitoring of the cells, with the images obtained providing additional information and helping in the identification of damaged cells or doublets. However, these methods are usually very low-throughput and time-consuming.

Microfluidic-based strategies [43, 44]: These provide a more integrated system for capture and library preparation. However, their capture efficiency is comparatively low, and for that reason they are not that well-suited when dealing with rare cell types or very small amounts of input. Still, the reagent volumes used are small, which make them a cost-effective alternative.

Microwell array-based methods [45, 46]: These allow for mRNA capture beads and single cells to be sealed in a microwell array for efficient cell lysis and transcript capture. Importantly, these microarrays can be scalable, low cost, and allow for massively parallel scRNA-seq.

Droplet-based methods [39, 41, 42]: These involve the co-encapsulation of each individual cell inside a droplet, together with a bead that contains a set of unique cell barcodes that attach to all the transcripts originating from each individual cell. In this way, all droplets can be pooled, undergo library preparation and be sequenced together, and the reads can be subsequently assigned to the cell of origin. Droplet-based strategies typically have high throughput, and library preparation costs are pretty low. However, coverage is usually low as well, with only a few thousand different transcripts detected.

Split-pool barcoding methods [47, 48]: These approaches allow for labeling of the cellular origin of RNA through combinatorial barcoding, and thus *scale exponentially.* Split-pool barcoding approaches have been successfully performed on fixed cells and nuclei, and do not require the physical partitioning of single cells into individual compartments, as the individual cells themselves serve as compartments. Additional levels of indexing in the split-pool barcoding strategy can provide greater complexity; thus, these approaches have the advantage of being low cost and high throughput.

1.3 Getting to Know the Performance Parameters

Prior to the enumeration of the basic principles for proper design of scRNA-seq experiments, let's briefly describe some of the main parameters that need to be considered when it comes to single-cell transcriptomic studies [7, 26, 49].

- *Coverage*: Number of reads that include a given nucleotide in the reference genome or transcriptome.

- *Sequencing depth*: Number of reads sequenced per transcript molecule. Importantly, sequencing depth is inversely correlated with throughput, so that large sample sizes impose a practical limit on sequencing depth.
- *Throughput*: Number of cells profiled. Technological advances (robotics and automation, microfluidics, reverse emulsion and hydrogel droplets, among others) have allowed for the substantial increase in the throughput of scRNA-seq technologies, from a handful to hundreds of thousands of cells sequenced per experiment [50].
- *Complexity*: Number of distinct molecules that can be captured from each cell (indeed, library complexity refers to the number of unique biological molecules that are represented in a sequencing library). Unlike throughput, complexity is difficult to characterize and optimize in scRNA-seq experiments. This is because individual cells have a variable RNA content, and because of varied, unpredictable cellular features can affect the recovery of their mRNA. This makes it impossible to estimate the performance of an assay using a replicate experiment. However, estimates of complexity can be obtained by assuming the mRNA content of cells is typically predicted to be between 10^5 to 10^6 molecules per cell [51].
- *Sensitivity*: Fraction of transcript molecules detected per cell. As already touched upon, we have to remember that scRNA-seq inherently has a sampling limitation in that it directly measures only some of the cells in a population, and only a fraction of the RNA molecules in each cell.

Among these performance parameters, sequencing depth and throughput deserve especial mention. The sequencing depth should be determined based on cDNA yield. We need to consider aspects of the experiment like the size and complexity of the transcriptome being assayed (are there repetitive regions?), the error rate of the sequencing platform, and, in particular, the biological question under investigation. Consider increasing the sequencing depth (number of reads) needed for a specific RNA-seq experiment if the question addressed involves:

- Identification of lowly expressed genes.
- Identification of very small fold changes between different conditions.
- Cell type classification in a mixed population.
- Quantification of alternative splicing/distinct transcript isoforms.
- Detection of chimeric transcripts.
- Detection of novel transcripts, transcription start and end sites.

- Performance of de novo transcript assembly.
- Analysis of the regulatory relationships between genes in single cells.
- Use of a sequencing platform that has a moderate to high error rate.

The case of the identification and classification of cell types in a mixed population also deserves special mention, as cell type classification is particularly dependent on sequencing depth. Drastic undersampling interferes with distinguishing between different cell types. However, always consider that increasing the number of biological replicates, in addition to the throughput, may be better suited to aid in classification than sequencing a smaller number of cells at a greater depth.

1.4 Addressing scRNA-seq Challenges from the Experimental Design Perspective

In spite of the substantial advances made over the last decade, scRNA-seq comes with major analytical challenges and yields complex data output. Among the factors that contribute to the challenges of single-cell analysis are the relatively small number of sequencing reads obtained per cell (sequencing depth), the sparsity of the data and its associated noise, and cell population heterogeneity. Fortunately, proper experimental design can address some of these challenges.

In order to distinguish transcriptomic changes due to the condition being studied (e.g., treatment A vs. treatment B, developmental stage A vs. stage B) from those caused by irrelevant biological differences between organisms (normal biological variation), experimenters, environmental or technical conditions (batch effects), it is necessary to perform our scRNA-seq experiments with a sufficient number of replicates, appropriate controls, and with a well-planned experimental design [26, 52]. In the end, our goal is to observe a reproducible effect that can only be ascribed to our differential treatment conditions (i.e., avoiding confounders and bias).

- *Capturing enough variability—the importance of replicates*: Ideally, our experiment should include the minimum number of replicates that is sufficient to capture the breadth of the variability of the response of interest and allows us to identify and isolate potential sources of noise. With enough replicates, outlier samples can be detected and removed. Replicates in scRNA-seq are important for validating differences in cellular composition between conditions or the specificity of markers for a given subpopulation. When working with mice, biological replicates represent biological samples taken from different animals. Of course, the number of replicates should be considered in terms of overall cost. Economic limitations play an important role in experimental design for scRNA-seq, which remains a very expensive experiment. Importantly, scRNA-seq can still be a highly valuable discovery tool even when limited replicates are used, as long as we can validate our findings with sufficient statistical power. Thus, validation of the scRNA-seq results is critically important!

- *Avoiding bias—Normalization and blocking*: The main goal of a well-planned experiment is to get accurate and precise results. This means that we should always consider the following:
  - Identify the question of interest, and design the experiment so that we can address it directly.
  - Identify possible sources of variability beforehand and plan the experiment in a way that reduces the effect of these expected factors.
  - Protect against unknown sources of variation. For this, it is important that we implement variable randomization and/or blocking. Randomization when assigning samples into groups can help avoid unconscious selection bias, whereas blocking by some factors that are very likely to be responsible for gene expression variation (like sex or age) may help increase statistical sensitivity.
- *Dealing with technical variation—Controlling for noise and avoiding further batch effects*: Technical artifacts pose an important challenge for data interpretation but can be accounted for at very early steps.
  - Technical noise: scRNA-seq is noisier than bulk RNA-seq due to:

    Small amounts of starting material (between 10 and 50 pg RNA on average).

    Sparse sampling: The efficiency of the RNA capture and cDNA conversion rate is always imperfect, so any molecules missed will lead to loss of information for that particular cell.

    Amplification bias and dropouts: Because of the above factors, starting material must be amplified significantly, which can lead to amplification bias and further gene dropout (in which a gene is observed at a moderate expression level in one cell but not detected in another one, and not necessarily because it is not expressed). This can lead to significant distortion of the gene expression profiles and an inflation of the estimates of cell-to-cell variability. Pooled amplification can also lead to recombination artifacts in which the barcode labeling one cell can become associated with a transcript molecule from a different cell.

- Batch effects: Introduced when cells from one biological group are cultured, isolated, captured, and/or sequenced separately from cells in a second condition. Implement good experimental designs and prevent batch effects from confounding your analysis, also at the library preparation and sequencing level: single-cell libraries derived from different experimental conditions should be distributed in equal fractions across the same set of lanes. For this, sample multiplexing using oligonucleotide-tagged antibodies can be very helpful [53]. Finally, since technical variability introduced during the sequencing protocol is lower than that arising from tissue preparation protocols, technical replicates accounting for library preparation alone are rarely undertaken.

- *Cellular heterogeneity*: While in some cases it can represent technical noise and in others it can be biologically meaningful, it is important to be aware that differences attributed to cell size, transcriptional bursts, or other factors may also be observable. In order to allow for robust and nonconfounded interpretation, it is essential to identify whether we are dealing with natural variation within the same cell type versus a functional state transition. This can be somewhat addressed by increasing the number of replicates or cells within a sample.

Finally, the power of a scRNA-seq experiment crucially depends on three of the parameters described above: throughput, sequencing depth, and complexity. These parameters, which can be experimentally controlled, need to be set according to the goal of the study.

- *Size of the dataset (throughput or number of cells)*: This is important for profiling cell composition with high sensitivity. Possible biases that might occur during isolation of the single-cell sample due to cell size or other factors need to be considered. One should always incorporate an estimate of the success rate (how many cells were informative compared to the number of cells I started with in the beginning?), as a number of samples will likely yield little to no material (RNA degradation, low amplification efficiency).

- *Sequencing depth and complexity*: It is important to sequence each cell with sufficient depth. If transcripts are counted with UMIs, then the sequencing depth should be adjusted such that every transcript is sequenced at least three to four times. This ensures that even lowly expressed genes can be quantified.

2 Materials

2.1 Single-Cell Dissociation of the Murine Intestinal Epithelium

1. 70% ethanol.
2. Surgical toolkit.
3. Styrofoam board and pins.
4. PBS.
5. Petri dish, 150 mm × 25 mm.
6. 10 mL syringe.
7. 20 G × 11/2 Needle.
8. 15 mL conical tubes.
9. 50 mL conical tubes.
10. 10 mM EDTA.
11. Collagenase/Dispase.
12. 40 μM cell strainer.
13. 5 mM Y27632.
14. Advanced DMEM/F12.
15. 1 M HEPES.
16. 100× Penicillin/Streptomycin/Glutamine.
17. FBS (fetal bovine serum).
18. SYTOX Blue.
19. Antibodies: anti-mouse CD31 antibody, anti-mouse CD45 antibody, anti-mouse EpCAM antibody.
20. FACS medium: Advanced DMEM/F12 with 10% FBS, 1× Penicillin/Streptomycin/Glutamine, 10 mM HEPES, 5 μM Y27632.

2.2 Epithelial Cell Enrichment and Sample Collection by FACS

1. FACS medium.
2. Stained epithelial single-cell suspension and unstained control.
3. Disposable 2 mL pipettes.
4. FACS tubes.
5. 1.5 mL collection tubes.
6. RNA lysis buffer.

2.3 scRNA-seq Using the 10× Genomics Chromium Platform

2.3.1 Viability Assay

1. Countess II FL Automated Cell Counter.
2. Trypan Blue.
3. Countess Cell Counting Chamber Slides.

2.3.2 GEM Preparation, Reverse Transcription, and Library Prep

For scRNA-Seq Library Preparation

1. Chromium™ Single Cell 3′ GEM, Library & Gel Bead Kit v3.
2. Chromium™ Chip B Single Cell Kit.
3. SPRIselect Reagent.
4. T-100 Thermal Cycler.

For cDNA and Library QC

5. 2200 TapeStation (Agilent).
6. High Sensitivity D5000 ScreenTapes and Reagents.
7. High Sensitivity D1000 ScreenTapes and Reagents.
8. Qubit 2.0.
9. Qubit dsDNA High Sensitivity Kit.

2.3.3 Sequencing

1. 2 N sodium hydroxide.
2. 1 M Tris–HCl (pH 8.0).
3. NovaSeq 6000 (Illumina).
4. NovaSeq S2 100 Cycle Reagent Kit (Illumina).

3 Methods

3.1 Single-Cell Dissociation of Intestinal Epithelium

1. Euthanize mice. Pin down limbs to immobilize on Styrofoam board with ventral side facing up. Disinfect with 70% ethanol and, with blunt scissors, make a central incision slightly above the pelvis and open the abdominal cavity.
2. Locate and liberate the stomach at its junction with the esophagus. Holding the stomach with forceps, begin to release the gastrointestinal tract by cutting away the liver, pancreas, spleen, and mesentery with surgical scissors.
3. Once the intestine has been removed from the animal, transfer to ice-cold PBS in a petri dish (*see* **Note 1**).
4. Isolate the intestine by trimming off the stomach and cecum. Trim away any remaining excess tissue adhering to the serosal surface of the intestine.
5. Flush out the luminal contents of the intestine using a 10 mL syringe with a fine 20 G needle and PBS. PBS should flow freely with minimal resistance.
6. Once the tissue has been flushed, discard dirty PBS in the petri dish and replace with abundant fresh cold PBS.
7. Cut the intestine open longitudinally with scissors. Rinse the tissue several times (3×–5×) until the intestinal walls are devoid of any adherent luminal contents.

8. Transfer intestine into a 15 mL conical tube containing 10 mL pre-chilled 10 mM EDTA in PBS. Incubate on ice for 30 min.
9. Transfer tissue into a 15 mL conical tube with 10 mL pre-chilled PBS and manually shake for ~5 min using hard, quick, energetic bursts. After shaking, the tube will appear cloudy as epithelial cells are liberated from the tissue. Supernatant can be checked by microscopy for successful isolation of intact intestinal crypts and villi.
10. Remove residual intestine tissue and spin down tube containing the epithelium for 5 min at 250–300 rcf (4 °C).
11. Discard supernatant and resuspend cell pellet in 10 mL pre-chilled PBS to fully wash out EDTA. Spin cell suspension down for 5 min at 250–300 rcf (4 °C).
12. Discard supernatant and resuspend cell pellet in 5 mL pre-chilled PBS containing Dispase/Collagenase up to a final concentration of 1 mg/mL (collagenase 0.1 U/mL; dispase 0.8 U/mL). Incubate at 37 °C for approximately 5 min in a water bath. Monitor for single cell dissociation by microscopy every 1–2 min.
13. Quickly place tube on ice and add FBS to 5% (v/v) and mix well.
14. Strain suspension through a 40 μm strainer into a 50 mL conical tube.
15. Spin cells down 250–300 rcf for 5 min (4 °C), discard supernatant, and resuspend once more in FACS medium supplemented with 5 μM Y27632 (Rho kinase inhibitor) and 1 mM EDTA.
16. Centrifuge again at 250–300 rcf for 5 min (4 °C), discard supernatant, and resuspend cell pellet in a suitable amount (1–5 mL) of supplemented FACS medium. Save a small fraction of the resuspended cells for use as an unstained control, then add the antibodies to the rest for immunostaining on ice for 30 min (*see* **Note 2**).
17. Wash the stained cells 3×. Centrifuge at 250–300 rcf for 5 min (4 °C) in between washes, and discard supernatant.
18. After the final wash and centrifugation, resuspend cell pellet in 1–3 mL of FACS media with 1× SYTOX Blue. Keep cells on ice and proceed to FACS.

3.2 Epithelial Cell Enrichment and Sample Collection by FACS

While flow cytometry cores at many academic institutions are operated by a team of trained experts who are able to provide guidance and assistance to the users, below we outline a basic protocol for the FACS-mediated isolation of intestinal epithelial cells based on the surface expression of epithelial cell adhesion molecule (EpCAM) [54], while simultaneously depleting the sample of endothelial (CD31+) and hematopoietic (CD45+) cells (Fig. 3).

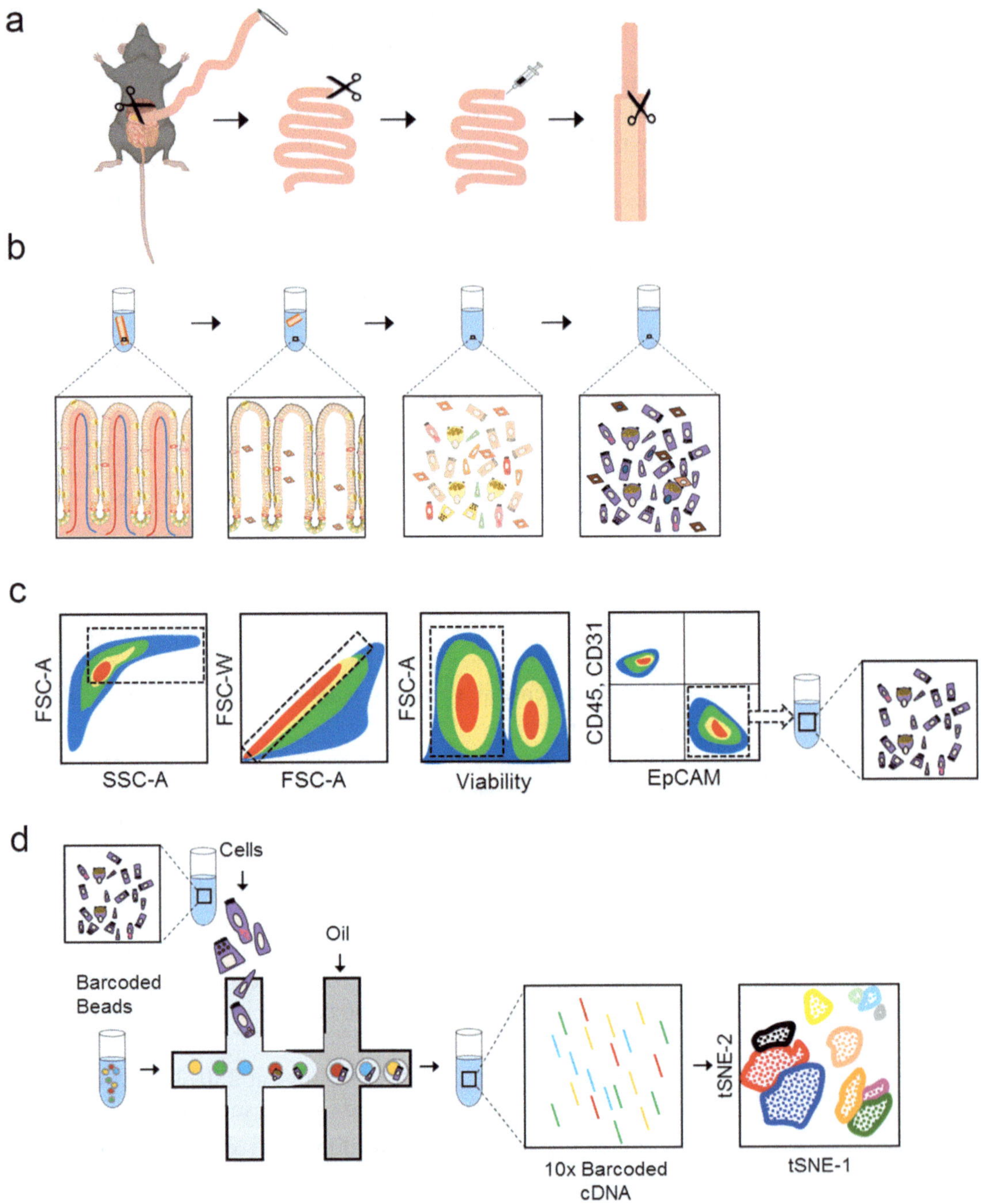

Fig. 3 Schema of single-cell dissociation of the murine small intestinal epithelium workflow. (**a**) Isolation of the murine small intestine. The intestine is removed en bloc from the mouse via an abdominal incision. (**b**) Single-cell dissociation of the intestinal epithelium. The epithelial sheet is extracted from the underlying tissue and dissociated into a single-cell suspension. (**c**) Single-cell isolation by FACS. Hierarchical FACS gating schema is shown. (**d**) 10× Chromium single-cell platform which enables the encapsulation of individual cells within droplets

1. Set up the flow cytometer. Use a FACS instrument consistent with the number and the spectral characteristics of the fluorophores contained in the sample. Ensure the cell sorting system is working and target the cells into the center of the collection tubes.
2. Set up the following panels and hierarchical gating system in the computer monitor (*x* and *y* axis). Adjust gates accordingly with the help of the negative (unstained) control and the stained populations: scatter, doublet/multicellular aggregate exclusion gate, viability, and epithelial cell discrimination (EpCAM+CD31−CD45−).
3. Sort viable epithelial cells out of stained sample (*see* **Note 3**).
4. Perform a post-sort check. Reload the sorted cells back to the cytometer and ensure sorting occurred properly by confirming high post-sort viability and purity. Prioritize the sorting of 10,000 cells back into 500 μL FACS media for scRNA-seq, and sort some of the remnant cells (at least 1000) straight into 300 μl RNA lysis buffer for validation purposes (*see* **Notes 4–6**). Quickly vortex the samples in RNA lysis buffer to ensure all cells come in contact with the lysis solution.
5. Transfer samples to ice (store at −80 °C for the sample in RNA lysis buffer if not proceeding to RNA extraction immediately after the sort).
6. Clean and shutdown FACS machine.
7. Proceed to scRNA-seq.

3.3 scRNA-seq Using 10× Genomics Chromium V3 Platform

Below is a simplified protocol for the scRNA-seq of the intestinal epithelium using 10× Genomics Chromium Platform V3 chemistry [55] for the generation of single cell 3′ gene expression libraries (Fig. 3d).

10× Genomics Chromium is a commonly used commercially available droplet-based method for single-cell transcriptomics, which partitions individual cells into nanoliter-scaled so-called gel beads-in-emulsion (GEMs). Each of these GEMs contains multiple copies of a single-stranded oligonucleotide that will allow for the capture of each individual transcriptome and the ultimate generation of cell-barcoded, full-length cDNA from all the polyadenylated mRNAs derived from every single cell, out of which a sequencing library can be generated.

The protocol can be divided into five different steps:

1. Determine cell viability. Cell viability immediately prior to sequencing should not drop below 80–90%.
2. Generate GEMs.
   (a) Prepare Single Cell Master Mix with desired cell recovery target.

(b) Load Chromium chip with Single Cell Master Mix, Gel Beads, and Partitioning Oil.

(c) Run Chromium controller.

(d) Transfer generated GEMs to new tube strip.

(e) Incubate GEMs in a thermal cycler with recommended protocol for RT.

3. Perform Post GEM-RT Cleanup and cDNA Amplification.

(a) Add Recovery Reagent to each sample. Wait 2 min and then remove Recovery Agent/Partitioning Oil.

(b) Perform GEM-RT Cleanup with Dynabeads.

(c) Amplify cDNA using cDNA Amplification Mix and recommended protocol on thermal cycler.

(d) Perform cDNA Cleanup and size selection with SPRIselect Reagent.

(e) Run cDNA quality control and quantification.

4. Construct Gene Expression Library.

(a) Add Fragmentation Mix to sample and 10× recommended protocol on thermal cycler for fragmentation, end repair, and A-tailing.

(b) Perform Post Fragmentation, End Repair, and A-tailing size selection with SPRIselect Reagent.

(c) Add Adaptor Ligation Mix and incubate in thermal cycler using recommended protocol.

(d) Perform Post Ligation Cleanup and size selection with SPRIselect Reagent.

(e) Prepare Sample Index PCR Mix. Incubate in thermal cycler with recommended protocol.

(f) Perform size selection with SPRIselect Reagent.

(g) Run diluted sample on Agilent Bioanalyzer High Sensitivity chip to determine average fragment size.

5. Sequence. The gene expression library is now ready for Illumina's pair-end sequencing. In general, libraries are first quantitated and denatured for sequencing on an Illumina Hi-Seq or Nova-Seq sequencer.

4 Notes

1. Work quickly to keep cells viable.
2. If attempting to use different sets of antibodies for immunostaining (different suppliers, or distinct fluorophore conjugates), empirically determine what concentrations are best to use. Pay attention to potential spectral overlap between fluorophores.

3. If low viability is observed, lower the concentration of EDTA, enzymes, or incubation times. Alternatively, Y27632 can be added at an earlier time point to prevent *anoikis*.
4. Single-cell isolation and FACS can be very easy and run smoothly, but it can also go awry at multiple points. Thus, before spending thousands of dollars in a scRNA-seq experiment that may not work, it is important to ensure that the cell isolation procedure worked. We recommend running a validation step in parallel to scRNA-seq. One easy way in which we can assess whether our sorted populations are the correct ones is through qPCR on the FACS-isolated samples to validate the expression of some expected markers. This entails RNA extraction, cDNA synthesis by reverse transcription, and qPCR analysis.
5. Beware the RNases! It is important to consider that RNA extraction is always complicated by the ubiquitous presence of hardy ribonuclease (RNase) enzymes in cells and tissues, which can rapidly degrade RNA. In order to account for this, it is necessary that all the equipment used for RNA extraction is thoroughly cleaned, kept separate from common lab equipment that handles other types of samples, and carefully treated with various harsh chemicals that destroy RNases. For the same reason, experimenters must take special care not to let their bare skin come in close contact to the samples, pieces of equipment, reagents, or surfaces where RNA extraction takes place.
6. The RNA Integrity Score (RIN). RNA is easily degraded and thus requires careful handling. For this reason, assessing the quality of the RNA used in RNA-seq experiments is critically important in order to get unbiased, comparable, robust and interpretable results that are free of additional confounding factors. In the case of scRNA-seq, different metrics exist at the level of sequencing analysis that allow us to quantitatively assess the status and integrity of our transcripts. However, for the purposes of FACS validation by qPCR, there are some quick tests that can be done in order to get an idea of the distribution of the RNA species in our sample, their physical integrity, and, after all, whether the sample can be used or not for further analyses. Currently, software tools that accompany instruments such as the Agilent Bioanalyzer are able to calculate the RNA Integrity Number (RIN) from a digital representation of the size distribution of the RNA molecules contained in our sample. Assessing RNA quality with a method like this is preferred since visual inspection by gel electrophoresis appears to be inconsistent, subjective, and thus less reliable. The automated approach facilitates the interpretation and reproducibility of RNA quality assessments, and provides a means by which

samples can be compared one to another in a standardized manner. Importantly, if you are unable to get good quality RNA out of your sorted cells in the beginning, it is best to optimize your protocol rather than proceeding with an expensive scRNA-seq experiment.

Acknowledgments

This work was supported by NIH (R03DK114656, U01DK103155, and P30CA013696, BWF CAMS, Louis V. Gerstner,Jr. Scholars Award, Lisa Dean Moseley Foundation, and the Irma T. Hirschl Career Award. C.C. was supported by a NYSTEM predoctoral training grant.

References

1. Gehart H, Clevers H (2019) Tales from the crypt: new insights into intestinal stem cells. Nat Rev Gastroenterol Hepatol 16(1):19–34. https://doi.org/10.1038/s41575-018-0081-y
2. Allaire JM, Crowley SM, Law HT, Chang SY, Ko HJ, Vallance BA (2018) The intestinal epithelium: central coordinator of mucosal immunity. Trends Immunol 39(9):677–696. https://doi.org/10.1016/j.it.2018.04.002
3. Middelhoff M, Westphalen CB, Hayakawa Y, Yan KS, Gershon MD, Wang TC, Quante M (2017) Dclk1-expressing tuft cells: critical modulators of the intestinal niche? Am J Physiol Gastrointest Liver Physiol 313(4): G285–G299. https://doi.org/10.1152/ajpgi.00073.2017
4. de Sousa EMF, de Sauvage FJ (2019) Cellular plasticity in intestinal homeostasis and disease. Cell Stem Cell 24(1):54–64. https://doi.org/10.1016/j.stem.2018.11.019
5. Barker N, van Es JH, Kuipers J, Kujala P, van den Born M, Cozijnsen M, Haegebarth A, Korving J, Begthel H, Peters PJ, Clevers H (2007) Identification of stem cells in small intestine and colon by marker gene Lgr5. Nature 449(7165):1003–1007. https://doi.org/10.1038/nature06196
6. Trapnell C (2015) Defining cell types and states with single-cell genomics. Genome Res 25(10):1491–1498. https://doi.org/10.1101/gr.190595.115
7. Tanay A, Regev A (2017) Scaling single-cell genomics from phenomenology to mechanism. Nature 541(7637):331–338. https://doi.org/10.1038/nature21350
8. Wang Z, Gerstein M, Snyder M (2009) RNA-Seq: a revolutionary tool for transcriptomics. Nat Rev Genet 10(1):57–63. https://doi.org/10.1038/nrg2484
9. Tang F, Barbacioru C, Wang Y, Nordman E, Lee C, Xu N, Wang X, Bodeau J, Tuch BB, Siddiqui A, Lao K, Surani MA (2009) mRNA-Seq whole-transcriptome analysis of a single cell. Nat Methods 6(5):377–382. https://doi.org/10.1038/nmeth.1315
10. Haber AL, Biton M, Rogel N, Herbst RH, Shekhar K, Smillie C, Burgin G, Delorey TM, Howitt MR, Katz Y, Tirosh I, Beyaz S, Dionne D, Zhang M, Raychowdhury R, Garrett WS, Rozenblatt-Rosen O, Shi HN, Yilmaz O, Xavier RJ, Regev A (2017) A single-cell survey of the small intestinal epithelium. Nature 551(7680):333–339. https://doi.org/10.1038/nature24489
11. Loo L, Simon JM, Xing L, McCoy ES, Niehaus JK, Guo J, Anton ES, Zylka MJ (2019) Single-cell transcriptomic analysis of mouse neocortical development. Nat Commun 10(1):134. https://doi.org/10.1038/s41467-018-08079-9
12. Mickelsen LE, Bolisetty M, Chimileski BR, Fujita A, Beltrami EJ, Costanzo JT, Naparstek JR, Robson P, Jackson AC (2019) Single-cell transcriptomic analysis of the lateral hypothalamic area reveals molecularly distinct populations of inhibitory and excitatory neurons. Nat Neurosci 22(4):642–656. https://doi.org/10.1038/s41593-019-0349-8
13. Zeisel A, Munoz-Manchado AB, Codeluppi S, Lonnerberg P, La Manno G, Jureus A, Marques S, Munguba H, He L, Betsholtz C,

Rolny C, Castelo-Branco G, Hjerling-Leffler J, Linnarsson S (2015) Brain structure. Cell types in the mouse cortex and hippocampus revealed by single-cell RNA-seq. Science 347 (6226):1138–1142. https://doi.org/10.1126/science.aaa1934

14. Mitra AK, Mukherjee UK, Harding T, Jang JS, Stessman H, Li Y, Abyzov A, Jen J, Kumar S, Rajkumar V, Van Ness B (2016) Single-cell analysis of targeted transcriptome predicts drug sensitivity of single cells within human myeloma tumors. Leukemia 30 (5):1094–1102. https://doi.org/10.1038/leu.2015.361
15. Sharma A, Cao EY, Kumar V, Zhang X, Leong HS, Wong AML, Ramakrishnan N, Hakimullah M, Teo HMV, Chong FT, Chia S, Thangavelu MT, Kwang XL, Gupta R, Clark JR, Periyasamy G, Iyer NG, DasGupta R (2018) Longitudinal single-cell RNA sequencing of patient-derived primary cells reveals drug-induced infidelity in stem cell hierarchy. Nat Commun 9(1):4931. https://doi.org/10.1038/s41467-018-07261-3
16. Lee MC, Lopez-Diaz FJ, Khan SY, Tariq MA, Dayn Y, Vaske CJ, Radenbaugh AJ, Kim HJ, Emerson BM, Pourmand N (2014) Single-cell analyses of transcriptional heterogeneity during drug tolerance transition in cancer cells by RNA sequencing. Proc Natl Acad Sci U S A 111(44):E4726–E4735. https://doi.org/10.1073/pnas.1404656111
17. Raj A, van Oudenaarden A (2008) Nature, nurture, or chance: stochastic gene expression and its consequences. Cell 135(2):216–226. https://doi.org/10.1016/j.cell.2008.09.050
18. Papalexi E, Satija R (2018) Single-cell RNA sequencing to explore immune cell heterogeneity. Nat Rev Immunol 18(1):35–45. https://doi.org/10.1038/nri.2017.76
19. Farrell JA, Wang Y, Riesenfeld SJ, Shekhar K, Regev A, Schier AF (2018) Single-cell reconstruction of developmental trajectories during zebrafish embryogenesis. Science 360(6392). https://doi.org/10.1126/science.aar3131
20. Chan TE, Stumpf MPH, Babtie AC (2017) Gene regulatory network inference from single-cell data using multivariate information measures. Cell Syst 5(3):251–267.e253. https://doi.org/10.1016/j.cels.2017.08.014
21. Moris N, Pina C, Arias AM (2016) Transition states and cell fate decisions in epigenetic landscapes. Nat Rev Genet 17(11):693–703. https://doi.org/10.1038/nrg.2016.98
22. Rizvi AH, Camara PG, Kandror EK, Roberts TJ, Schieren I, Maniatis T, Rabadan R (2017) Single-cell topological RNA-seq analysis reveals insights into cellular differentiation and development. Nat Biotechnol 35 (6):551–560. https://doi.org/10.1038/nbt.3854
23. Treutlein B, Brownfield DG, Wu AR, Neff NF, Mantalas GL, Espinoza FH, Desai TJ, Krasnow MA, Quake SR (2014) Reconstructing lineage hierarchies of the distal lung epithelium using single-cell RNA-seq. Nature 509 (7500):371–375. https://doi.org/10.1038/nature13173
24. Hwang B, Lee JH, Bang D (2018) Single-cell RNA sequencing technologies and bioinformatics pipelines. Exp Mol Med 50(8):96. https://doi.org/10.1038/s12276-018-0071-8
25. Streets AM, Huang Y (2014) How deep is enough in single-cell RNA-seq? Nat Biotechnol 32(10):1005–1006. https://doi.org/10.1038/nbt.3039
26. Grun D, van Oudenaarden A (2015) Design and analysis of single-cell sequencing experiments. Cell 163(4):799–810. https://doi.org/10.1016/j.cell.2015.10.039
27. Adam M, Potter AS, Potter SS (2017) Psychrophilic proteases dramatically reduce single-cell RNA-seq artifacts: a molecular atlas of kidney development. Development 144 (19):3625–3632. https://doi.org/10.1242/dev.151142
28. Koyanagi M, Takahashi J, Arakawa Y, Doi D, Fukuda H, Hayashi H, Narumiya S, Hashimoto N (2008) Inhibition of the Rho/ROCK pathway reduces apoptosis during transplantation of embryonic stem cell-derived neural precursors. J Neurosci Res 86(2):270–280. https://doi.org/10.1002/jnr.21502
29. Kivioja T, Vaharautio A, Karlsson K, Bonke M, Enge M, Linnarsson S, Taipale J (2011) Counting absolute numbers of molecules using unique molecular identifiers. Nat Methods 9(1):72–74. https://doi.org/10.1038/nmeth.1778
30. Islam S, Zeisel A, Joost S, La Manno G, Zajac P, Kasper M, Lonnerberg P, Linnarsson S (2014) Quantitative single-cell RNA-seq with unique molecular identifiers. Nat Methods 11(2):163–166. https://doi.org/10.1038/nmeth.2772
31. Jindal A, Gupta P, Jayadeva, Sengupta D (2018) Discovery of rare cells from voluminous single cell expression data. Nat Commun 9 (1):4719. https://doi.org/10.1038/s41467-018-07234-6
32. Nguyen A, Khoo WH, Moran I, Croucher PI, Phan TG (2018) Single cell RNA sequencing

of rare immune cell populations. Front Immunol 9:1553. https://doi.org/10.3389/fimmu.2018.01553

33. Shalek AK, Satija R, Adiconis X, Gertner RS, Gaublomme JT, Raychowdhury R, Schwartz S, Yosef N, Malboeuf C, Lu D, Trombetta JJ, Gennert D, Gnirke A, Goren A, Hacohen N, Levin JZ, Park H, Regev A (2013) Single-cell transcriptomics reveals bimodality in expression and splicing in immune cells. Nature 498 (7453):236–240. https://doi.org/10.1038/nature12172
34. Hayashi T, Ozaki H, Sasagawa Y, Umeda M, Danno H, Nikaido I (2018) Single-cell full-length total RNA sequencing uncovers dynamics of recursive splicing and enhancer RNAs. Nat Commun 9(1):619. https://doi.org/10.1038/s41467-018-02866-0
35. Ziegenhain C, Vieth B, Parekh S, Reinius B, Guillaumet-Adkins A, Smets M, Leonhardt H, Heyn H, Hellmann I, Enard W (2017) Comparative analysis of single-cell RNA sequencing methods. Mol Cell 65(4):631–643.e634. https://doi.org/10.1016/j.molcel.2017.01.023
36. Svensson V, Natarajan KN, Ly LH, Miragaia RJ, Labalette C, Macaulay IC, Cvejic A, Teichmann SA (2017) Power analysis of single-cell RNA-sequencing experiments. Nat Methods 14(4):381–387. https://doi.org/10.1038/nmeth.4220
37. Prakadan SM, Shalek AK, Weitz DA (2017) Scaling by shrinking: empowering single-cell 'omics' with microfluidic devices. Nat Rev Genet 18(6):345–361. https://doi.org/10.1038/nrg.2017.15
38. Picelli S, Faridani OR, Bjorklund AK, Winberg G, Sagasser S, Sandberg R (2014) Full-length RNA-seq from single cells using Smart-seq2. Nat Protoc 9(1):171–181. https://doi.org/10.1038/nprot.2014.006
39. Macosko EZ, Basu A, Satija R, Nemesh J, Shekhar K, Goldman M, Tirosh I, Bialas AR, Kamitaki N, Martersteck EM, Trombetta JJ, Weitz DA, Sanes JR, Shalek AK, Regev A, McCarroll SA (2015) Highly parallel genome-wide expression profiling of individual cells using nanoliter droplets. Cell 161 (5):1202–1214. https://doi.org/10.1016/j.cell.2015.05.002
40. Keren-Shaul H, Kenigsberg E, Jaitin DA, David E, Paul F, Tanay A, Amit I (2019) MARS-seq2.0: an experimental and analytical pipeline for indexed sorting combined with single-cell RNA sequencing. Nat Protoc 14 (6):1841–1862. https://doi.org/10.1038/s41596-019-0164-4
41. Zheng GX, Terry JM, Belgrader P, Ryvkin P, Bent ZW, Wilson R, Ziraldo SB, Wheeler TD, McDermott GP, Zhu J, Gregory MT, Shuga J, Montesclaros L, Underwood JG, Masquelier DA, Nishimura SY, Schnall-Levin M, Wyatt PW, Hindson CM, Bharadwaj R, Wong A, Ness KD, Beppu LW, Deeg HJ, McFarland C, Loeb KR, Valente WJ, Ericson NG, Stevens EA, Radich JP, Mikkelsen TS, Hindson BJ, Bielas JH (2017) Massively parallel digital transcriptional profiling of single cells. Nat Commun 8:14049. https://doi.org/10.1038/ncomms14049
42. Klein AM, Mazutis L, Akartuna I, Tallapragada N, Veres A, Li V, Peshkin L, Weitz DA, Kirschner MW (2015) Droplet barcoding for single-cell transcriptomics applied to embryonic stem cells. Cell 161 (5):1187–1201. https://doi.org/10.1016/j.cell.2015.04.044
43. Kimmerling RJ, Lee Szeto G, Li JW, Genshaft AS, Kazer SW, Payer KR, de Riba Borrajo J, Blainey PC, Irvine DJ, Shalek AK, Manalis SR (2016) A microfluidic platform enabling single-cell RNA-seq of multigenerational lineages. Nat Commun 7:10220. https://doi.org/10.1038/ncomms10220
44. Streets AM, Zhang X, Cao C, Pang Y, Wu X, Xiong L, Yang L, Fu Y, Zhao L, Tang F, Huang Y (2014) Microfluidic single-cell whole-transcriptome sequencing. Proc Natl Acad Sci U S A 111(19):7048–7053. https://doi.org/10.1073/pnas.1402030111
45. Gierahn TM, Wadsworth MH II, Hughes TK, Bryson BD, Butler A, Satija R, Fortune S, Love JC, Shalek AK (2017) Seq-Well: portable, low-cost RNA sequencing of single cells at high throughput. Nat Methods 14 (4):395–398. https://doi.org/10.1038/nmeth.4179
46. Yuan J, Sims PA (2016) An automated microwell platform for large-scale single cell RNA-Seq. Sci Rep 6:33883. https://doi.org/10.1038/srep33883
47. Cao J, Packer JS, Ramani V, Cusanovich DA, Huynh C, Daza R, Qiu X, Lee C, Furlan SN, Steemers FJ, Adey A, Waterston RH, Trapnell C, Shendure J (2017) Comprehensive single-cell transcriptional profiling of a multicellular organism. Science 357 (6352):661–667. https://doi.org/10.1126/science.aam8940
48. Rosenberg AB, Roco CM, Muscat RA, Kuchina A, Sample P, Yao Z, Graybuck LT, Peeler DJ, Mukherjee S, Chen W, Pun SH, Sellers DL, Tasic B, Seelig G (2018) Single-cell profiling of the developing mouse brain

and spinal cord with split-pool barcoding. Science 360(6385):176–182. https://doi.org/10.1126/science.aam8999

49. Sims D, Sudbery I, Ilott NE, Heger A, Ponting CP (2014) Sequencing depth and coverage: key considerations in genomic analyses. Nat Rev Genet 15(2):121–132. https://doi.org/10.1038/nrg3642
50. Svensson V, Vento-Tormo R, Teichmann SA (2018) Exponential scaling of single-cell RNA-seq in the past decade. Nat Protoc 13 (4):599–604. https://doi.org/10.1038/nprot.2017.149
51. Marinov GK, Williams BA, McCue K, Schroth GP, Gertz J, Myers RM, Wold BJ (2014) From single-cell to cell-pool transcriptomes: stochasticity in gene expression and RNA splicing. Genome Res 24(3):496–510. https://doi.org/10.1101/gr.161034.113
52. Leek JT, Scharpf RB, Bravo HC, Simcha D, Langmead B, Johnson WE, Geman D, Baggerly K, Irizarry RA (2010) Tackling the widespread and critical impact of batch effects in high-throughput data. Nat Rev Genet 11 (10):733–739. https://doi.org/10.1038/nrg2825
53. Stoeckius M, Zheng S, Houck-Loomis B, Hao S, Yeung BZ, Mauck WM III, Smibert P, Satija R (2018) Cell Hashing with barcoded antibodies enables multiplexing and doublet detection for single cell genomics. Genome Biol 19(1):224. https://doi.org/10.1186/s13059-018-1603-1
54. Magness ST, Puthoff BJ, Crissey MA, Dunn J, Henning SJ, Houchen C, Kaddis JS, Kuo CJ, Li L, Lynch J, Martin MG, May R, Niland JC, Olack B, Qian D, Stelzner M, Swain JR, Wang F, Wang J, Wang X, Yan K, Yu J, Wong MH (2013) A multicenter study to standardize reporting and analyses of fluorescence-activated cell-sorted murine intestinal epithelial cells. Am J Physiol Gastrointest Liver Physiol 305(8):G542–G551. https://doi.org/10.1152/ajpgi.00481.2012
55. Chromium Single Cell 3′ Reagents Kits User Guide (v3 Chemistry) (2019). 10× Genomics

Chapter 9

Single-Cell Studies of Intestinal Stem Cell Heterogeneity During Homeostasis and Regeneration

Maxim Norkin, Claudia Capdevila, Ruben I. Calderon, Tianhong Su, Maria Trifas, Paloma Ordóñez-Morán, and Kelley S. Yan

Abstract

Single-cell RNA-sequencing (scRNA-seq) provides a unique opportunity to study heterogeneous cell populations within tissues, including the intestinal epithelium, to gain detailed molecular insights into their biology. Many new putative markers of intestinal stem cells and their progeny have been described using single-cell transcriptomics, which has contributed to the identification of novel subpopulations of mature cell types and insight into their developmental trajectories. This approach has revealed tremendous cellular heterogeneity within the intestinal epithelium that is concordant with its diverse and multifaceted functions. We discuss the function of these subpopulations during tissue homeostasis, as well as putative subpopulations with inducible regenerative potential following tissue injury.

Key words Single-cell RNA-sequencing, Stem cell, Intestine, Epithelium, Lineage, Homeostasis, Regeneration, Differentiation, Plasticity

1 Introduction

The renewal of the intestinal epithelium is characterized by both tight regulation and an immense turnover capacity. This rapid renewal is important given the relatively harsh conditions to which the epithelial cells are exposed during their lifetime. In addition to its remarkable proliferative activity, the intestinal epithelium is extremely diverse in its cellular composition, reflecting a high degree of heterogeneity within its major lineages [1, 2] (Fig. 1). For example, the enteroendocrine cell (EEC) lineage is thought to be comprised of at least ten different cell types [1, 3]. Additionally, there is controversy over the heterogeneity of its stem cell compartment that supports tissue turnover and regeneration. Multiple studies highlight the regenerative function of

Maxim Norkin and Claudia Capdevila contributed equally to this work.

Paloma Ordóñez-Morán (ed.), *Intestinal Stem Cells: Methods and Protocols*, Methods in Molecular Biology, vol. 2171,
https://doi.org/10.1007/978-1-0716-0747-3_9,

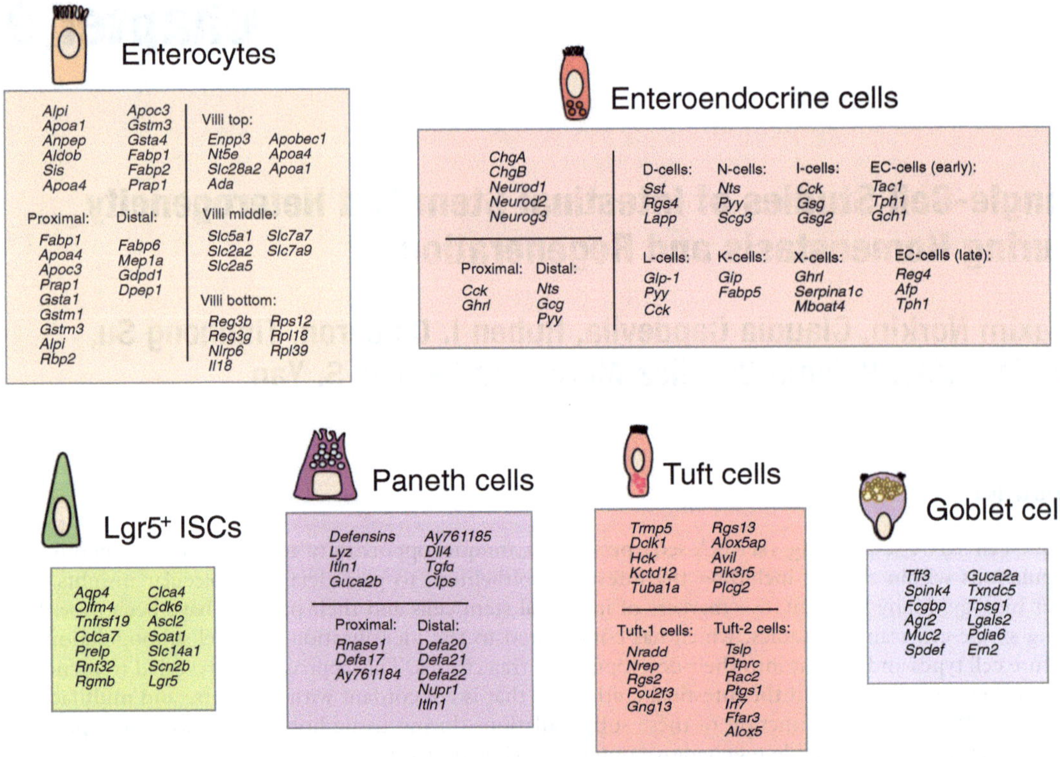

Fig. 1 Markers of intestinal stem cells and their progeny. Summary of putative marker genes of different intestinal cell types based on transcriptional profiling. Enterocytes: left upper corner—enterocyte specific markers, left bottom corner—proximal and distal markers of enterocytes, right—genes with variable expression in enterocytes along the intestinal villus axis. EECs: left upper corner—EEC specific markers, left bottom corner—proximal and distal markers of EECs, right—specific markers of EEC subpopulations. Lgr5$^+$ ISCs: stem cell specific markers. Paneth cells: top—specific genes, bottom—proximal and distal markers of corresponding populations. Tuft cells: up—specific markers, bottom—specific markers of tuft subpopulations. Goblet cells: specific markers

reserve stem cells that are deployed during times of tissue injury using distinct mechanisms from those employed during homeostasis [1, 4–14]. Moreover, numerous progenitor cells and intermediates along the developmental trajectories of individual lineages remain to be established.

Single-cell RNA-sequencing (scRNA-seq) is a rapidly evolving method that is becoming an indispensable tool to investigate the cellular composition of tissues. Continuous changes in cellular identity can also be analyzed by real time-resolved single-cell transcriptomics, which can be used for cellular processes that display transient activation of a marker gene [3]. Early studies using scRNA-seq in the intestine were performed on mouse epithelial cells [15, 16]. More recent reports have delineated the mRNA expression in human epithelial cells. The cell types identified in human tissues largely correlated with the previous scRNA-seq analyses performed on the mouse [17]. Overall, scRNA-seq studies have revealed new insights into the cell types and mechanisms that underlie intestinal homeostasis and epithelial regeneration. Here, we summarize recent findings using this technology as well as new questions raised by these studies (Figs. 1 and 2).

2 Cellular Heterogeneity During Intestinal Homeostasis

The constant tissue renewal of the intestinal epithelium during homeostasis is fueled by continuously dividing intestinal stem cells (ISCs) that reside at the base of the crypts [18–20]. These crypt-base columnar ISCs have been molecularly defined by the expression of the R-spondin receptor *Lgr5*, a leucine rich repeat containing G protein-coupled receptor [18]. Single-cell transcriptomic analysis reveals that Lgr5-eGFP$^+$ ISCs isolated from *Lgr5-eGFP-IRES-CreERT2* mice are composed of distinct clusters of both cycling and non-cycling Wnt- and R-spondin-dependent ISC as well as nascent transit-amplifying (TA) cell populations, consistent with multiple reports suggesting heterogeneity within the ISC pool [10, 21, 22]. Many markers of ISCs were initially identified based on microarray and bulk RNA-seq profiling of cells isolated using Lgr5 reporter expression. Subsequently, transcriptional profiling using scRNA-seq technology helped to support these markers and also allowed the identification of new ones, with varying degrees of characterization and validation (Fig. 1): *Aqp4*, *Olfm4*, *Tnfrsf19*, *Cdca7*, *Prelp*, *Rnf32*, *Cdk6*, *Rgmb*, *Clca4*, *Scn2b*, *Slc14a1*, *Ascl2*, and *Soat1* [2, 15, 16, 23–25].

Under homeostatic conditions, Lgr5$^+$ ISCs regenerate the intestinal epithelium [18]. In contrast, a distinct, label-retaining cell at the +4 cell position in intestinal crypts has been reported and historically proposed to serve as a quiescent/slowly cycling, reserve stem cell population that comes into play upon loss of Lgr5$^+$ ISCs [4–6, 8, 26–28]. Even though multiple markers have been described for the +4 cell [4–8], more recent studies show that many of these overlap with Lgr5$^+$ ISCs, or are expressed very broadly throughout the crypt [29–31]. While it remains unclear if there are other dedicated reserve stem cell populations that coexist within the crypt alongside the Lgr5$^+$ ISCs, there is increasing evidence of diverse cell types that are capable of exhibiting cellular plasticity to become activated upon injury conditions to rapidly regenerate the epithelium following loss of Lgr5$^+$ ISCs [32]. Indeed, some recent observations suggest that it is not a single type of quiescent stem cell but rather a functionally distinct, non-*Lgr5*-expressing cell type that reverts to a stem cell state in an injury-inducible fashion [1, 4, 8].

Globally, along the crypt–villus axis, the progenitor cells arising from Lgr5$^+$ ISCs and their immediate TA progeny reside in the

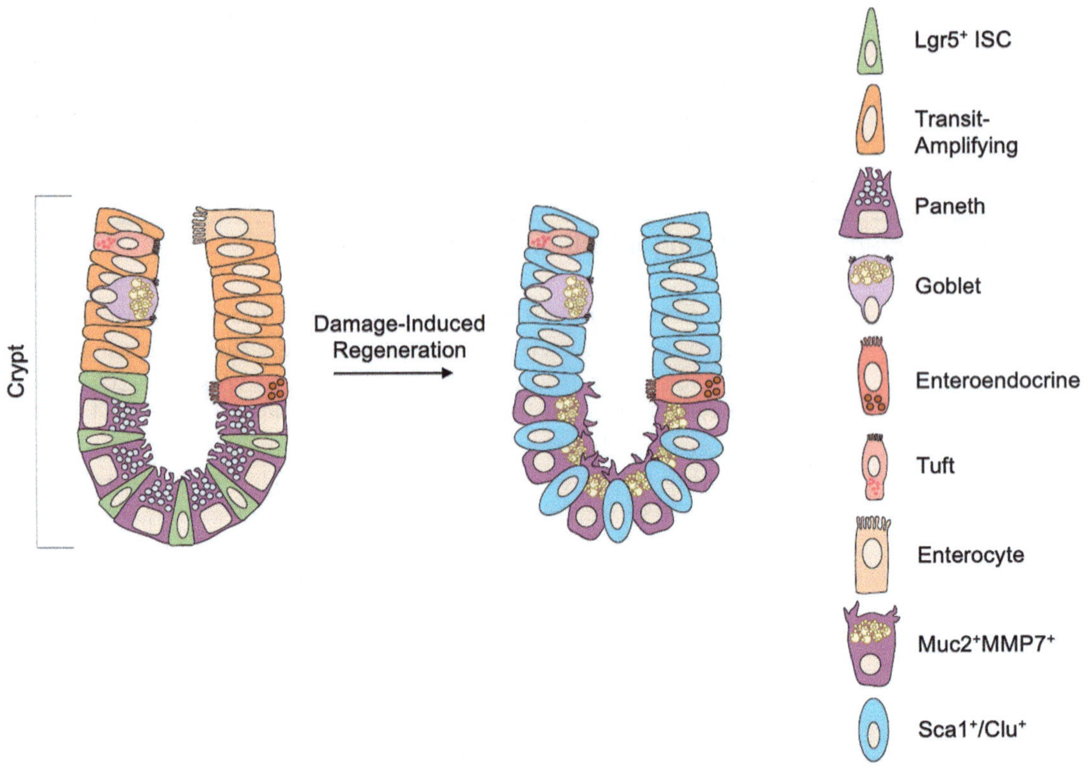

Fig. 2 Injury-induced regeneration of the intestinal epithelium. Injury-induced regeneration of the intestinal epithelium is mediated by $Sca1^+$ and/or Clu^+ cells (blue), which are rare under homeostatic conditions. $Sca1^+$/Clu^+ cells repopulate the crypts of the small intestine and regenerate the epithelium following damage-induced loss of $Lgr5^+$ ISCs

crypts close to the stem cells and move into the villi upon differentiation. Absorptive enterocytes are the most abundant cell lineage in the intestinal epithelium. Additionally, there are multiple secretory cell lineages including mucus-producing goblet cells, Paneth cells, EECs, and rare tuft cells [20]. All these differentiated cell lineages contribute to specific functions of the intestine. The advantage of scRNA-seq methodology is that it can be applied to study specific gene expression in individual subpopulations and it can help to infer the hierarchical relationships of individual lineages (Fig. 1).

Tuft cells are a rare cell type involved in chemosensory function. Tuft cells express proteins known to be involved in taste signal transduction and also in an immune response against parasite infection [33, 34]. Tuft cells express such markers as *Trpm5*, *Dclk1*, *Rgs13*, *Alox5ap*, *Avil*, *Hck*, *Kctd12*, *Tuba1a*, *Pik3r5*, and *Plcg2* [2, 15, 16, 35, 36]. Recent findings using scRNA-seq demonstrated the existence of two distinct subtypes of tuft cells: tuft-1 cells are highly enriched in genes related to neuronal development, while tuft-2 cells show upregulation of genes related to immunity. Specifically, tuft-1 cells are enriched for *Nradd*, *Gng13*, *Nrep*, *Rgs2*, and *Pou2f3* [2] (Fig. 1). Conversely, tuft-2 cells showed higher expression levels of the TH2-promoting cytokine thymic stromal lymphopoietin (*Tslp*) and *Ptprc* (pan-immune cell marker CD45), as well as enrichment in transcripts for *Rac2*, *Ptgs1*, *Irf7*, *Ffar3*, and *Alox5*.

Enterocytes are the most abundant cell type in the intestinal epithelium, making up to 80% of intestinal epithelial cells [37]. Enterocytes are predominantly located in the villus region and function in the hydrolysis, absorption, and transport of nutrients [38]. Many enterocyte markers were identified using bulk RNA-seq [39] and scRNA-seq [2, 15, 16, 41]; among them are *Alpi*, *Apoa1*, *Anpep*, *Aldob*, *Sis*, *Apoa4*, *Prap1*, *Apoc3*, *Gstm3*, *Gsta4*, *Fabp1*, and *Fabp2*. Recently, scRNA-seq was used to identify enterocyte progenitor populations [15, 16], to examine enterocyte regional diversity throughout the gut [2], and to investigate the spatial allocation of distinct functional classes of enterocytes along the intestinal crypt–villus axis [40, 41]. It has been shown that earlier enterocyte progenitors express high transcript levels of ribosomal proteins (*Rn45s*, *Rps19*, and *Rps12*), *Dmbt1*, *Reg3g*, *Ube2c* and low levels of those for enterocyte-specific genes (*Prap1*, *Apoa1*, *Apoa4*, *Apoc3*, etc.) [2]. Late progenitor cells lose the expression of ribosomal proteins with concurrent elevation of enterocyte-specific transcripts. Finally, mature enterocytes are characterized by further upregulation of enterocyte-specific mRNAs.

A study of regional enterocyte markers reveals that *Fabp1*, *Apoa4*, *Apoc3*, *Gsta1*, *Gstm3*, *Gstm1*, *Alpi*, *Prap1*, and *Rbp2* are highly expressed in the proximal part of the intestine, while *Fabp6*, *Mep1a*, *Dpep1*, and *Gdpd1* are predominantly expressed in the distal small intestine [2]. This is consistent with differential absorptive functions along the longitudinal proximal-to-distal gut axis. Moreover, a combination of high-throughput scRNA-seq and bulk RNA-seq of laser-microdissected crypt–villus sections revealed that enterocyte transcriptional signatures differ along the crypt–villus axis consistent with their zonation [41]. Enterocytes at the bottom of the crypt showed enrichment in ribosomal/proliferation signatures (*Rps12*, *Rpl18*, and *Rpl39*) and antimicrobial program peptides (*Reg3b*, *Reg3g*, *Nlrp6*, and *Il18*) [41]. Enterocytes in the middle of the crypt–villus axis are enriched in the genes responsible for the processing and absorption of various nutrients, especially amino acids (*Slc7a7* and *Slc7a9*) and carbohydrates (*Slc5a1*, *Slc2a2*, and *Slc2a5*) [41]. Enterocytes at the top of the villus exhibited a distinct expression program: enrichment in cell adhesion signature genes (*Egfr*, *Klf4*, *Fos*, *Junb*), purine catabolism genes (*Enpp3*, *Nt5e*, *Slc28a2*, and *Ada*), and apolipoproteins/cholesterol processing genes (*Apobec1*, *Apoa4*, and *Apoa1*) (Fig. 1) [41].

Goblet cells are secretory cells present in both the small intestine and colon. They produce and secrete mucus into the intestinal lumen, which facilitates the migration of chyme through the gut and contributes to a physical barrier that prevents microorganisms and toxins from direct contact with the mucosa. In addition, goblet cells are also shown to be expanded in response to parasite infection [2, 42, 43]. Several scRNA-seq experiments revealed multiple markers of goblet cells: *Tff3*, *Spink4*, *Fcgbp*, *Agr2*, *Muc2*, *Txndc5*, *Tpsg1*, *Spdef*, *Guca2a*, *Lgals2*, *Pdia6*, and *Ern2* [2, 15, 16]. *Spdef* has been shown to have an essential role in goblet cell differentiation [44, 45] (Fig. 1).

Paneth cells are intermingled with $Lgr5^+$ ISCs at the crypt base and function in antimicrobial defense and metabolic regulation of ISCs. They interact with ISCs via signaling pathways such as Notch, Wnt, and EGF [46–48]. While the first Paneth cell markers (lysozyme and defensins) were reported decades ago [49, 50], recent RNA-seq studies on sorted $CD24^+$ Paneth cells reveal additional transcripts differentially enriched in Paneth cells relative to ISCs: *Dll4*, *Tgfα*, *Wnt11*, *Clps*, *AY761185*, *Itln1*, and *Guca2b* [46]. Recent scRNA-seq studies additionally showed that Paneth cells also exhibit regional diversity, and the expression profiles of these cells are different in distal and proximal part of the small intestine: for example, *Rnase1*, *Defa17*, and *AY761184* are highly expressed in the duodenum, while *Defa20*, *Defa21*, *Defa22*, *Nupr1*, and *Itln1* show the highest expression in ileum [2]. Other Paneth-specific markers that show uniform expression throughout the gut include *Lyz1*, *Ang4*, *Clps*, and *Habp2* (Fig. 1) [2]. It still remains unclear how and why Paneth cells are heterogeneous, and the functional implications on their interactions with ISCs and on the other Paneth cell functions.

EECs produce hormones that regulate digestion and metabolism, and participate in chemosensation, including nutrient detection. EECs represent a very small fraction of epithelial cells that are widely dispersed throughout the intestinal epithelium and have been elusive to molecular characterization due to their marked cellular heterogeneity. While EECs are located throughout the entire intestinal epithelium, they also exhibit regionalized patterns of gene expression. Recently, scRNA-seq studies have identified new putative markers for the EECs subpopulations and further underscored the heterogeneity of this lineage. At least ten different EEC subtypes have been identified either in vivo or via in vitro intestinal organoid culture [2, 3, 15, 16, 19, 51]. *Neurog3*, *Neurod1*, and *Neurod2* are known EEC markers specific to EEC progenitor populations [2]. In addition, the EEC markers *ChgA* and *ChgB* were shown to be predominantly expressed in enterochromaffin (EC) cells [2, 3]. Many more putative novel markers were identified for each of the EEC subpopulations, using sequencing technology, including scRNA-seq: *Gip*, *Fabp5* (K-cells); *Reg4*, *Afp*, *Tph1* (EC-cells (late)); *Sst*, *Lapp*, *Rgs4* (D-cells); *Ghrl*, *Serpina1c*, *Mboat4* (X-cells); *Cck*, *Gsg*, *Gsg*-2 (I-cells); *Glp-1*, *Pyy*, *Cck* (L-cells); *Nts*, *Pyy*, *Scg3* (N-cells); *Tac1*, *Tph1*, *Gch1* (EC cells (early)) [2, 3,

15, 16, 19, 51, 52]. Some of the markers such as *Cck* and *Gcg* are shown to be expressed in several EEC subtypes simultaneously. Furthermore, while some markers are predominantly expressed in the proximal small intestine (*Cck* and *Ghrl*), others are more localized in the distal region (*Nts*, *Gcg*, and *Pyy*). *Sct* is expressed in all subtypes of EECs and is detected in both proximal and distal parts of the gastrointestinal tract [3, 53] (Fig. 1). These EEC lineage markers and their distribution need further confirmatory experimental validation and biological interpretation. The diverse repertoire of EEC subsets found by scRNA-seq likely reflects the multifaceted functions of these cells that orchestrate chemosensation, digestion, metabolism, and communication with other organs via their hormone products.

3 Cellular Heterogeneity During Tissue Repair

Multiple distinct types of cells capable of regenerating the intestinal epithelium after tissue injury or $Lgr5^+$ ISC loss have been described [32]. Importantly, these include progenitor cells of both secretory and absorptive cell lineages, as well as more differentiated cells that reside within the intestinal crypt. Functional studies using lineage tracing in mice using $Alpi^+$ absorptive enterocytes [54] and secretory lineage markers [8, 9, 55, 56] support the possibility that there are multiple populations capable of reconstituting the intestinal epithelium following injury, suggesting a model of cellular plasticity in which perturbation of homeostasis enables even lineage-committed cells to regain regenerative capability. For instance, recent comparative transcriptomics using scRNA-seq to examine cells isolated from various reporter mice revealed that lineage-committed EECs also exhibit regenerative potential in the context of irradiation injury [1]. In this setting, $Lgr5^+$ ISCs are rapidly ablated, and $Prox1^+$ crypt cells of the enteroendocrine lineage can come into play to regenerate the epithelium. Notably, *Prox1* is expressed in quiescent crypt cells near the +4 cell position and identifies cells of the tuft and enteroendocrine lineages, suggesting a shared lineage progenitor [1]. Thus, despite being already-committed secretory populations, these cells are capable of regenerating the epithelium following tissue injury.

scRNA-seq has recently provided additional insight into potential mechanisms for injury-induced regeneration of the intestinal epithelium. Recent reports of injury-induced regeneration have identified and characterized two additional subpopulations of epithelial cells that occupy the crypt compartment following tissue damage. Various injury models, including helminth infection, result in expansion of crypt cells that adopt a fetal-like transcriptional program, in which the crypt cells are functionally Wnt-independent and express the protein Sca1 [12, 14, 57,

58]. In another report, a rare population that selectively expands upon damage is characterized by high expression of clusterin (Clu) (Fig. 2) [13]. The inter-relationship between these populations remains undefined.

In contrast to Wnt pathway-dependent Lgr5$^+$ ISCs, Sca1$^+$ cells rise, peak, and return to basal expression levels in different injury/perturbation models including parasitic helminth infection, irradiation, and Lgr5$^+$ ISC ablation by diphtheria toxin, in a manner that is inversely proportional to the loss of Lgr5$^+$ ISCs [12]. These Sca1$^+$ cells are capable of forming regenerative, primarily undifferentiated, fetal-like proliferative crypts devoid of canonical adult Lgr5$^+$ ISC markers shortly after injury [12]. By the time Lgr5$^+$ ISCs re-emerge and repopulate the regenerating epithelium, Sca1 expression decreases. Notably, Sca1$^+$ cells in the adult small intestine are ultimately derived from Lgr5$^+$ ISCs; yet, they can arise and regenerate the epithelium independent of Lgr5$^+$ ISCs themselves [12]. Importantly, this report provides evidence of epithelial cells co-opting a fetal-like transcriptional program for repair of damaged adult tissue [12].

Cells characterized by high expression levels of clusterin (Clu) and referred to as "revival" stem cells (revSCs) were recently reported to selectively expand following irradiation damage [13]. RevSCs were identified by scRNA-seq analysis of the regenerating intestinal epithelium 3 days after irradiation, and were characterized as a preexisting Lgr5$^-$, Clu$^+$, YAP-dependent injury-induced quiescent cell type that is most prominently found following tissue damage [13]. Immunofluorescent analysis of transgenic Clu reporter mice confirmed limited numbers of Clu$^+$ cells within the epithelium during homeostatic conditions. Indeed, crypts populated by Clu$^+$ cells did not harbor Lgr5$^+$ ISCs [13]. However, lineage tracing experiments initiated under homeostatic conditions demonstrated that those rare Clu$^+$ cells were multipotent and capable of giving rise to Lgr5$^+$ ISCs [13]. Following irradiation damage, Clu$^+$ cells repopulated the small intestine and colon epithelium [13]. Consistent with these data, ablation of Clu$^+$ cells displayed no detrimental phenotype under homeostasis; however, impaired regeneration of the epithelium was observed in the context of irradiation, acute and chronic colitis [13]. Similarly, both Sca1$^+$ and Clu$^+$ cells are derived from Lgr5$^+$ progeny, although the degree of overlap between these and other regenerative populations remains unknown. Is this repair capacity restricted to certain discrete cell populations, or do these gene signatures represent a common intermediate state where lineage-committed cells transiently converge as they go backward in the lineage hierarchy and reacquire regenerative potential (Fig. 2)? Further studies are needed in order to address these questions.

4 Conclusion

scRNA-seq has enabled the description of intestinal epithelial cell populations and their interrelationships with unprecedented granularity. One caveat to using this technology to study cellular heterogeneity is that cellular categorization ultimately requires experimental functional validation, which currently remains a bottleneck. In addition to providing more accurate definitions for mature populations and describing potential inferred lineage trajectories, scRNA-seq has demonstrated regional differences in cell type prevalence, as well as substantial compartmentalization along the crypt–villus axis. Taken together, these studies highlight the high degree of cellular diversity in intestinal tissue architecture, which is concordant with the great diversity of its biological functions. Moreover, these studies underscore the notion that cellular identity is malleable and influenced by environmental cues, as cells can adopt different functions in different biological contexts, further expanding the potential cellular heterogeneity within the tissue.

Acknowledgments

K.S.Y. is supported by NIH R03DK114656 and U01DK103155, BWF CAMS, Louis V. Gerstner,Jr. Scholars Award, Lisa Dean Moseley Foundation, and the Irma T. Hirschl Career Award. C. C. is supported by a NYSTEM predoctoral training grant and T.S. is supported by a Berrie Foundation fellowship.

References

1. Yan KS, Gevaert O, Zheng GXY, Anchang B, Probert CS, Larkin KA, Davies PS, Cheng ZF, Kaddis JS, Han A, Roelf K, Calderon RI, Cynn E, Hu X, Mandleywala K, Wilhelmy J, Grimes SM, Corney DC, Boutet SC, Terry JM, Belgrader P, Ziraldo SB, Mikkelsen TS, Wang F, von Furstenberg RJ, Smith NR, Chandrakesan P, May R, Chrissy MAS, Jain R, Cartwright CA, Niland JC, Hong YK, Carrington J, Breault DT, Epstein J, Houchen CW, Lynch JP, Martin MG, Plevritis SK, Curtis C, Ji HP, Li L, Henning SJ, Wong MH, Kuo CJ (2017) Intestinal enteroendocrine lineage cells possess homeostatic and injury-inducible stem cell activity. Cell Stem Cell 21(1):78–90 e76. https://doi.org/10.1016/j.stem.2017.06.014
2. Haber AL, Biton M, Rogel N, Herbst RH, Shekhar K, Smillie C, Burgin G, Delorey TM, Howitt MR, Katz Y, Tirosh I, Beyaz S, Dionne D, Zhang M, Raychowdhury R, Garrett WS, Rozenblatt-Rosen O, Shi HN, Yilmaz O, Xavier RJ, Regev A (2017) A single-cell survey of the small intestinal epithelium. Nature 551(7680):333–339. https://doi.org/10.1038/nature24489
3. Gehart H, van Es JH, Hamer K, Beumer J, Kretzschmar K, Dekkers JF, Rios A, Clevers H (2019) Identification of enteroendocrine regulators by real-time single-cell differentiation mapping. Cell 176(5):1158–1173 e1116. https://doi.org/10.1016/j.cell.2018.12.029
4. Sangiorgi E, Capecchi MR (2008) Bmi1 is expressed in vivo in intestinal stem cells. Nat Genet 40(7):915–920. https://doi.org/10.1038/ng.165
5. Tian H, Biehs B, Warming S, Leong KG, Rangell L, Klein OD, de Sauvage FJ (2011) A

reserve stem cell population in small intestine renders Lgr5-positive cells dispensable. Nature 478(7368):255–259. https://doi.org/10.1038/nature10408

6. Montgomery RK, Carlone DL, Richmond CA, Farilla L, Kranendonk ME, Henderson DE, Baffour-Awuah NY, Ambruzs DM, Fogli LK, Algra S, Breault DT (2011) Mouse telomerase reverse transcriptase (mTert) expression marks slowly cycling intestinal stem cells. Proc Natl Acad Sci U S A 108(1):179–184. https://doi.org/10.1073/pnas.1013004108
7. Powell AE, Wang Y, Li Y, Poulin EJ, Means AL, Washington MK, Higginbotham JN, Juchheim A, Prasad N, Levy SE, Guo Y, Shyr Y, Aronow BJ, Haigis KM, Franklin JL, Coffey RJ (2012) The pan-ErbB negative regulator Lrig1 is an intestinal stem cell marker that functions as a tumor suppressor. Cell 149 (1):146–158. https://doi.org/10.1016/j.cell.2012.02.042
8. Yan KS, Chia LA, Li X, Ootani A, Su J, Lee JY, Su N, Luo Y, Heilshorn SC, Amieva MR, Sangiorgi E, Capecchi MR, Kuo CJ (2012) The intestinal stem cell markers Bmi1 and Lgr5 identify two functionally distinct populations. Proc Natl Acad Sci U S A 109 (2):466–471. https://doi.org/10.1073/pnas.1118857109
9. van Es JH, Sato T, van de Wetering M, Lyubimova A, Yee Nee AN, Gregorieff A, Sasaki N, Zeinstra L, van den Born M, Korving J, Martens ACM, Barker N, van Oudenaarden A, Clevers H (2012) Dll1+ secretory progenitor cells revert to stem cells upon crypt damage. Nat Cell Biol 14 (10):1099–1104. https://doi.org/10.1038/ncb2581
10. Buczacki SJ, Zecchini HI, Nicholson AM, Russell R, Vermeulen L, Kemp R, Winton DJ (2013) Intestinal label-retaining cells are secretory precursors expressing Lgr5. Nature 495 (7439):65–69. https://doi.org/10.1038/nature11965
11. Asfaha S, Hayakawa Y, Muley A, Stokes S, Graham TA, Ericksen RE, Westphalen CB, von Burstin J, Mastracci TL, Worthley DL, Guha C, Quante M, Rustgi AK, Wang TC (2015) Krt19 (+)/Lgr5(−) cells are radioresistant cancer-initiating stem cells in the colon and intestine. Cell Stem Cell 16(6):627–638. https://doi.org/10.1016/j.stem.2015.04.013
12. Nusse YM, Savage AK, Marangoni P, Rosendahl-Huber AKM, Landman TA, de Sauvage FJ, Locksley RM, Klein OD (2018) Parasitic helminths induce fetal-like reversion in the intestinal stem cell niche. Nature 559 (7712):109–113. https://doi.org/10.1038/s41586-018-0257-1
13. Ayyaz A, Kumar S, Sangiorgi B, Ghoshal B, Gosio J, Ouladan S, Fink M, Barutcu S, Trcka D, Shen J, Chan K, Wrana JL, Gregorieff A (2019) Single-cell transcriptomes of the regenerating intestine reveal a revival stem cell. Nature 569(7754):121–125. https://doi.org/10.1038/s41586-019-1154-y
14. Pont AR, Yan KS (2018) Intestinal crypts assume the fetal position in response to injury. Cell Stem Cell 23(2):158–159. https://doi.org/10.1016/j.stem.2018.07.013
15. Grun D, Lyubimova A, Kester L, Wiebrands K, Basak O, Sasaki N, Clevers H, van Oudenaarden A (2015) Single-cell messenger RNA sequencing reveals rare intestinal cell types. Nature 525(7568):251–255. https://doi.org/10.1038/nature14966
16. Grun D, Muraro MJ, Boisset JC, Wiebrands K, Lyubimova A, Dharmadhikari G, van den Born M, van Es J, Jansen E, Clevers H, de Koning EJP, van Oudenaarden A (2016) De novo prediction of stem cell identity using single-cell transcriptome data. Cell Stem Cell 19(2):266–277. https://doi.org/10.1016/j.stem.2016.05.010
17. Fujii M, Matano M, Toshimitsu K, Takano A, Mikami Y, Nishikori S, Sugimoto S, Sato T (2018) Human intestinal organoids maintain self-renewal capacity and cellular diversity in niche-inspired culture condition. Cell Stem Cell 23(6):787–793 e786. https://doi.org/10.1016/j.stem.2018.11.016
18. Barker N, van Es JH, Kuipers J, Kujala P, van den Born M, Cozijnsen M, Haegebarth A, Korving J, Begthel H, Peters PJ, Clevers H (2007) Identification of stem cells in small intestine and colon by marker gene Lgr5. Nature 449(7165):1003–1007. https://doi.org/10.1038/nature06196
19. Gehart H, Clevers H (2019) Tales from the crypt: new insights into intestinal stem cells. Nat Rev Gastroenterol Hepatol 16(1):19–34. https://doi.org/10.1038/s41575-018-0081-y
20. Clevers H (2013) The intestinal crypt, a prototype stem cell compartment. Cell 154 (2):274–284. https://doi.org/10.1016/j.cell.2013.07.004
21. Barriga FM, Montagni E, Mana M, Mendez-Lago M, Hernando-Momblona X, Sevillano M, Guillaumet-Adkins A, Rodriguez-Esteban G, Buczacki SJA, Gut M, Heyn H, Winton DJ, Yilmaz OH, Attolini CS, Gut I, Batlle E (2017) Mex3a marks a slowly dividing subpopulation of Lgr5+ intestinal stem cells. Cell Stem Cell 20(6):801–816

e807. https://doi.org/10.1016/j.stem.2017.02.007

22. von Furstenberg RJ, Buczacki SJ, Smith BJ, Seiler KM, Winton DJ, Henning SJ (2014) Side population sorting separates subfractions of cycling and non-cycling intestinal stem cells. Stem Cell Res 12(2):364–375. https://doi.org/10.1016/j.scr.2013.10.012
23. van der Flier LG, Haegebarth A, Stange DE, van de Wetering M, Clevers H (2009) OLFM4 is a robust marker for stem cells in human intestine and marks a subset of colorectal cancer cells. Gastroenterology 137(1):15–17. https://doi.org/10.1053/j.gastro.2009.05.035
24. van der Flier LG, van Gijn ME, Hatzis P, Kujala P, Haegebarth A, Stange DE, Begthel H, van den Born M, Guryev V, Oving I, van Es JH, Barker N, Peters PJ, van de Wetering M, Clevers H (2009) Transcription factor achaete scute-like 2 controls intestinal stem cell fate. Cell 136(5):903–912. https://doi.org/10.1016/j.cell.2009.01.031
25. Muñoz J, Stange DE, Schepers AG, van de Wetering M, Koo BK, Itzkovitz S, Volckmann R, Kung KS, Koster J, Radulescu S, Myant K, Versteeg R, Sansom OJ, van Es JH, Barker N, van Oudenaarden A, Mohammed S, Heck AJR, Clevers H (2012) The Lgr5 intestinal stem cell signature: robust expression of proposed quiescent '+4' cell markers. EMBO J 31 (14):3079
26. Potten CS, Kovacs L, Hamilton E (1974) Continuous labelling studies on mouse skin and intestine. Cell Tissue Kinet 7(3):271–283
27. Potten CS (1975) Kinetics and possible regulation of crypt cell populations under normal and stress conditions. Bull Cancer 62(4):419–430
28. Potten CS (1977) Extreme sensitivity of some intestinal crypt cells to X and gamma irradiation. Nature 269(5628):518–521. https://doi.org/10.1038/269518a0
29. Munoz J, Stange DE, Schepers AG, van de Wetering M, Koo BK, Itzkovitz S, Volckmann R, Kung KS, Koster J, Radulescu S, Myant K, Versteeg R, Sansom OJ, van Es JH, Barker N, van Oudenaarden A, Mohammed S, Heck AJ, Clevers H (2012) The Lgr5 intestinal stem cell signature: robust expression of proposed quiescent '+4′ cell markers. EMBO J 31 (14):3079–3091. https://doi.org/10.1038/emboj.2012.166
30. Li N, Yousefi M, Nakauka-Ddamba A, Jain R, Tobias J, Epstein JA, Jensen ST, Lengner CJ (2014) Single-cell analysis of proxy reporter allele-marked epithelial cells establishes intestinal stem cell hierarchy. Stem Cell Reports 3 (5):876–891. https://doi.org/10.1016/j.stemcr.2014.09.011
31. Itzkovitz S, Lyubimova A, Blat IC, Maynard M, van Es J, Lees J, Jacks T, Clevers H, van Oudenaarden A (2011) Single-molecule transcript counting of stem-cell markers in the mouse intestine. Nat Cell Biol 14(1):106–114. https://doi.org/10.1038/ncb2384
32. Santos AJM, Lo YH, Mah AT, Kuo CJ (2018) The intestinal stem cell niche: homeostasis and adaptations. Trends Cell Biol 28 (12):1062–1078. https://doi.org/10.1016/j.tcb.2018.08.001
33. Gerbe F, Jay P (2016) Intestinal tuft cells: epithelial sentinels linking luminal cues to the immune system. Mucosal Immunol 9 (6):1353–1359. https://doi.org/10.1038/mi.2016.68
34. Howitt MR, Lavoie S, Michaud M, Blum AM, Tran SV, Weinstock JV, Gallini CA, Redding K, Margolskee RF, Osborne LC, Artis D, Garrett WS (2016) Tuft cells, taste-chemosensory cells, orchestrate parasite type 2 immunity in the gut. Science 351(6279):1329–1333. https://doi.org/10.1126/science.aaf1648
35. Ting H-A, von Moltke J (2019) The immune function of tuft cells at gut mucosal surfaces and beyond. J Immunol 202(5):1321. https://doi.org/10.4049/jimmunol.1801069
36. Gerbe F, Brulin B, Makrini L, Legraverend C, Jay P (2009) DCAMKL-1 expression identifies tuft cells rather than stem cells in the adult mouse intestinal epithelium. Gastroenterology 137(6):2179–2180. https://doi.org/10.1053/j.gastro.2009.06.072
37. de Santa BP, van den Brink GR, Roberts DJ (2003) Development and differentiation of the intestinal epithelium. Cell Mol Life Sci 60 (7):1322–1332. https://doi.org/10.1007/s00018-003-2289-3
38. van der Flier LG, Clevers H (2009) Stem cells, self-renewal, and differentiation in the intestinal epithelium. Annu Rev Physiol 71:241–260. https://doi.org/10.1146/annurev.physiol.010908.163145
39. Kazakevych J, Sayols S, Messner B, Krienke C, Soshnikova N (2017) Dynamic changes in chromatin states during specification and differentiation of adult intestinal stem cells. Nucleic Acids Res 45(10):5770–5784. https://doi.org/10.1093/nar/gkx167
40. Gassler N, Newrzella D, Bohm C, Lyer S, Li L, Sorgenfrei O, van Laer L, Sido B, Mollenhauer J, Poustka A, Schirmacher P, Gretz N (2006) Molecular characterisation of non-absorptive and absorptive enterocytes in human small intestine. Gut 55

(8):1084–1089. https://doi.org/10.1136/gut.2005.073262
41. Moor AE, Harnik Y, Ben-Moshe S, Massasa EE, Rozenberg M, Eilam R, Bahar Halpern K, Itzkovitz S (2018) Spatial reconstruction of single enterocytes uncovers broad zonation along the intestinal villus axis. Cell 175(4):1156–1167 e1115. https://doi.org/10.1016/j.cell.2018.08.063
42. Artis D, Wang ML, Keilbaugh SA, He W, Brenes M, Swain GP, Knight PA, Donaldson DD, Lazar MA, Miller HR, Schad GA, Scott P, Wu GD (2004) RELMbeta/FIZZ2 is a goblet cell-specific immune-effector molecule in the gastrointestinal tract. Proc Natl Acad Sci U S A 101(37):13596–13600. https://doi.org/10.1073/pnas.0404034101
43. Biton M, Levin A, Slyper M, Alkalay I, Horwitz E, Mor H, Kredo-Russo S, Avnit-Sagi T, Cojocaru G, Zreik F, Bentwich Z, Poy MN, Artis D, Walker MD, Hornstein E, Pikarsky E, Ben-Neriah Y (2011) Epithelial microRNAs regulate gut mucosal immunity via epithelium-T cell crosstalk. Nat Immunol 12(3):239–246. https://doi.org/10.1038/ni.1994
44. Noah TK, Kazanjian A, Whitsett J, Shroyer NF (2010) SAM pointed domain ETS factor (SPDEF) regulates terminal differentiation and maturation of intestinal goblet cells. Exp Cell Res 316(3):452–465. https://doi.org/10.1016/j.yexcr.2009.09.020
45. Gregorieff A, Stange DE, Kujala P, Begthel H, van den Born M, Korving J, Peters PJ, Clevers H (2009) The ets-domain transcription factor Spdef promotes maturation of goblet and paneth cells in the intestinal epithelium. Gastroenterology 137(4):1333–1345 e1331-1333. https://doi.org/10.1053/j.gastro.2009.06.044
46. Sato T, van Es JH, Snippert HJ, Stange DE, Vries RG, van den Born M, Barker N, Shroyer NF, van de Wetering M, Clevers H (2011) Paneth cells constitute the niche for Lgr5 stem cells in intestinal crypts. Nature 469 (7330):415–418. https://doi.org/10.1038/nature09637
47. Pellegrinet L, Rodilla V, Liu Z, Chen S, Koch U, Espinosa L, Kaestner KH, Kopan R, Lewis J, Radtke F (2011) Dll1- and dll4-mediated notch signaling are required for homeostasis of intestinal stem cells. Gastroenterology 140 (4):1230–1240 e1231-1237. https://doi.org/10.1053/j.gastro.2011.01.005
48. Yilmaz OH, Katajisto P, Lamming DW, Gultekin Y, Bauer-Rowe KE, Sengupta S, Birsoy K, Dursun A, Yilmaz VO, Selig M, Nielsen GP, Mino-Kenudson M, Zukerberg LR, Bhan AK, Deshpande V, Sabatini DM (2012) mTORC1 in the Paneth cell niche couples intestinal stem-cell function to calorie intake. Nature 486(7404):490–495. https://doi.org/10.1038/nature11163
49. Ghoos Y, Vantrappen G (1971) The cytochemical localization of lysozyme in Paneth cell granules. Histochem J 3(3):175–178. https://doi.org/10.1007/bf01002560
50. Porter EM, Liu L, Oren A, Anton PA, Ganz T (1997) Localization of human intestinal defensin 5 in Paneth cell granules. Infect Immun 65 (6):2389–2395
51. Basak O, Beumer J, Wiebrands K, Seno H, van Oudenaarden A, Clevers H (2017) Induced quiescence of Lgr5+ stem cells in intestinal organoids enables differentiation of hormone-producing enteroendocrine cells. Cell Stem Cell 20(2):177–190 e174. https://doi.org/10.1016/j.stem.2016.11.001
52. Habib AM, Richards P, Cairns LS, Rogers GJ, Bannon CA, Parker HE, Morley TC, Yeo GS, Reimann F, Gribble FM (2012) Overlap of endocrine hormone expression in the mouse intestine revealed by transcriptional profiling and flow cytometry. Endocrinology 153 (7):3054–3065. https://doi.org/10.1210/en.2011-2170
53. Egerod KL, Engelstoft MS, Grunddal KV, Nohr MK, Secher A, Sakata I, Pedersen J, Windelov JA, Fuchtbauer EM, Olsen J, Sundler F, Christensen JP, Wierup N, Olsen JV, Holst JJ, Zigman JM, Poulsen SS, Schwartz TW (2012) A major lineage of enteroendocrine cells coexpress CCK, secretin, GIP, GLP-1, PYY, and neurotensin but not somatostatin. Endocrinology 153(12):5782–5795. https://doi.org/10.1210/en.2012-1595
54. Tetteh PW, Basak O, Farin HF, Wiebrands K, Kretzschmar K, Begthel H, van den Born M, Korving J, de Sauvage F, van Es JH, van Oudenaarden A, Clevers H (2016) Replacement of lost Lgr5-positive stem cells through plasticity of their enterocyte-lineage daughters. Cell Stem Cell 18(2):203–213. https://doi.org/10.1016/j.stem.2016.01.001
55. Jadhav U, Saxena M, O'Neill NK, Saadatpour A, Yuan GC, Herbert Z, Murata K, Shivdasani RA (2017) Dynamic reorganization of chromatin accessibility signatures during dedifferentiation of secretory precursors into Lgr5+ intestinal stem cells. Cell Stem Cell 21(1):65–77 e65. https://doi.org/10.1016/j.stem.2017.05.001
56. Yu S, Tong K, Zhao Y, Balasubramanian I, Yap GS, Ferraris RP, Bonder EM, Verzi MP, Gao N (2018) Paneth cell multipotency induced by notch activation following injury. Cell Stem

Cell 23(1):46–59. e45. https://doi.org/10.1016/j.stem.2018.05.002

57. Huels DJ, Medema JP (2018) Think about the environment: cellular reprogramming by the extracellular matrix. Cell Stem Cell 22(1):7–9. https://doi.org/10.1016/j.stem.2017.12.006
58. Yui S, Azzolin L, Maimets M, Pedersen MT, Fordham RP, Hansen SL, Larsen HL, Guiu J, Alves MRP, Rundsten CF, Johansen JV, Li Y, Madsen CD, Nakamura T, Watanabe M, Nielsen OH, Schweiger PJ, Piccolo S, Jensen KB (2018) YAP/TAZ-dependent reprogramming of colonic epithelium links ECM remodeling to tissue regeneration. Cell Stem Cell 22 (1):35–49 e37. https://doi.org/10.1016/j.stem.2017.11.001

Part III

Organoids and Applications

Chapter 10

Large-Scale Production of Recombinant Noggin and R-Spondin1 Proteins Required for the Maintenance of Stem Cells in Intestinal Organoid Cultures

David L. Hacker and Paloma Ordóñez-Morán

Abstract

The presence of the proteins mouse R-Spondinl (mRSpol) and mouse Noggin (mNoggin) in a 3D-organoid culture allows for the maintenance of intestinal stem cells. Here, we describe a transient gene expression method for the production of these proteins from human embryo kidney 293 (HEK293) cells cultivated in suspension using orbitally shaken bioreactors. Plasmid DNA was delivered into cells using the cationic polymer polyethylenimine (PEI). The 7-day production cultures were performed in the presence of valproic acid (VPA), an enhancer of recombinant gene expression. Both proteins were secreted from the transfected cells. mRSpo1 was produced as a secreted Fc fusion protein (mRSpo1-Fc) and purified by protein A-based affinity chromatography. mNoggin was produced as a secreted histidine-tagged protein (mNoggin-His) and purified by immobilized metal affinity chromatography (IMAC). This transient transfection system supports a high production efficiency.

Key words HEK293, Transfection, mR-Spondin1, mNoggin, Recombinant protein, Orbital shaking, Polyethylenimine, Affinity chromatography, Intestinal organoids, Stem cells

1 Introduction

Stem cells are essential to maintain homeostasis. In the intestine, stem cells are surrounded by an environment which provides signals that govern cell maintenance, proliferation and differentiation. In these last years, 3D-organoid technology has become a powerful tool to study these epithelial cells. The organoid systems maintain a 3D self-organized tissue and allow long-term intestinal epithelial culture. Organoids are embedded in a gelatinous protein mixture, often Matrigel, that is secreted by mouse sarcoma cells and contains structural proteins in combination with several growth stimuli (the Wnt agonist R-Sspondin1, epidermal growth factor, and the bone morphogenetic protein (BMP) inhibitor Noggin). The 3D culture requires the presence of these growth factors for the maintenance of stem cells in vitro [1].

Paloma Ordóñez-Morán (ed.), *Intestinal Stem Cells: Methods and Protocols*, Methods in Molecular Biology, vol. 2171,
https://doi.org/10.1007/978-1-0716-0747-3_10,

Here, we describe a protocol for a highly efficient production of mRSpo1-Fc and mNoggin-His. To this end, we use large-scale transient gene expression of mammalian cells in suspension culture because it allows rapid access to recombinant proteins [2, 3]. The two main host cell lines for this technology are HEK293 and Chinese hamster ovary (CHO) cells [3]. With this approach, milligram to gram quantities of a protein can be produced within days of plasmid DNA delivery using linear 25-kDa PEI. Improved transient protein yields were observed by increasing the cell density at the time of transfection and by adding VPA, a histone deacetylase inhibitor, to the culture post-transfection [4, 5]. We describe the production of mRSpo1-Fc and mNoggin-His in 2-L cultures of HEK293E cells, a variant of HEK293 cells that overexpresses the Epstein–Barr virus nuclear antigen 1 (EBNA1). The cultures were performed in 5-L glass bottles on an orbital shaker [6, 7]. Simple methods for the benchtop purification of the two proteins by gravity-flow affinity chromatography are described.

2 Materials

2.1 Routine Cell Culture

1. HEK293E cells adapted to cultivation in serum-free suspension.
2. Cylindrical and square-shaped glass bottles with nominal volumes of 500 mL to 5 L.
3. Ex-cell 293 medium without L-glutamine and phenol red.
4. FreeStyle 293 expression medium.
5. 50× L-glutamine and phenol red: dissolve 29.23 g glutamine and 250 mg phenol red in 1 L water. Sterilize by filtration, transfer into sterile 50-mL tubes, and store frozen at −20 °C.
6. Trypan blue solution (0.4%).
7. Phosphate-buffered saline (PBS).
8. Neubauer hemocytometer.

2.2 Plasmids

1. mNoggin-His expression vector (Fig. 1) (*see* **Note 1**).
2. mRSpo1-Fc expression vector (Fig. 1).

   The expression plasmid for mRSpo1-Fc (mIgG2a) was a kind gift from Calvin Kuo (Stanford University, USA) (*see* **Note 2**).

2.3 Plasmid DNA Preparation

1. LB agar petri plates with 100 μg/mL ampicillin.
2. LB medium with 100 μg/mL ampicillin.
3. Nucleobond AX 500 chromatography column.
4. Resuspension buffer: 50 mM Tris–HCl, 10 mM EDTA, 100 μg/mL RNase A, pH 8.0.

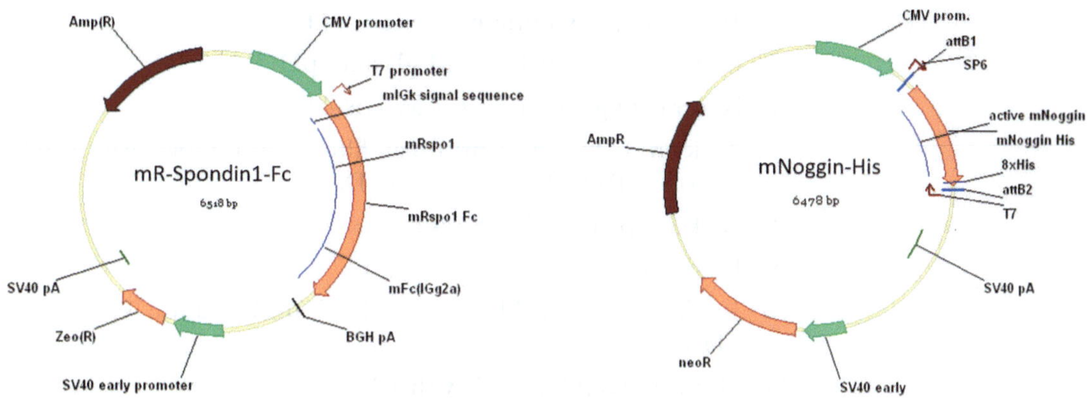

Fig. 1 Maps of expression vectors for mRSpo1-Fc and mNoggin-His

5. Lysis buffer: 200 mM NaOH, 1% SDS.
6. Neutralization buffer: 2.8 M potassium acetate, pH 5.1.
7. Equilibration buffer: 100 mM Tris, 15% ethanol, 900 mM KCl, 0.15% Triton X-100, adjusted to pH 6.3 with H_3PO_4.
8. Wash buffer: 100 mM Tris, 15% ethanol, 1.15 M KCl, adjusted to pH 6.3 with H_3PO_4.
9. Elution buffer: 100 mM Tris, 15% ethanol, 1 M KCl, adjusted to pH 8.5 with H_3PO_4.
10. 95% ethanol.
11. 70% ethanol.
12. TE: 10 mM Tris–HCl, pH 7.4, and 1 mM EDTA sterilized by filtration.
13. Virkon disinfection tablets.

2.4 Transfection

1. RPMI 1640 medium containing 25 mM HEPES and 4 mM glutamine.
2. 10% Pluronic F-68: dissolve 100 g Pluronic F-68 in 1 L water, filter sterilize, and store at room temperature.
3. 0.5 M valproic acid (VPA): dissolve 72.1 g of VPA in 1 L water, filter sterilize, and store at −20 °C in sterile 50-mL tubes.
4. Linear 25 kDa polyethylenimine (PEI): dissolve 1 g of PEI in 800 mL water. Adjust the pH to 3 with 1N HCl. When the PEI is in solution, adjust the pH to 7 with 1N NaOH. Adjust the volume to 1 L, filter sterilize, and store at −20 °C in sterile 50-mL tubes.
5. Plasmid DNA in TE at a concentration of 1–2 mg/mL.

2.5 Protein Purification

1. MabSelect resin.
2. Econo-Pac chromatography column.
3. PBS.

4. 10× TBS (Tris-buffered saline at pH 7.5).
5. Elution solution: 100 mM glycine (pH 3).
6. Neutralizing solution: 1.4 M Tris–HCl (pH 8.0).
7. Amicon Ultra-15 centrifugal filters with molecular weight cut-off (MWCO) of 10 and 3 kDa.
8. Nickel sepharose Excel resin.
9. Imidazole.
10. Buffer A: 150 mM NaCl and 25 mM sodium phosphate, pH 7.2.
11. 10 mM wash: buffer A with 10 mM imidazole.
12. 25 mM wash: buffer A with 25 mM imidazole.
13. 50 mM wash: buffer A with 50 mM imidazole.
14. Elution solution: buffer A with 250 mM imidazole.
15. SnakeSkin dialysis tubing (22 mm diameter with 3.5 kDa MWCO).
16. Precast NuPAGE 4–12% Bis-Tris gel.
17. 20× MES running buffer.
18. InstantBlue gel staining solution.
19. Millex-HV syringe-driven filter with low protein binding membrane with 0.45 μm pore size.

2.6 Equipment

1. Laminar flow hood.
2. Two incubator shakers maintained at 37 °C, one for mammalian cell cultivation in the presence of CO_2 and one for bacteria cultivation.
3. Floor-model centrifuge.
4. Tabletop centrifuge.
5. Inverted phase contrast microscope.
6. NanoDrop spectrophotometer.
7. Mini-polyacrylamide gel electrophoresis (PAGE) gel box with power supply.
8. Benchtop roller apparatus.
9. Magnetic stirrer.

3 Methods

3.1 Plasmid Production

1. *E. coli* strain DH5α is separately transformed with each plasmid by the standard $CaCl_2$ method and spread onto LB agar plates with 100 μg/mL ampicillin (*see* **Note 3**).
2. Incubate the plates overnight (16 h) at 37 °C.

3. With a pipette tip, transfer a single colony from each plate to a 14 mL culture tube containing 3 mL LB broth with 100 μg/mL ampicillin.
4. Incubate at 37 °C for 6 h with agitation at 220 rpm.
5. Use the 3 mL culture to inoculate a 5-L Erlenmeyer flask containing 1 L of LB broth with 100 μg/mL ampicillin.
6. Incubate 12–16 h at 37 °C with agitation at 220 rpm.
7. Transfer the culture to two 500-mL centrifuge bottles.
8. Centrifuge at 1000 × *g* for 20 min at 4 °C and decant the medium into an Erlenmeyer flask. Retain the cell pellets and dispose of the medium after treatment with Virkon disinfectant.
9. Resuspend each cell pellet in 12 mL of resuspension buffer by pipetting the cells with a 10-mL pipette.
10. Transfer the resuspended cells into a 50-mL centrifuge tube.
11. Add 12 mL of lysis buffer to the bacterial suspension. Close the cap and mix gently by inverting the tube 6–8 times.
12. Let the solution stand at room temperature (20–25 °C) for 2–3 min.
13. Add 12 mL of prechilled neutralization buffer to the suspension. Close the cap and mix gently by inverting the tube 6–8 times until a homogeneous suspension containing an off-white flocculate is formed.
14. Let the tube stand in ice for 5 min.
15. Centrifuge the suspension at 1800 × *g* for 30 min at 4 °C. Repeat this step if the supernatant contains residual particles after the first centrifugation.
16. Attach the NucleoBond AX 500 column to a support stand and equilibrate the column with 6 mL of equilibration buffer. Allow the column to empty by gravity flow and discard the flow-through.
17. Load the cleared lysate onto the NucleoBond column. Allow the column to empty by gravity flow.
18. Wash the column twice with 32 mL of wash buffer. Collect the flow-through in a beaker and then discard.
19. Elute the plasmid DNA with 15 mL of elution buffer. Collect the eluate in a clean 50-mL centrifuge tube.
20. Add 11 mL of isopropanol at room temperature to precipitate the plasmid DNA. Mix well and centrifuge at 2500 × *g* for 30 min at 4 °C.
21. In a laminar flow hood, carefully pour the supernatant into a waste container.

22. To the pellet, add 15 mL of 70% ethanol. Vortex briefly and centrifuge at 2500 × *g* for 20 min at room temperature.
23. In a laminar flow hood, carefully decant the 70% ethanol. Allow the pellet to air-dry in the flow hood at room temperature (*see* **Note 4**).
24. To the pellet add 0.7 mL of sterile TE and incubate at 37 °C for 2–3 h on an orbital shaker.
25. Determine the plasmid yield using a Nanodrop spectrophotometer. Record the concentration and the A_{260}/A_{280} ratio (*see* **Note 5**).

3.2 Routine Cell Cultivation

1. HEK293E cells are subcultivated every 3–4 days (*see* **Note 6**) in a 1-L square-shaped glass bottle by inoculation in 300 mL Ex-cell 293 medium containing L-glutamine and phenol red at an initial cell density of 0.3×10^6 cells/mL (*see* **Note 7**).
2. At the end of the subcultivation period, determine the cell density and viability by Trypan Blue staining by mixing 20 μL cells culture, 40 μL PBS, and 20 μL 0.4% Trypan blue solution.
3. After transferring the stained solution to a Neubauer haemocytometer chamber, determine the cell density and viability with an inverted phase contrast microscope.
4. Transfer a culture volume corresponding to 9×10^7 cells into a 50-mL conical centrifuge tube and centrifuge at 500 × *g* for 5 min in a standard tabletop centrifuge.
5. Remove the medium by aspiration and resuspend the cell pellet in 20 mL of Ex-cell 293 medium.
6. Transfer the cells to a 1-L square-shaped bottle containing 280 mL of prewarmed Ex-cell 293 medium.
7. Attach the bottle to a platform mounted on an orbital shaker with a rotational diameter of 5 cm using double-sided adhesive tape and agitate at 110 rpm at 37 °C in a 4.5% CO_2 atmosphere without humidity, keeping the cap of the bottle opened about one quarter of a turn (*see* **Note 8**).

3.3 Cell Expansion for Transfection

1. One day before transfection, determine the cell density and viability as described in Subheading 3.2.
2. For a 2-L transfection, transfer 12×10^8 cells (about 300 mL) into a 500-mL conical centrifuge bottle.
3. Centrifuge the cells for 10 min at 500 × *g* at room temperature.
4. Remove the medium by aspiration and gently resuspend the cell pellet in 50 mL of prewarmed Ex-cell 293 medium.

5. Transfer the cells into a 2-L cylindrical glass bottle with 550 mL of prewarmed Ex-cell 293 medium. The starting cell density of the culture is about 2×10^6 cells/mL.
6. Place the bottle on an orbital shaker as described in Subheading 3.2, **step** 7 and incubate overnight.

3.4 Large-Scale Transfection

1. Determine the cell density and viability of the culture in the 2-L bottle as described in Subheading 3.2 (*see* **Notes 9** and **10**).
2. Transfer a total of 2×10^9 cells (about 500 mL) from the overnight culture into one or two 500-mL conical centrifuge bottle(s) and centrifuge at $500 \times g$ for 10 min at room temperature.
3. Remove the medium by aspiration and resuspend the cells in a total volume of 100 mL by addition of prewarmed RPMI 1640 medium containing 0.1% Pluronic F-68.
4. Transfer the cells to a 250-mL square-shaped glass bottle.
5. Add 3 mg of plasmid DNA to the culture and mix gently by swirling the bottle (*see* **Note 11**).
6. Add 6 mL PEI solution (1 mg/mL) to the culture and gently mix by swirling the bottle.
7. Attach the bottle to the platform of the orbital shaker and agitate as described in Subheading 3.2, **step** 7.
8. After 1 h of incubation, transfer the cells to a 5-L cylindrical bottle containing 2 L of prewarmed Ex-cell 293 medium, if mRSpo1-Fc is being produced, or 2 L of FreeStyle 293 medium, if mNoggin-His is being produced (*see* **Note 12**).
9. Add 15 mL of 0.5 M VPA to achieve a final VPA concentration of 3.75 mM.
10. Incubate the culture as described in Subheading 3.2, **step** 7 for 7–8 days (Fig. 2).

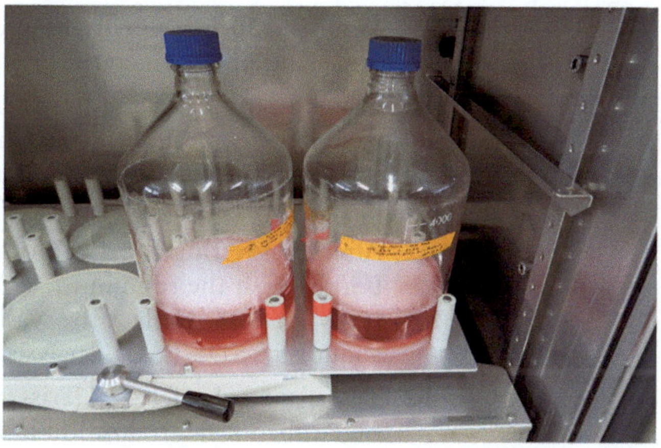

Fig. 2 Image of incubator shaker for the agitation of 5-L glass bottles. The shaker platform is equipped with supports to hold 5-L glass bottles

3.5 Conditioned Medium Harvest

1. At the end of the production phase of 7–8 days, harvest the conditioned medium by transferring the culture into four 500-mL conical centrifuge bottles (*see* **Note 13**).
2. Centrifuge at 1500 × *g* for 30 min at 4 °C.
3. Recover the supernatant by decanting into a 2-L glass bottle.
4. Remove any additional cell debris by passing the medium through a 1-L filtration unit with a 0.2 μm membrane, collecting the filtrate in a sterile 2-L glass bottle.
5. Store the filtered medium at 4 °C until needed for protein purification.

3.6 Purification of Recombinant mRSpo1-Fc

1. In a 50-mL centrifuge tube, wash 4 mL of MabSelect with 40 mL of PBS. Mix the tube gently by hand and allow the tube to stand at room temperature. When the resin has sedimented, decant the PBS (*see* **Notes 14** and **15**).
2. Repeat the washing step two more times.
3. After the last wash, transfer the MabSelect resin into the 2-L bottle containing the medium from the mRSpo1-Fc production. After transferring, rinse the remaining resin from the 50-mL tube with PBS and add to the 2-L bottle.
4. Install the 2-L bottle on a benchtop rotator in a refrigerator or cold room at 4 °C.
5. Rotate the bottle at least 4 h at about 4 rpm (*see* **Note 16**).
6. Remove the bottle from the rotator and let it stand at 4 °C until the MabSelect resin has sedimented.
7. Decant about 1.6 L of the conditioned medium into a clean 2-L glass bottle making sure that the MabSelect is not too disturbed.
8. Transfer the remaining medium with the MabSelect to a 500-mL conical centrifuge bottle and let it stand at 4 °C to allow the MabSelect to sediment.
9. Mount a 20-mL gravity-flow chromatography column on a ring stand in a refrigerator or cold room.
10. Using a 10-mL pipette, transfer the MabSelect from the bottom of the conical bottle into the column. Collect the medium that flows through the column in a clean bottle. Save an aliquot of the flow-through for SDS-PAGE analysis.
11. After transferring the 4 mL of MabSelect from the centrifuge bottle, wash the resin in the column with 20 column volumes (CVs) of PBS. Collect the wash in a clean bottle and save an aliquot for SDS-PAGE analysis.
12. Transfer 5 mL of 1.4 M Tris–HCl (pH 8) to a 50-mL centrifuge tube for neutralization of the eluate.

13. Elute the mRSpo1-Fc from the MabSelect with 20 mL (5 CVs) of 100 mM glycine (pH 3), collecting the eluate in the 50-mL centrifuge tube prepared in **step 12**.
14. Thoroughly mix the eluate in the 50-mL tube and save a sample for SDS-PAGE.
15. Wash the resin with 5 CVs of 100 mM glycine (pH 3) and then 5 CVs PBS, transfer into a 50-mL centrifuge tube, and store at 4 °C in PBS plus 20% ethanol.
16. Measure the protein concentration in the eluate with a Nanodrop spectrophotometer.
17. Analyze the flow-through, wash and elution fractions by SDS-PAGE (Fig. 3a).
18. Transfer 2 L of cold PBS into a 3-L beaker for the dialysis of the recombinant protein and place the beaker on a magnetic stirrer in a refrigerator or cold room at 4 °C.
19. Transfer the eluted protein into dialysis tubing (MWCO 3.5 kDa) and dialyze in 2 L of cold PBS at 4 °C, stirring gently for at least 3 h.
20. Make two changes of the PBS for dialysis at intervals of at least 3 h (*see* **Note 17**).
21. Remove the dialysis bag from the PBS after the third dialysis. Open the bag and transfer the protein to a 50-mL centrifuge tube.

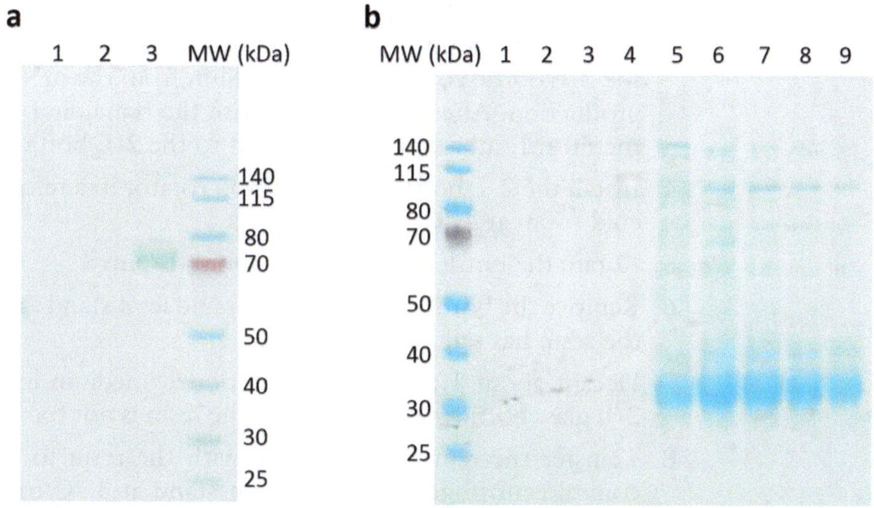

Fig. 3 SDS-PAGE analysis of protein purification. (**a**) Chromatography fractions from mRSpo1-Fc purification with protein A. Lane 1: flow-through fraction; lane 2: wash fraction; lane 3: elution fraction. (**b**) Chromatography fractions from mNoggin-His purification by IMAC. Lane 1: flow-through fraction; lane 2: 25 mM imidazole wash; lane 3: 50 mM imidazole wash; lanes 4–9: elution with 250 mM imidazole solution. Acrylamide gels were stained with Coomassie Blue

22. Measure the protein concentration with a NanoDrop spectrophotometer.
23. If the protein concentration is too low for the desired application, then the solution can be concentrated with an Amicon Ultra-15 centrifugal filter (MWCO 10 kDa). Washing the filter first with 15 mL of PBS by centrifugation at 1500 × *g* for 30 min.
24. After washing, discard the PBS from the two chambers of the centrifugal filter and then transfer the protein solution to the top chamber. Centrifuge at 1500 × *g* for 20 min repeatedly until the desired reduction in volume is achieved.
25. Transfer the concentrated protein solution from the centrifugal filter using a needle and syringe.
26. Attach the syringe to a syringe-driven filter unit with a 0.45 μm membrane. Filter the protein solution into a sterile 15-mL centrifuge tube.
27. Measure the protein concentration with a Nanodrop spectrophotometer.
28. Store the protein at 4 °C if used immediately or divide the solution into appropriate aliquots and store at −80 °C.

3.7 Purification of Recombinant mNoggin-His

1. In a 50-mL centrifuge tube, wash 4 mL of nickel sepharose Excel with 40 mL of buffer A. Mix the tube gently by hand and allow the tube to stand at room temperature. When the resin has sedimented, decant the buffer (*see* **Note 18**).
2. Repeat the washing step two more times.
3. After the last wash, the resin in buffer A can be transferred into the 2-L bottle containing the medium from the mNoggin-His production. After transferring, rinse the remaining resin from the 50-mL tube with PBS and add to the 2-L bottle.
4. Install the 2-L bottle on a benchtop rotator in a refrigerator or cold room at 4 °C.
5. Rotate the bottle at least 4 h at about 4 rpm.
6. Remove the bottle from the rotator and let it stand at 4 °C until the resin has sedimented.
7. Decant about 1.6 L of the conditioned medium into a clean 2-L glass bottle making sure that the resin is not too disturbed.
8. Transfer the remaining medium with the resin to a 500-mL conical centrifuge bottle and let it stand at 4 °C to allow the resin to sediment.
9. Mount a 20-mL gravity-flow chromatography column on a ring stand in a refrigerator or cold room at 4 °C.

10. Using a 10-mL pipette, transfer the resin from the bottom of the conical bottle into the column. Collect the medium that flows through the column in a clean glass bottle. Save an aliquot of the flow-through for SDS-PAGE analysis.
11. After transferring the 4 mL of resin from the centrifuge bottle, wash the resin in the column with 10 CVs of 10 mM wash solution. Collect the wash in a clean bottle. Save an aliquot of the wash for SDS-PAGE analysis.
12. Wash the resin with 5 CVs of 25 mM wash solution and collect as before. Keep an aliquot for SDS-PAGE analysis.
13. Wash the resin with 5 CVs of 50 mM wash solution and collect as before. Keep an aliquot for SDS-PAGE analysis.
14. Elute mNoggin-His from the resin 6 times with 3 mL of elution buffer, collecting each 3-mL fraction in a 15-mL centrifuge tube. Save an aliquot of each for SDS-PAGE analysis.
15. Wash the resin with 10 CVs of buffer A, transfer into a 50-mL centrifuge tube, and store at 4 °C in buffer A plus 20% ethanol.
16. Analyze the flow-through, wash and elution fractions by SDS-PAGE (Fig. 3b).
17. Pool the wash and elution fractions containing mNoggin-His and transfer into dialysis tubing (MWCO 3.5 kDa).
18. Dialyze in 2 L of cold TBS at 4 °C on a magnetic stirrer with two changes of the PBS as described in Subheading 3.6.
19. Remove the dialysis bag after the third dialysis. Open the bag and transfer the protein solution to a 50-mL centrifuge tube.
20. Measure the protein concentration with a NanoDrop spectrophotometer.
21. If the protein concentration is too low for the desired application, then the solution can be concentrated with an Amicon Ultra-15 centrifugal filter (MWCO 3 kDa). Wash the filter with 15 mL of PBS by centrifugation at 1500 × *g* for 30 min as described in Subheading 3.6, **step 23**.
22. Discard the PBS from the two chambers of the centrifugal filter. Transfer the protein solution to the top chamber and centrifuge at 1500 × *g* for 20 min repeatedly until the desired reduction in volume is achieved (*see* **Note 19**).
23. Filter the protein solution with a syringe-driven filter as described in Subheading 3.6, **step 26**.
24. Measure the protein concentration with a Nanodrop spectrophotometer.
25. Store the protein at 4 °C if used immediately or divide the solution into appropriate aliquots and store at −80 °C.
26. The proteins are ready to use to treat the intestinal organoids (Fig. 4) (*see* **Note 20**).

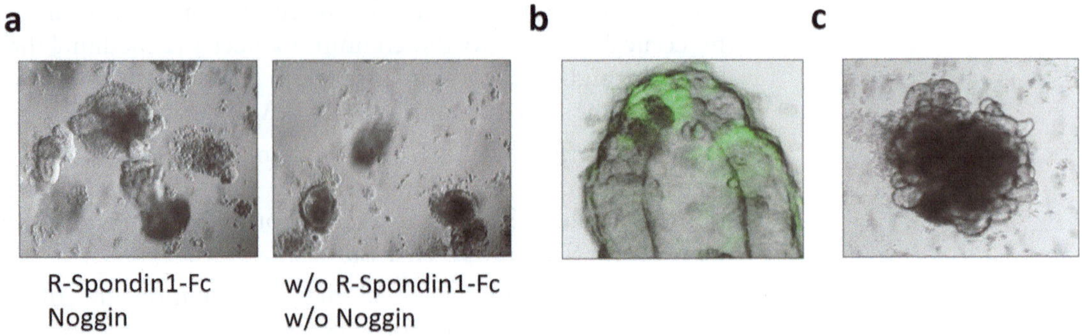

Fig. 4 Microscopic images of intestinal organoid culture. (**a**) Cultures with (left panel) and without (right panel) growth factors. (**b**) Maintenance of stem cells (Lgr5-GFP-positive cells) in the presence of growth factors. (**c**) Intestinal organoid in the presence of growth factors

4 Notes

1. An 8× polyhistidine (His) tag was added to the C-terminus of mNoggin for Ni-NTA enrichment. A linker sequence was added between mNoggin and the His tag in order to prevent the polyhistidine tag from affecting the activity of the protein. The oligonucleotides used to generate the linker are the following: 5′ CCG GCG GCC GCG GAG GCG GAT CTG GAG GCG GAT CTG GCG GAG GAT CCC CC 3′ and 5′ GGG GGA TCC TCC GCC AGA TCC GCC TCC AGA TCC GCC TCC GCG GCC G 3′. We cloned the mNoggin cDNA and linker first into pENTR1a (attL1-mNoggin_linker_8xHis-attL2) and afterward it was cloned into pDEST12.2 (CMVprom-mNoggin_linker_8xHis-SV40pA) using the Gateway recombination cloning technology.
2. We produce and purify mR-Spondin1, but other Spondins can replace this protein for the intestinal organoid culture. A Wnt activity assay should be done before using the different Spondins in the culture (Fig. 5).
3. Since a significant amount of plasmid DNA (1.5 mg/L) is necessary for TGE at large scale, it is important to maximize plasmid yields by choosing a vector with a high copy number origin of replication. This will also make the recovery of plasmid DNA easier, since the ratio of plasmid DNA to contaminants such as genomic DNA, RNA, and protein will be less.
4. The plasmid DNA needs to be prepared sterilely since we do not use antibiotics in the cell culture medium. The most efficient way to achieve sterility is to precipitate the DNA in alcohol.
5. Only DNA preparations with A_{260}/A_{280} ratios ≥ 1.8 are used for transfection.

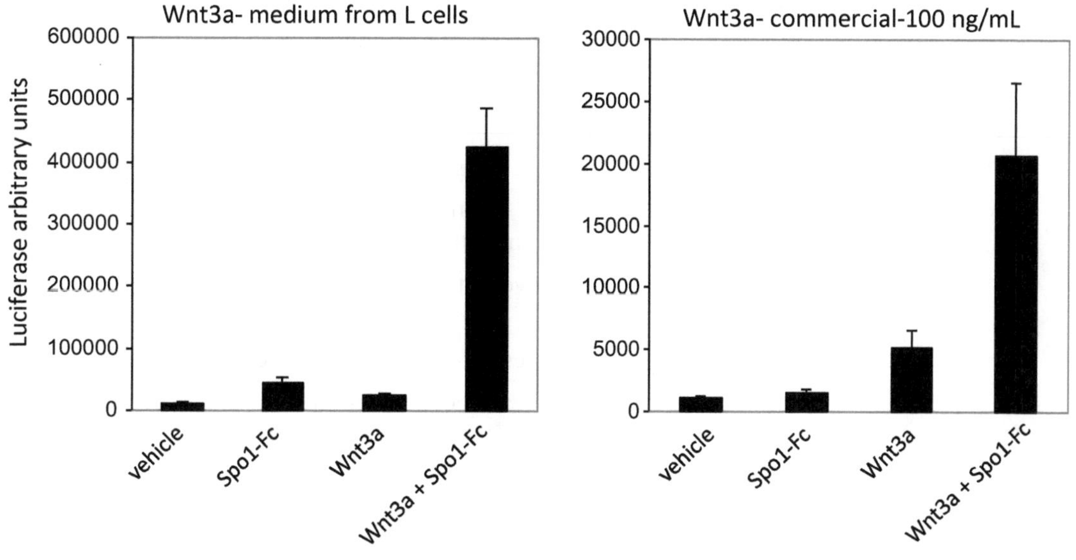

Fig. 5 Wnt activity assay of recombinant mRSpo1-Fc

6. To assure reproducibility of transfection results, we do not recommend keeping HEK293E cells in culture longer than 3 months (20–25 passages). We also recommend maintaining the cells in exponential growth phase at all times.
7. For routine cell culture maintenance, we keep a culture of 100 mL in a 250-mL square-shaped glass bottle.
8. The agitation conditions described here depend on a shaking diameter of 5 cm for the shaker platform. For incubator shakers with a lower shaking diameter, the shaking speed needs to be increased. For a shaking diameter of 2.5 cm, for example, the bottles need to be agitated at about 150 rpm.
9. The protocol described here has been established for 2-L transfections. For some users, it may not be feasible to transfect at this volume or it may be necessary to do small-scale test transfections. For optimizing our protocol, we used small-scale transfections of 5–10 mL in TubeSpin 50 bioreactors. Protocols for the cultivation and transfection of HEK293E cells in this type of vessel have been previously published [7].
10. We do not use cultures in which the cell viability is less than 95% or if the cells have not doubled overnight.
11. The method described here does not involve precomplex formation with DNA and PEI prior to addition to the culture. It is very important to minimize the time delay between addition of DNA and PEI and to mix the culture well after each component is added.

12. Nickel-affinity chromatography of a secreted histidine-tagged protein from Ex-Cell293 medium may be difficult due to stripping of the nickel by a component of the medium. This can be avoided by producing a secreted his-tagged protein in another medium that supports nickel-affinity chromatography such Freestyle293 or Pro293s. In addition, it is possible to purify the protein from Ex-cell293 with Ni Sepharose Excel resin as described in Subheading 3.7.
13. The optimal harvest time should be determined for each protein.
14. Other protein A resins can be used besides MabSelect.
15. The purification can be performed on a fast protein liquid chromatography (FPLC) apparatus using a prepacked protein A column.
16. We usually do this step overnight at 4 °C.
17. We usually do the first or second dialysis overnight.
18. The purification can be performed on a FPLC apparatus using a prepacked Ni sepharose Excel column.
19. To remove contaminating proteins from the mNoggin-His solution after affinity chromatography, it is advisable to perform size-exclusion chromatography. However, this will result in the loss of some mNoggin-His.
20. Long-term maintenance of intestinal organoids is possible in the presence of mRSpo1-Fc mNoggin-His, and epidermal growth factor (EGF) (Fig. 4).

References

1. Gjorevski N, Ordóñez-Morán P (2017) Intestinal stem cell niche insights gathered from both in vivo and novel in vitro models. Stem Cells Int 2017:8387297
2. Pham PL, Kamen A, Durocher Y (2006) Large-scale transfection of mammalian cells for the fast production of recombinant protein. Mol Biotechnol 34:225–237
3. Hacker DL, Kiseljak D, Rajendra Y, Thurnheer S, Baldi L, Wurm FM (2013) Polyethyleneimine-based transient gene expression processes for suspension-adapted HEK-293E and CHO-DG44 cells. Protein Expr Purif 92:67–76
4. Backliwal G, Hildinger M, Hasija V, Wurm FM (2008) High-density transfection with HEK-293 cells allows doubling of transient titers and removes need for a priori DNA complex formation with PEI. Biotechnol Bioeng 99:721–727
5. Backliwal G, Hildinger M, Kuettel I, Delegrange F, Hacker DL, Wurm FM (2008) Valproic acid: a viable alternative to sodium butyrate for enhancing protein expression in mammalian cell cultures. Biotechnol Bioeng 101:182–189
6. Muller N, Girard P, Hacker DL, Jordan M, Wurm FM (2005) Orbital shaker technology for the cultivation of mammalian cells in suspension. Biotechnol Bioeng 89:400–406
7. Hacker DL, Durrer L, Quinche S (2019) CHO and HEK293 cultivation and transfection in single-use orbitally shaken bioreactors. Methods Mol Biol 1850:123–131

Chapter 11

Primary Intestinal Epithelial Organoid Culture

Tomohiro Mizutani and Hans Clevers

Abstract

The intestinal epithelium is known as one of the most regenerative tissues in our body. The lining of the intestine is composed of a single layer of epithelial cells generated by rapidly renewing stem cells residing at the crypt bottoms, resulting in a flow of cells to the villus tips. The stereotypical crypt–villus architecture makes the intestine an ideal model for stem cell research. Based on recent advances in research of stem cell niche signals in vivo, we have established an intestinal epithelial stem cell culture method. Under this culture condition, single Lgr5+ intestinal stem cells (ISCs) or isolated whole crypts efficiently expand into three-dimensional spherical structures recapitulating the intestinal crypt–villus organization. These organoids can be passaged weekly and maintained for years in culture. Moreover, they can be cryopreserved. As intestinal organoids recapitulate many aspects of the epithelial biology and are amenable to most, if not all, current experimental manipulations, they are widely used to study stem cell biology, cell fate determination, gene function, and disease mechanism.

Key words Intestinal epithelial stem cells, Lgr5, Crypt isolation, Organoid culture, Small intestine, Colon, Mouse

1 Introduction

The intestinal epithelium is one of the most regenerative tissues in our body. The lining of the small intestine and large intestine (colon) is composed of a single layer of intestinal epithelial cells and is completely regenerated every 4–5 days in rodents [1]. The intestinal epithelium consists of two parts: the ISC-containing crypts and the villi that project into the gut lumen. The ISCs, marked by the Leucine-rich repeat-containing G protein-coupled receptor 5 (Lgr5), reside at the bottom of the intestinal crypts and divide every 24 h to generate rapidly dividing transit-amplifying (TA) cells [2]. TA cells subsequently differentiate into all types of differentiated cells and migrate on the villi. The renewal kinetics and the highly stereotypical crypt–villus architecture makes this an ideal model for stem cell research.

By combining several stem cell niche signals, we established a mouse small intestinal epithelial stem cell culture method (also

Paloma Ordóñez-Morán (ed.), *Intestinal Stem Cells: Methods and Protocols*, Methods in Molecular Biology, vol. 2171, https://doi.org/10.1007/978-1-0716-0747-3_11,

called organoid culture) [3]. In this culture condition, a single Lgr5 + ISC or an isolated whole crypt is embedded in Matrigel®, a laminin- and collagen-rich extracellular matrix mimicking the basement membrane. The cells efficiently expand as three-dimensional spherical structures resembling intestinal crypt–villus structures. The stem cell niche factors for mouse intestinal organoids are: epidermal growth factor (EGF), the bone morphogenetic protein (BMP) signal inhibitor Noggin, and the Wnt agonist/ligand of Lgr5 R-spondin1 for mouse small intestinal organoids. Later, we also developed mouse colon organoid culture method by adding exogenous Wnt3a [4].

Intestinal organoids established from whole intestinal crypts or single Lgr5 ISCs contain Lgr5+ ISCs and all known types of differentiated cells. These organoids can be passaged at 1:5 ratio weekly and can be maintained in culture for years. Once organoids are established, the cells can be cryopreserved and thawed when needed. Their structural features, cell type composition and karyotype remain unchanged. Importantly, several organoid transplantation experiments have revealed that Lgr5+ ISCs retain their stem cell activity during in vitro organoid culture [5, 6]. Thus, organoids recapitulate stem cell maintenance and the differentiation hierarchy [7]. The organoid culture system has allowed to genetically manipulate and clone intestinal epithelial stem cells by transfection of DNA and small interfering RNA, retroviral/lentiviral transduction and CRISPR-Cas9 mediated genome editing [8–12].

Intestinal organoids are readily amenable to standard techniques for imaging (confocal microscopy, scanning and transmission electron microscopy [3, 13–15]), molecular analyses (single cell) RNA sequencing, whole genome Sequencing, ChIP/ATAC sequencing, etc.) or proteomic analyses [16–21]. It has thus been shown that the culture system recapitulates the hierarchical crypt–villus architecture; preserves gene functions, cell fate determination, and phenotypical characteristics; and is amenable to most experimental manipulations. Thus, organoids can be used for studying stem cell function, cell fate determination and differentiation, gene function, drug/nutrient absorption, interaction with nonepithelial cell types such as immune cells, and infectious agents [22–26].

Here, we describe our improved protocol for the isolation and establishment of long-term culture of mouse small intestinal and colonic organoids. Moreover, we provide details and discuss pitfalls in long-term maintenance of organoid culture, cryopreservation, thawing, preparation for immunohistochemical analysis, and whole-mount immunostaining.

2 Materials

2.1 Isolation of Intestinal Crypts

1. Dulbecco's phosphate-buffered saline without Ca^{2+} and Mg^{2+} (DPBS).
2. 10% (v/v) fetal bovine serum (FBS) in DPBS.
3. Intestinal crypt isolation buffer: 2.5 mM ethylenediaminetetraacetic acid (EDTA) in DPBS (Table 1).
4. Chelation stock buffer 5× [27] (Table 2) (*see* **Note 1**).
5. Colonic crypt isolation buffer: 5 mM EDTA in Chelation buffer (Table 3).
6. Basal medium (DMEM+): 100 mg/mL streptomycin and 100 U/mL penicillin in DMEM.
7. 70 μm cell strainer.
8. 15 and 50 mL tube coated with 10% (v/v) FBS in DPBS (FBS-coated tubes).
9. Coverslip 24 × 50 mm.
10. Murine small intestinal and colonic tissue (*see* **Note 2**).

2.2 Primary Organoid Culture from Isolated Intestinal Crypts

1. Basal culture medium (AdDF+++): 10 mM HEPES, 2 mM GlutaMAX, 100 U/mL penicillin, and 100 μg/mL streptomycin in advanced DMEM/F12.

Table 1
Preparation of intestinal crypt isolation buffer

Intestinal crypt isolation buffer	
DPBS	20 mL
EDTA [0.5 M]	100 μL [2.5 mM]

Table 2
Preparation of chelation stock buffer 5×

Chelation stock buffer 5×	500 mL of distilled water
Na_2HPO_4	1.97 g [28 mM]
KH_2PO_4	2.7 g [40 mM]
NaCl	14 g [480 mM]
KCl	0.3 g [8 mM]
Sucrose	37.5 g [220 mM]
D-sorbitol	25 g [274 mM]

Table 3
Preparation of colonic crypt isolation buffer

Colonic crypt isolation buffer	
Chelation stock buffer 5×	4 mL
Distilled water	16 mL
DL-dithiothreitol [100 mM]	100 μL [0.5 mM]
EDTA [0.5 M]	200 μL [5 mM]

Table 4
Organoid culture media components

Reagents	Final concentration	Small intestine ENR	Colon WENR
AdDMF+++		+	+
Wnt 3a-conditioned medium	50% (v/v)	–	+
EGF	50 ng/mL	+	+
Noggin Noggin-conditioned medium	100 ng/mL 10% (v/v)	+	+
R-spondin1 R-spondin1-conditioned medium	500 ng/mL 10% (v/v)	+	+
B27 supplement	1×	+	+
N-acetyl-L-cysteine	1 mM	+	+

2. Mouse ENR medium: 50 ng/mL recombinant mouse EGF, 100 ng/mL recombinant mouse Noggin, 500 ng/mL recombinant human R-spondin-1 (*see* **Note 3**), 1× B27 supplement, and 1 mM *N*-acetyl-L-cysteine in AdDF+++ (Table 4).
3. Mouse WENR medium: 50% (v/v) Wnt-3a-conditioned medium (*see* **Note 4**), 50 ng/mL recombinant mouse EGF, 100 ng/mL recombinant mouse Noggin, 500 ng/mL recombinant human R-spondin-1, 1× B27 supplement, and 1 mM *N*-acetyl-L-cysteine in AdDF+++ (*see* **Notes 5** and **6**) (Table 4).
4. Tissue culture plates: 48-well/24-well, suspension culture, flat bottom.
5. Matrigel®, basement membrane matrix, growth factor reduced (GFR), phenol red-free.
6. Cultrex® Reduced Growth Factor BME, Type 2 PathClear® (*see* **Note 7**).
7. Y-27632 (Rho kinase inhibitor).

2.3 Passaging Intestinal Organoids

1. A narrow-tip Pasteur pipette (*see* **Note 8**).
2. TrypLE™ Express Enzyme (1X), phenol red.

2.4 Cryopreservation and Thawing of Organoids

1. Recovery™ Cell Culture Freezing Medium.
2. Freezing medium: 10% (v/v) dimethyl sulfoxide (DMSO) and 40% (v/v) fetal bovine serum (FBS) in AdDF+++ (*see* **Note 9**).
3. 1.5 mL cryotubes.
4. CoolCell® Cell Freezing Container.

2.5 Preparation for Immunohistochemistry

1. 4% paraformaldehyde (PFA).
2. 25%, 50%, 70%, 96%, and 100% ethanol.
3. Eosin.
4. *N*-butanol.
5. Paraffin.
6. Tissue-Tek® Base Mold.

2.6 Whole-Mount Immunostaining

1. 4% formalin.
2. Triton X-100.
3. Normal donkey serum.
4. VECTASHIELD® Antifade Mounting Medium.
5. Slide glass.
6. Coverslip 18 × 18 mm.
7. Vaseline.
8. Nail polish.

3 Methods

3.1 Isolation of Intestinal Crypts

1. Before starting, thaw Matrigel/Cultrex BME at 4 °C. Prewarm a culture plate by placing in a CO_2 incubator (5% CO_2, 37 °C). Place all medium and buffer bottle on ice (*see* **Note 10**). The culture medium must be warmed to room temperature before use.
2. Sacrifice a mouse and harvest a part of small intestine or colon.
3. Place the intestinal segment into a 100-mm dish containing 5 mL cold DPBS.
4. Flush the intestine with ice-cold DPBS using a 10-mL syringe to remove luminal contents (*see* **Note 11**).
5. Open the intestine longitudinally using surgical scissors, open the segment and gently scrape off luminal contents and villi using a coverslip.

Table 5
Crypt isolation conditions

	Small intestinal crypt isolation	Colonic crypt isolation
Isolation buffer	Intestinal crypt isolation buffer	Colonic crypt isolation buffer
Buffer	DPBS 20 mL	Chelation buffer 20 mL
EDTA concentration	2.5 mM [100 μL]	5 mM [200 μL]
Incubation duration	30 min	60 min
Temperature	4 °C	4 °C

6. Wash three times with ice-cold DPBS in a 100-mm dish using forceps (*see* **Note 12**).
7. Cut the opened sample into 10-mm pieces and transfer to a 50 mL tube containing 20-mL ice-cold DPBS using forceps.
8. Prewet a 10 mL serological pipette with DPBS and gently pipet the intestinal pieces up and down several times.
9. Allow the pieces to settle down to the bottom of the tube by gravity and gently discard the supernatant with a 10 mL serological pipette.
10. Add 15 mL ice-cold DPBS and repeat the washing step by pipetting the pieces with a 10 mL serological pipette.
11. Repeat steps 8–10 another about 10 times until the supernatant is clear enough.
12. Transfer the pieces to another 50-mL tube containing 20 mL of crypt isolation buffer/colonic crypt isolation buffer for small intestinal and colonic tissue respectively (Table 5).
13. Incubate the tissue pieces for 30–60 min at 4 °C on a tube roller (Table 5).
14. After incubation, let the tissue pieces settle by gravity, then gently pipet off the supernatant.
15. Resuspend the tissue pieces in 10 mL ice-cold DPBS containing 10% FBS and vigorously shake the tube until the crypts are released and can be seen in the supernatant (*see* **Note 13**).
16. Gently aspirate the supernatant with the pipette and filter it through a 70 μm filter to collect the supernatant in a new FBS-coated 50 mL tube.
17. Repeat steps 15–16 three more times to collect fresh crypts.
18. Check the contents of the tube under a light microscope to determine whether they contain healthy crypt fractions (Fig. 1a) (*see* **Note 14**).

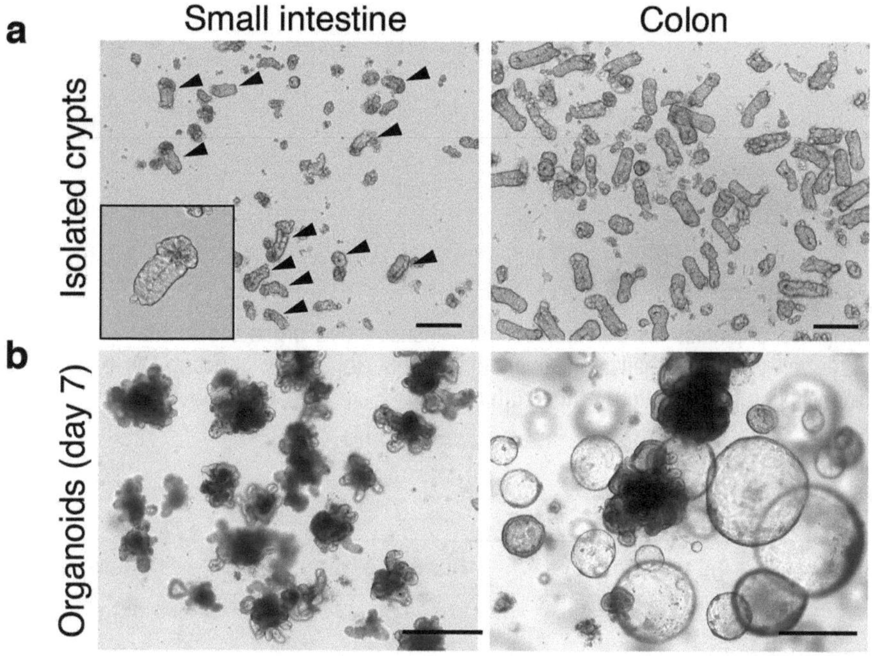

Fig. 1 Representative images of isolated mouse intestinal crypts and cultured organoids. (**a**) Isolated crypts from mouse small intestine and colon can be checked under a microscope. Black arrowheads indicate healthy small intestinal crypts showing shiny finger-like structures. Small intestinal crypts show large secretory granules of Paneth cells (inset) Scale bar, 100 μm. (**b**) Cultured small intestinal organoids have budding crypt-like structures. Colonic organoids show cystic structure and contain apoptotic cells inside. Scale bar, 500 μm

19. Centrifuge the supernatants containing the crypts at 100 × *g* for 5 min at 4 °C.
20. Resuspend the pellet in 10 mL cold DMEM+ and centrifuge at 100 × *g* for 5 min at 4 °C to remove single cells.
21. Resuspend the pellet in 10 mL DMEM+ and take 25 μL of the crypt resuspension to count crypt numbers under a microscope (*see* **Note 15**).

3.2 Primary Organoid Culture from Isolated Intestinal Crypts

1. Transfer the desired amount of crypt suspension into a new FBS-coated 15 mL tube and centrifuge at 100 × *g* for 5 min at 4 °C. Discard the supernatant (*see* **Note 16**).
2. Using a P200 pipette, resuspend the crypt pellet in cold Matrigel/Cultrex BME (200 crypts in 20 μL of Matrigel/Cultrex BME). Carefully pipet up and down to prevent bubbling.
3. Plate 20 μL of the mixture into the center of each well of a prewarmed 48-well plate (Fig. 2a) (*see* **Note 17**).
4. Gently invert the plate to make the gels hanging from the bottom of the plate (Fig. 2b) (*see* **Note 18**).

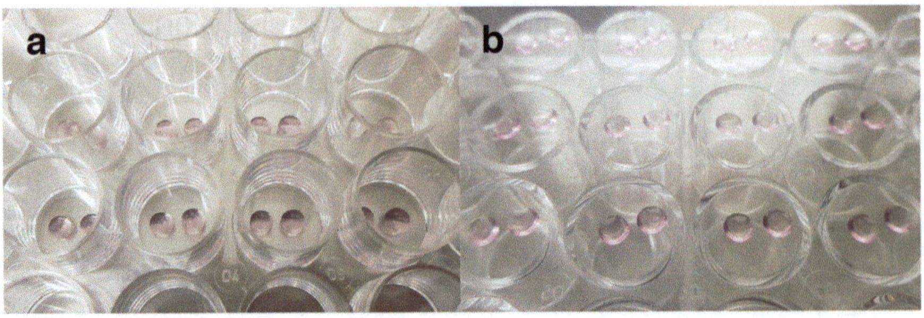

Fig. 2 Seeding the organoids as gel droplets. (**a**) Resuspended organoids in Matrigel (20 μL) plated in prewarmed 24-well plate as small hemispherical droplets. (**b**) Gently inverting the plates to prevent the organoids from sinking to the bottom of the culture plate

Table 6
Volume of gel and culture medium

Culture plate	Gel volume (μL)	Number of droplets	Culture medium volume
48-well	20–25	1 droplet	250 μL/well
24-well	40–45	2–3 droplets	500 μL/well
12-well	100–120	5–7 droplets	1 mL/well
6-well	200–240	12–14 droplets	2 mL/well

5. Place the culture plate into a CO_2 incubator (5% CO_2 at 37 °C) for 30 min to solidify the gel.
6. Add 250 μL mouse ENR/WENR culture medium to each well and culture for 5–7 days (Table 6).
7. To prevent anoikis, add 10 μmol/L Y-27632 to the culture medium for the first 2 days.
8. Exchange the culture medium every 2–3 days.
9. Passage the cultured organoids with 1:5 split ratio after 5–10 days before organoids overgrow or cell debris in their lumen accumulates (Fig. 1b) (*see* **Note 19**).

3.3 Passaging Intestinal Organoids

1. Before passaging, thaw Matrigel/Cultrex BME at 4 °C. Pre-warm a culture plate by placing in a CO_2 incubator (5% CO_2, 37 °C). Place DMEM+ bottle on ice. The culture medium must be warmed to room temperature before use.
2. Remove the culture medium from the wells and add 500 μL of ice-cold DMEM+ to each well with a P1000 pipette.
3. Scrape and suspend the gel cultures in cold medium and transfer into a 15 mL tube.
4. Add 10 mL of cold medium to the tube and gently pipet approximately 10 times to dissolve the gel (*see* **Note 20**).

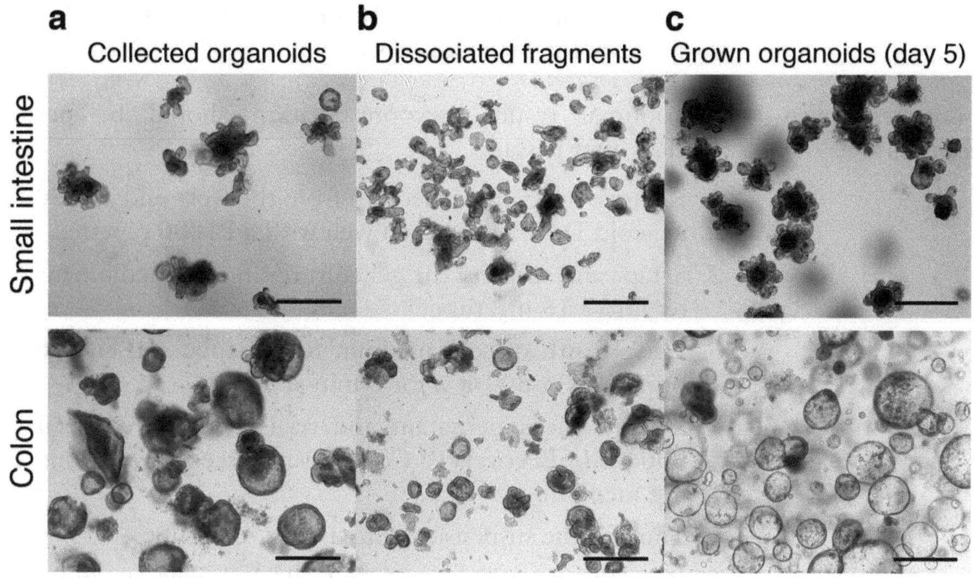

Fig. 3 Mouse intestinal organoid passage. (**a**) Collected intestinal organoids can be checked under a microscope. (**b**) After shearing with a narrow-tip Pasteur pipette, intestinal organoids are dissociated into crypt fragments. Colonic organoids are sheared into small cell clusters and fragmented organoids. (**c**) Passaged cell fragments grow into organoids within 5–7 days. Scale bar, 500 μm

5. Centrifuge the organoids at 300 × *g* for 5 min at 4 °C.
6. Discard the supernatant and resuspend the cell pellet in 2 mL cold medium (Fig. 3a).
7. Pipet up and down 10–20 times with a narrow-tip Pasteur pipette to shear the organoids. Dissociation of organoids can be assessed under a microscope. Organoids should break into individual crypt fragments (Fig. 3b) (*see* **Note 21**).
8. Add 10 mL of cold medium to the tube and centrifuge at 300 × *g* for 5 min at 4 °C (*see* **Note 22**).
9. Discard the supernatant without disturbing the pellet.
10. Resuspend the cell pellet in cold Matrigel/Cultrex BME. The gel volume can be calculated from the desired split ratio, and the optimal size of the culture plate and the gel volume are referred in Table 6.
11. Plate 20 μL of the mixture into the center of each well of a prewarmed 48-well plate.
12. Gently invert the plate to make the gels hanging from the bottom of the plate.
13. Place the culture plate into a CO_2 incubator (5% CO_2 at 37 °C) for 30 min to solidify the gel.
14. Add 250 μL mouse ENR/WENR culture medium to each well and culture for 5–7 days (Fig. 3c).

15. To prevent anoikis, add 10 μmol/L Y-27632 to the culture medium for the first 2 days.

3.4 Cryopreservation of Organoids

For optimal results, cryopreservation should be performed 2–4 days after passage.

1. Remove the culture medium from the wells and add 500 μL of ice-cold DMEM+ to each well with a P1000 pipette.
2. Scrape and suspend the gel cultures in cold medium and transfer into a 15 mL tube.
3. Add 10 mL of cold medium to the tube and centrifuge the organoids at 300 × *g* for 5 min at 4 °C.
4. Discard the supernatant and resuspend the cell pellet with 500 μL of Recovery™ Cell Culture Freezing Medium/Freezing medium.
5. Transfer the suspension into a 1 mL cryotube.
6. Place the cryotubes in a CoolCell® cell freezing container and store at −80 °C.
7. After overnight freezing, transfer the tubes into liquid N_2. Samples can be stored in liquid N_2 for years.

3.5 Thawing the Cryopreserved Organoids

1. Before thawing the organoids, thaw Matrigel/Cultrex BME at 4 °C. Prewarm a culture plate by placing in a CO_2 incubator (5% CO_2, 37 °C). Place DMEM+ bottle on ice. The culture medium must be warmed to room temperature before use.
2. Thaw the vial of frozen organoids at 37 °C in a water bath for 1–2 min, just before all ice crystals are melt.
3. Transfer the suspension of organoids into a 15 mL tube and dilute by adding 10 mL of basal medium slowly to avoid osmotic shock.
4. Centrifuge the organoids at 300 × *g* for 3 min at 4 °C.
5. Discard the supernatant without disturbing the pellet.
6. Resuspend the cell pellet in cold Matrigel/Cultrex BME (*see* **Note 23**).
7. Plate the mixture into the center of each well of a prewarmed 48-well plate.

3.6 Preparation for Immunohistochemistry

1. Remove the culture medium from the wells and add 500 μL of ice-cold DMEM to each well with a P1000 pipette.
2. Scrape and suspend the gel cultures in cold medium and transfer into a 15 mL tube.
3. Add 10 mL of cold medium to the tube and gently pipet almost 10 times to dissolve the gel.
4. Centrifuge the organoids at 300 × *g* for 5 min at 4 °C.

5. Discard the supernatant, resuspend the cell pellet in 5 mL of 4% PFA, and incubate for 60 min at 4 °C on a tube roller.
6. After fixation, let the organoids settle down to the bottom of the tube and remove the supernatant carefully.
7. Wash the organoids with PBS three times.
8. Dehydrate the organoids using 25%, 50%, and 70% ethanol each for 15 min (*see* **Note 24**).
9. Dehydrate the organoids in 96% ethanol and stain the organoids by adding 1% eosin dissolved in ethanol 96% for 30 min.
10. Dehydrate the organoids in 100% ethanol three times for 30 min.
11. Clear the organoids with *n*-butanol three times for 30 min.
12. Immerse the organoids in paraffin at 65 °C three times for 30 min and embed in Tissue-Tek® Base Mold.

3.7 Whole-Mount Immunostaining

1. Remove the culture medium from the wells and add 500 μL of cold DMEM to each well with a P1000 pipette.
2. Scrape and suspend the gel cultures in cold DMEM and transfer into a 15 mL tube.
3. Add 10 mL of cold medium to the tube and gently pipet almost 10 times to dissolve the gel.
4. Centrifuge the organoids at 300 × *g* for 5 min at 4 °C.
5. Discard the supernatant, resuspend the cell pellet in 1 mL of 4% formalin, transfer to a 1.5 mL tube. Be sure to prewet the tip of P1000 pipette with 10% (v/v) FBS in PBS to prevent the organoids from sticking to the wall of the tip.
6. Incubate for 2 h at R/T on a tube roller.
7. Centrifuge the fixed organoids at 300 × *g* for 5 min.
8. Wash organoids with 2% (v/v) normal Donkey Serum in PBS.
9. Centrifuge at 300 × *g* for 5 min and discard the supernatant as much as possible.
10. Incubate organoids with 2% (v/v) normal Donkey Serum in PBS for 30 min.
11. Centrifuge at 300 × *g* for 5 min and discard the supernatant as much as possible.
12. Incubate organoids with 0.5% Triton + 2% (v/v) normal Donkey Serum in PBS for 30 min for permeabilization.
13. Centrifuge at 300 × *g* for 5 min and discard the supernatant as much as possible.
14. Wash organoids with 2% (v/v) normal Donkey Serum in PBS.
15. Centrifuge at 300 × *g* for 5 min and discard the supernatant as much as possible.

16. Add 100 μL of primary antibody and incubate O/N at 4 °C (*see* **Note 25**).
17. Wash organoids with 2% (v/v) normal Donkey Serum in PBS. Repeat twice.
18. Centrifuge at 300 × *g* for 5 min and discard the supernatant as much as possible.
19. Repeat steps 15–17 for second and third antibodies.
20. Wash organoids with 2% (v/v) normal Donkey Serum in PBS. Repeat twice.
21. Centrifuge at 300 × *g* for 5 min and discard the supernatant as much as possible.
22. Suspend with one drop of mounting medium.
23. Put a coverslip under the slide glass and place a tiny droplet of Vaseline at each corner of the coverslip on the slide glass.
24. Remove the coverslip and apply suspended sample on the slide glass in the center of Vaseline drops.
25. Carefully put the cover glass on and close with nail polish.
26. The organoids are analyzed by confocal fluorescence microscopy (Fig. 4).

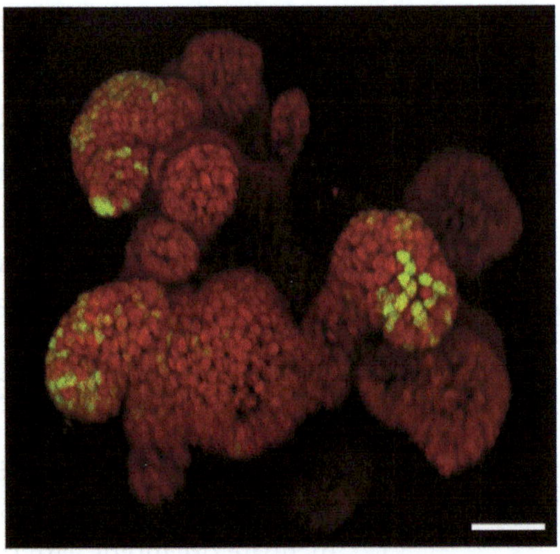

Fig. 4 A representative whole-mount confocal image of mouse intestinal organoid. A three-dimensional intestinal organoid imaged by whole-mount immunofluorescent staining and confocal microscopy. Intestinal organoids were established from Lgr5-EGFP-ires-CreERT2 mouse intestinal crypts. Lgr5-GFP stem cells (green) are localized at the tip of budding crypt-like structures. Counterstain is TO-PRO-3 (red). Scale bar, 50 μm. (Adapted with permission from Sato et al. [3], Springer Nature)

4 Notes

1. After dissolved in distilled water, pass it through 0.22 μm filter and store at 4 °C.
2. Mouse intestinal tissue should be processed immediately after harvesting. However, it's possible to preserve it in ice-cold medium (AdDMF+++) for one day.
3. Recombinant Noggin and R-spondin1 can be replaced with 10% (v/v) Noggin conditioned medium from Noggin-Fc-producing HEK293T cell line (stably transfected with mouse Noggin-Fc expression vector) established in our lab (available on request) [28] and 10% (v/v) R-spondin1-conditioned medium from R-spondin1-Fc-producing HEK293T cell line (a gift from Calvin Kuo, Stanford University, available as Cultrex® R-spondin1 Cells [29]), respectively.
4. Wnt-3a-conditioned medium is prepared using L-Wnt3a cells (L cells stably transfected with the expression plasmid; pcDNA3.1/Zeo(−)-mouse Wnt3a) established in our lab (available on request) [30]. The TOP flash assay can be used to test the transcriptional activity of Wnt with HEK293 STF cell line (ATCC® CRL-3249™). Collect 40 mL aliquots of fresh Wnt-3a-conditioned medium in 50 mL conical tubes and store at 4 °C for up to 2 months. Do not freeze the conditioned medium to avoid decreasing the Wnt activity.
5. Recently commercially produced medium for organoid culture (IntestiCult™ Organoid Growth Medium (Mouse)) is also available.
6. Culture medium containing growth factors can be stored at 4 °C for 2 weeks.
7. Matrigel® can be used without dilution. As Cultrex® BME is provided with higher concentration, Cultrex® BME can be diluted with basal medium to optimize at a final concentration of 8–10 mg/mL.
8. To make a glass tip narrower, rotate the tip of a Pasteur pipette in the Bunsen burner flame for a few seconds (Fig. 5a). Alternatively, a P1000 pipette tip with a P10 tip fixed on the tip can be used (Fig. 5b).
9. Commercially available freezing media or conventional homemade freezing media can be used.
10. All the medium for isolation should be put on ice to avoid damaging the tissue.
11. Insert P200 pipette tip or a blunt end 18G needle attached on a 10 mL syringe to flush DPBS through a lumen of the intestine.
12. After three times washing in a petri dish, most of the residual contents are cleaned.

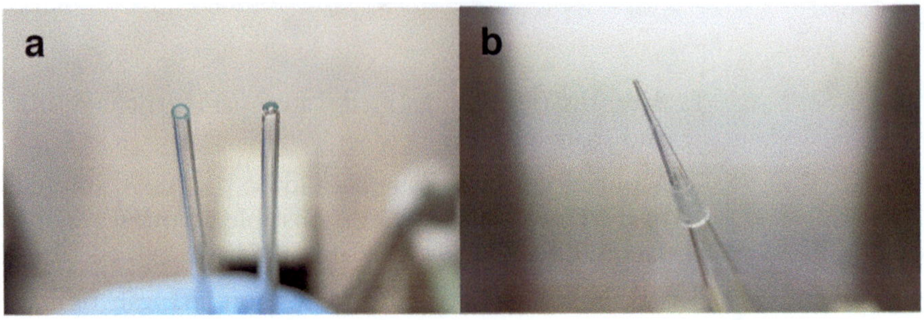

Fig. 5 A narrow-tip Pasteur pipette for shearing organoids. (**a**) After rotation the tip of a Pasteur pipette (Left) in the Bunsen burner flame, the glass tip pore is narrowed (Right) for efficient shearing organoids. (**b**) A P10 tip fixed on the P1000 tip can be used alternatively

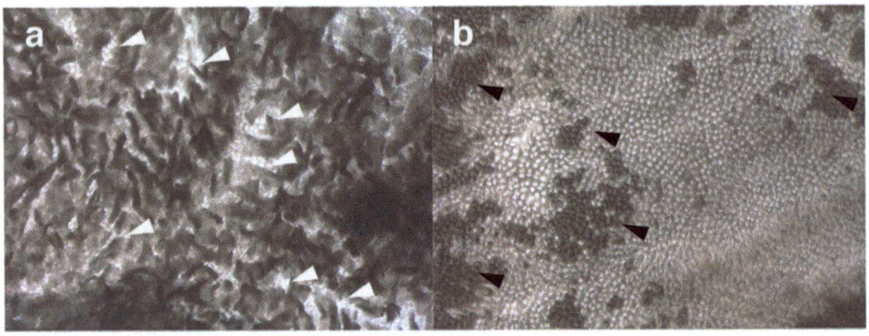

Fig. 6 Intestinal tissue fragments after crypts released. (**a**) Small intestinal tissue after crypt dissociation shows small pores of crypts (white arrowheads) between projecting villus structure. (**b**) Vacant pores can be seen in colonic tissue under a stereomicroscope. Remaining crypts can be observed as dark colored patchy areas (black arrowheads). If most of the crypts remain in the tissue, chelation steps are expected not working well

13. To prevent the liberated crypts from adhering to the inside of tubes and pipettes, 10% FBS containing DPBS is used after chelation step.
14. Put the tissue fragments on a petri dish and see under a light microscope to check if all the crypts are liberated. If you see most of the crypts are not liberated, chelation step must not work properly (Fig. 6). Repeat steps 12–15 to incubate tissue fragments in crypt isolation buffer for an additional 30 min until released crypts can be observed.
15. Healthy crypts show shiny finger-like structures. Dark colored cell aggregates might be liberated part of villi and are expected not to grow as organoids (Fig. 1a).
16. Carefully aspirate supernatant as much as possible not to dilute Matrigel. It is recommended to use a P200 pipette not to lose the pellet.

17. Better to plate less than 25 μL of the cell mixture. Making bigger droplet cause less organoid growth in the center of the gel droplet. When plating to bigger plates, keep the volume of droplets (20–25 μL) and cell density (Fig. 2) (Table 6).
18. Flipping the plate is optional, but this prevents the crypts sink and attach to the bottom.
19. Depending on the quality of Wnt activity, colonic organoids grow a bit slower than small intestinal organoids. They are recommended to be passaged with 1:1–3 split ratio.
20. The remaining gel can be dissolved by pipetting with an ice-cold medium. No need to use Cell Recovery Solution, specifically designed for Matrigel digestion.
21. The organoids can be dissociated by TrypLE express cell dissociation enzyme. The organoid pellet is incubated in 1 mL of TrypLE express containing 10 μmol/L Y-27632 for 5 min at 37 °C and dissociated by pipetting 10 times with P1000 tip.
22. For removing dead cells and debris, centrifuge at 100 × *g* for 5 min at 4 °C.
23. Normally, the frozen organoids harvested from 1 well can be plated into 1 well with 1:1 split ratio.
24. Organoids can be stored in 70% ethanol at 4 °C for weeks before you continue processing.
25. The appropriate incubation conditions need to be optimized for each antibody.

Acknowledgments

The authors would like to thank Joep Beumer for advice on whole-mount imaging technique and critical reading of the manuscript, and Jeroen Korving and Harry Begthel for their advice on immunohistochemistry protocol.

References

1. Leblond CP, Stevens CE (1948) The constant renewal of the intestinal epithelium in the albino rat. Anat Rec 100:357–377
2. Barker N, van Es JH, Kuipers J et al (2007) Identification of stem cells in small intestine and colon by marker gene Lgr5. Nature 449:1003–1007
3. Sato T, Vries RG, Snippert HJ et al (2009) Single Lgr5 stem cells build crypt-villus structures in vitro without a mesenchymal niche. Nature 459:262–265
4. Sato T, Stange DE, Ferrante M et al (2011) Long-term expansion of epithelial organoids from human colon, adenoma, adenocarcinoma, and Barrett's epithelium. Gastroenterology 141:1762–1772
5. Yui S, Nakamura T, Sato T et al (2012) Functional engraftment of colon epithelium expanded in vitro from a single adult Lgr5+ stem cell. Nat Med 18:618–623
6. Fukuda M, Mizutani T, Mochizuki W et al (2014) Small intestinal stem cell identity is maintained with functional Paneth cells in heterotopically grafted epithelium onto the colon. Genes Dev 28:1752–1757

7. Sato T, Clevers H (2013) Growing self-organizing mini-guts from a single intestinal stem cell: mechanism and applications. Science 340(6137):1190–1194
8. Koo B-K, Stange DE, Sato T, Karthaus W, Farin HF, Huch M, van Es JH, Clevers H (2011) Controlled gene expression in primary Lgr5 organoid cultures. Nat Methods 9 (1):81–83
9. Schwank G, Koo B-K, Sasselli V et al (2013) Functional repair of CFTR by CRISPR/Cas9 in intestinal stem cell organoids of cystic fibrosis patients. Cell Stem Cell 13:653–658
10. Li X, Nadauld L, Ootani A et al (2014) Oncogenic transformation of diverse gastrointestinal tissues in primary organoid culture. Nat Med 20(7):769–777
11. Drost J, van Jaarsveld RH, Ponsioen B et al (2015) Sequential cancer mutations in cultured human intestinal stem cells. Nature 521:43–47
12. Matano M, Date S, Shimokawa M, Takano A, Fujii M, Ohta Y, Watanabe T, Kanai T, Sato T (2015) Modeling colorectal cancer using CRISPR-Cas9-mediated engineering of human intestinal organoids. Nat Med 21 (3):256–262. https://doi.org/10.1038/nm.3802
13. Basak O, Beumer J, Wiebrands K, Seno H, van Oudenaarden A, Clevers H (2017) Induced quiescence of Lgr5+ stem cells in intestinal organoids enables differentiation of hormone-producing enteroendocrine cells. Cell Stem Cell 20:177–190.e4
14. Schuijers J, van der Flier LG, van Es J, Clevers H (2014) Robust cre-mediated recombination in small intestinal stem cells utilizing the olfm4 locus. Stem Cell Reports 3(2):234–241
15. Mustata RC, Vasile G, Fernandez-Vallone V, Strollo S, Lefort A, Libert F, Monteyne D, Pérez-Morga D, Vassart G, Garcia M-I (2013) Identification of Lgr5-independent spheroid-generating progenitors of the mouse fetal intestinal epithelium. Cell Reports 5:421–432
16. Grün D, Lyubimova A, Kester L, Wiebrands K, Basak O, Sasaki N, Clevers H, van Oudenaarden A (2015) Single-cell messenger RNA sequencing reveals rare intestinal cell types. Nature 525:251–255
17. Haber AL, Biton M, Rogel N et al (2017) A single-cell survey of the small intestinal epithelium. Nat Publ Group 551:333–339
18. Lindeboom RG, van Voorthuijsen L, Oost KC et al (2018) Integrative multi-omics analysis of intestinal organoid differentiation. Mol Syst Biol 14:e8227
19. Behjati S, Huch M, van Boxtel R et al (2014) Genome sequencing of normal cells reveals developmental lineages and mutational processes. Nat Publ Group 513:422–425
20. Gonneaud A, Jones C, Turgeon N, Lévesque D, Asselin C, Boudreau F, Boisvert F-M (2016) A SILAC-based method for quantitative proteomic analysis of intestinal organoids. Sci Rep 6:38195
21. Cristobal A, van den Toorn HWP, van de Wetering M, Clevers H, Heck AJR, Mohammed S (2017) Personalized proteome profiles of healthy and tumor human colon organoids reveal both individual diversity and basic features of colorectal cancer. Cell Reports 18:263–274
22. Beumer J, Artegiani B, Post Y, Reimann F, Gribble F, Nguyen TN, Zeng H, van den Born M, van Es JH, Clevers H (2018) Enteroendocrine cells switch hormone expression along the crypt-to-villus BMP signalling gradient. Nat Cell Biol 20:909–916
23. Gehart H, van Es JH, Hamer K, Beumer J, Kretzschmar K, Dekkers JF, Rios A, Clevers H (2019) Identification of enteroendocrine regulators by real-time single-cell differentiation mapping. Cell 176:1158–1173.e16
24. Nozaki K, Mochizuki W, Matsumoto Y, Matsumoto T, Fukuda M, Mizutani T, Watanabe M, Nakamura T (2016) Co-culture with intestinal epithelial organoids allows efficient expansion and motility analysis of intraepithelial lymphocytes. J Gastroenterol 51:206–213
25. Zhang Y-G, Wu S, Xia Y, Sun J (2014) Salmonella-infected crypt-derived intestinal organoid culture system for host-bacterial interactions. Physiol Rep 2:e12147–e12111
26. Heo I, Dutta D, Schaefer DA et al (2018) Modelling Cryptosporidium infection in human small intestinal and lung organoids. Nat Microbiol 3(7):814–823
27. Booth C, O'Shea JA, Freshney RI (2002) Isolation and culture of intestinal epithelial cells. Culture of epithelial cells, 2nd edn. Wiley-Liss, New York, pp 303–335
28. Farin HF, van Es JH, Clevers H (2012) Redundant sources of Wnt regulate intestinal stem cells and promote formation of Paneth cells. Gastroenterology 143(6):1518–1529.e7. https://doi.org/10.1053/j.gastro.2012.08.031
29. Kim KA (2005) Mitogenic influence of human R-spondin1 on the intestinal epithelium. Science 309:1256–1259
30. Barker N, Huch M, Kujala P et al (2010) Lgr5 (+ve) stem cells drive self-renewal in the stomach and build long-lived gastric units in vitro. Cell Stem Cell 6:25–36

Chapter 12

In Vivo Human PSC-Derived Intestinal Organoids to Study Stem Cell Maintenance

Simon Vales, Holly M. Poling, Nambirajan Sundaram, Michael A. Helmrath, and Maxime M. Mahe

Abstract

Human intestinal organoids (HIOs), derived from pluripotent stem cells, are a new tool to gain insights in gastrointestinal development, physiology, and associated diseases. Herein, we present a method for renal transplantation of HIOs in immunocompromised mice and subsequent analysis to study intestinal epithelial cell proliferation. In addition, we describe how to generate enteroids from transplanted HIOs. The method highlights the specific steps to successful engraftment and provides insight into the study of human intestinal stem cells.

Key words Human intestinal organoids, Pluripotent stem cells, Transplantation, In vivo model, Intestinal stem cell, Enteroid

1 Introduction

The intestinal tract is lined by a simple columnar epithelium shaped into finger-like protrusions called villi and surrounded by multiple invaginations or crypts. The intestinal epithelium is one of the fastest renewing tissues in humans [1]. The constant epithelial regeneration is driven by active intestinal stem cells (ISCs) located at the bottom of the crypts [2]. ISCs divide to produce fast-proliferative progenitor cells also called transit amplifying cells (TA cells). TA cells undergo additional cell divisions as they migrate toward the tip of the villi. These cells will ultimately give rise to either secretory (enteroendocrine, Paneth, and goblet cells) or absorptive (enterocytes, colonocytes) lineages that will contribute to the mature epithelium [3].

In the last decade, intestinal organoid models have been developed to study ISC maintenance and differentiation [4]. Methods based on pluripotent stem cell-directed differentiation have allowed the generation of human intestinal organoids (HIOs) [5]. Using a

Paloma Ordóñez-Morán (ed.), *Intestinal Stem Cells: Methods and Protocols*, Methods in Molecular Biology, vol. 2171, https://doi.org/10.1007/978-1-0716-0747-3_12,

stepwise differentiation process that mimics embryonic intestinal development, human pluripotent stem cells (hPSC) can be differentiated into definitive endoderm and then into hindgut spheres that will form tridimensional intestinal structures. After 28 days in culture, HIOs present a polarized epithelium surrounded by supporting mesenchymal cells. In addition, in vivo engraftment of HIOs further both tissue expansion and maturation. Transplanted HIOs (tHIO) develop into an intestinal tissue that includes an epithelium containing all main differentiated intestinal cell lineages (enterocytes, goblet cells, Paneth cells, tuft cells, and enteroendocrine cells) and a laminated mesenchyme forming submucosal and muscle layers [6].

This model system provides us with a way to study ISC and progenitor cells in a fully developed and mature tissue of human origin. Herein, we describe a method to transplant hPSC-derived intestinal organoids under the kidney capsule and generate crypt-derived enteroids from them.

2 Materials

2.1 Human Intestinal Organoid (HIO) Maintenance

All solutions should be prepared fresh using sterile cell culture grade reagents.

1. Human intestinal organoids derived from human pluripotent stem cell lines.
2. Matrigel® Growth Factor Reduced; Phenol red-free.
3. Intestinal growth medium: Advanced DMEM/F12 medium supplemented with 2 mM glutamine, 10 mM HEPES, 100 U/mL penicillin, 100 μg/mL streptomycin, 1× N2 supplement, 1× B27 supplement and filter sterilized with 0.22 μm filter (*see* **Note 1**).
4. Human recombinant Epidermal Growth Factor (EGF) (5000× stock; 500 μg/mL in sterile PBS/0.1% bovine serum albumin).

2.2 Murine Recipients, Surgical Equipment, and Reagents

Aseptic technique is essential for any survival surgery and requires that all surgical instruments and supplies be sterile. All surgical instruments should be washed and sterilized in an autoclave prior to use. All surgeries are performed under a HEPA-filtered laminar flow bioBubble to prevent microbial contamination of the surgical site.

1. Mice: Female or Male immunocompromised NOD-scid IL2R-gamma^{null} (NSG) mice are housed in microisolator systems in a barrier facility. The mice are used between 6 and 14 weeks of age (*see* **Note 2**).

2. Antibiotic diet: a modified chow diet (PicoLab Rodent Diet 20) is supplemented with 275 ppm Sulfamethoxazole and 1365 ppm Trimethoprim (*see* **Note 3**).
3. Antibacterial drugs: 100 mg/kg of piperacillin and tazobactam are diluted in sterile saline solution and used for any surgeries.
4. Buprenorphine.
5. Surgical Instruments: Suture tying forceps, ring forceps, dissecting scissors, Bishop-Harmon forceps, Halsey needle holder, sterilization tray, Vannas spring scissors.
6. Isoflurane and anesthesia system (*see* **Note 4**).
7. Sterile 4-0 coated absorbable suture.
8. Sterile 18G blunt fill needles.

2.3 Thymidine Analog Injection, Sample Preparation, and Staining

1. Ca^{2+} and Mg^{2+} free Phosphate buffered saline (PBS), pH 7.2–7.6.
2. PBT solution: 0.5% Triton X-100 is diluted in PBS.
3. Blocking solution: 10% normal donkey serum and 1% Bovine Serum Albumin (BSA) Fraction V are diluted in PBS.
4. Mounting medium: 70% glycerol diluted in PBS (*see* **Note 5**).
5. Ethanol solutions: 70, 75, 85, 90, 95, and 100% histology grade ethanol (v/v) diluted in Milli-Q purified water.
6. 5-ethynyl-2′-deoxyuridine (EdU) solution: EdU powder is diluted at 10 mg/mL in sterile PBS (*see* **Note 6**).
7. 4% paraformaldehyde.
8. HistoPrep Xylene.
9. Deionized water.
10. Click-iT EdU Alexa Fluor 488 Imaging Kit.
11. Citrate Buffer (pH 6).
12. Antibody diluent: 1% BSA is diluted in PBS.
13. Antibodies and fluorescent counterstain (Table 1).
14. Tissue Processor.

Table 1
List of primary and secondary antibodies and counterstain used for immunofluorescence staining

	Antigen	Dilution	Host
Primary	MKI67 CDH1	1:500 1:500	Rabbit Mouse
Secondary	Anti-rabbit Anti-mouse	1:1000 1:1000	Donkey Donkey
DAPI	Nuclei		

15. Embedding workstation stocked with paraffin.
16. Hybridization oven or equivalent.
17. Orbital shaker.
18. Tissue flotation bath.
19. Microtome and blades.
20. Slide staining set, including slide holders and chemical resistant reagent buckets.
21. Square bioassay dishes, as humidity chamber.
22. Tissue processing embedding cassettes.
23. Disposable base molds.
24. 1 mL TB syringe with 27G × ½ needle.
25. 75 × 25 × 1 mm positively charged microscope slides.
26. 24 × 50 mm, No. 1.5 Thickness glass coverslips.
27. Hydrophobic PAP pen.
28. Paper towels.

2.4 HIO-Derived Epithelial Organoid Generation and Culture

1. Dulbecco's phosphate buffered saline (DPBS).
2. Chelation buffer: 1% sorbitol, 1% sucrose, 1% bovine serum albumin fraction V (BSA), and 1× Gentamicin/Amphotericin solution are diluted in DPBS and filter-sterilized with a 0.22 μm filter.
3. 0.5 M ethylenediaminetetraacetic acid (EDTA) diluted in ultra-pure water.
4. Matrigel® Growth Factor Reduced; Phenol red-free.
5. 10 mM Y-27632 compound diluted (Stock solution) in sterile ultrapure water.
6. Intesticult® Organoid Growth medium (*see* **Note 7**).
7. Inverted tissue culture stereoscope.
8. Petri dishes.
9. Razor blades.
10. Minutien pins.
11. Dissecting microscope.
12. Silicone-coated glass petri dish.
13. Small scissors.
14. Curved and fine forceps.
15. Micropipettes and micropipette tips.
16. Orbital shaker.
17. 15 mL polypropylene conical tubes.
18. 150 μm nylon mesh.

19. Centrifuge.
20. 24-well tissue-culture treated plates.
21. Cell culture incubator at 37 °C, 5% CO_2.

3 Methods

3.1 Human Intestinal Organoid Generation and Maintenance

The generation of HIOs from pluripotent stem cells has been described in the following protocols [7, 8].

1. Culture HIOs in Matrigel® on a 24-well plate in a 37 °C, 5% CO_2 incubator.
2. Overlay HIOs with 500 μL of intestinal growth medium supplemented with 100 ng/mL of human recombinant EGF (1:5000 dilution of 500 μg/mL stock).
3. Change intestinal growth media supplemented with 100 ng/mL of human recombinant EGF every 4 days (*see* **Note 8**).
4. Bring the HIOs plate in the surgical room at the day of transplantation (*see* **Note 9**).

3.2 Renal Subcapsular Transplantation of Human Intestinal Organoids

A comprehensive method for the transplantation of HIOs into both the mesentery and kidney capsule of immunocompromised mice has been previously described [9]. In this section, we briefly outline the renal transplantation methodology, as it is the basis for subsequent analysis of the stem cell compartment, which will be examined in this chapter.

Proper surgical technique must be practiced, that is, asepsis, gentle tissue handling, minimal dissection of tissue, appropriate use of instruments, effective hemostasis, and correct use of suture materials and patterns. The person performing the procedures must be appropriately trained and working under a mentor or veterinarian to perform the procedure. During a surgical procedure, the person performing the procedures must wear clean scrubs, sterile surgical gown, mask, cap, and sterile gloves. Sterile surgical gown and gloves must be donned and maintained in an aseptic manner. All NSG mice are maintained on regular chow supplemented with antibiotics prior to transplantation.

1. Bring the mice to the operating suite where weighting and assessment of health status are performed.
2. Anesthetize the mouse in an anesthetic gas vaporizer delivering an isoflurane–O_2 mixture (*see* **Note 10**).
3. Shave the left flank of the abdomen between the last rib and the iliac crest, and the spine and the lower third of the abdominal wall. Remove loose fur (*see* **Note 11**).

4. Prepare the surgical site using povidone–iodine with a cotton swab. Repeat the procedure with new cotton swabs three times.
5. Repeat the procedure using 70% isopropyl alcohol with cotton-swabs. Repeat the procedures with new cotton swabs three times.
6. Place ophthalmic ointment on the eyes to prevent drying of the cornea and administer buprenorphine (0.05 mg/kg) subcutaneously (*see* **Note 12**).
7. Restrain the mouse on lateral recumbency, left kidney facing upward, and secure the mouse to a nosecone vaporizing isoflurane–O_2 mixture (*see* **Note 13**).
8. Monitor respiratory rate and effort, along with the surgical plane of anesthesia. Confirm the loss of pedal reflex by pinching the toe with forceps.
9. Use straight forceps and fine scissors to make an 8–10 mm left posterior subcostal skin incision just below the last rib.
10. Use fine scissors to make a subsequent 8–10 mm retroperitoneal muscle incision.
11. Identify automatically the kidney using ring forceps and mobilize it into the wound.
12. Stabilize the kidney in the wound by placing a 7–0 silk suture loop with an untied square knot around the incision. Secure one ear of the suture with a needle holder to hold the knot and leave the other ear free.
13. Lift the kidney caudal pole through the abdominal incision and tie the silk suture by gently pulling the free ear until the kidney remain still (*see* **Note 14**).
14. Use a cotton-swab to dry out the renal capsule.
15. Grasp the capsule under a surgical stereoscope with fine forceps and make a 2–3 mm incision with Vannas spring scissors in the capsule over the lateral aspect of the kidney.
16. Create a subcapsular pocket by gently sliding back and forth straight suture tying forceps under the kidney capsule (*see* **Note 15**). Allowing the forceps to gently open when inside the capsule helps create enough space for the HIO.
17. Grab one HIO using straight suture tying forceps and insert it in the subcapsular pocket (*see* **Note 16**).
18. Cut the 7-0 silk suture and return the kidney within the abdominal cavity.
19. Flush the abdominal cavity with 2–3 mL of piperacillin/tazobactam solution to help prevent bacterial infection.
20. Close the incision in double layers with continuous over and over sutures using 4-0 VICRYL RAPIDE® suture (*see* **Note 17**).

21. Allow mice to recover in a warm and dry incubator (30 °C) and monitor at least every 15 min until they resume activity and are able to maintain a sternal or sitting position.
22. After recovery, place mice back into cages with regular bedding and provide ad lib Bactrim diet and water.
23. Evaluate animals 12 h later, and then daily throughout the remainder of the experiment. Appetite, attitude, and hydration should be noted as an indication of recovery from the surgery. Supplemental fluids or/and analgesics should be administered postoperatively as needed.

3.3 Sample Collection and Proliferative Cell Staining

3.3.1 Tissue Preparation

1. Inject mice intraperitoneally with a single dose of EdU (50 mg/kg) using a sterile 27G × ½ needle attached to a 1 mL TB syringe 24 h prior harvest (*see* **Notes 18** and **19**).
2. Sacrifice mice in accordance with the approved animal protocol in place.
3. Dissect out the kidney with the transplanted HIO (tHIO).
4. Cut the tHIO in half removing any accumulated mucous.
5. Transfer tissue into an appropriately labeled tissue processing embedding cassette.
6. Fix samples in 4% paraformaldehyde (PFA) diluted in PBS at 4 °C overnight.
7. Wash samples in PBS for 15–60 min with gentle agitation. Repeat the procedure three times.
8. Place samples in a tissue processor and proceed overnight to a standard tissue processing protocol:
   (a) 70% Ethanol × 2 for 30 min each.
   (b) 75% Ethanol for 1 h.
   (c) 90% Ethanol for 1 h.
   (d) 95% Ethanol for 1 h.
   (e) 100% Ethanol × 2 for 1 h each.
   (f) 100% Ethanol for 20 min.
   (g) Xylene × 3 for 1 h each.
   (h) Paraffin × 3 for 1 h each.
9. Remove samples from tissue processor and embed them into appropriately sized disposable base molds.
10. Discard the lid of the cassette and affix the bottom portion on top of the mold creating a tissue block.
11. Remove the disposable base mold from the paraffin tissue block.
12. Section into the block slightly, exposing the desired tissue face.

13. Rehydrate the tissue soaking in a bath of PBS for 1 h at room temperature (RT), before chilling on ice.
14. Section the tissue into ribbons 5 μm thick and float sections until wrinkles disappear in a tissue flotation bath.
15. Mount sections on to microscope slides.
16. Bake the microscope slides at 60 °C for 1 h.
17. Store microscope slides at room temperature until further use.

3.3.2 Tissue Section Staining

1. Warm microscope slides 30 min at 65 °C.
2. Rehydrate microscope slides by immersing them in the following series of baths:
   (a) Xylene × 3 for 15 min each.
   (b) 100% EtOH × 2 for 2 min each.
   (c) 95% EtOH for 2 min.
   (d) 85% EtOH for 2 min.
   (e) 70% EtOH for 2 min.
   (f) Deionized water for 5 min.
3. Perform heat induced epitope retrieval in a pH 6 citrate buffer.
4. Permeabilize tissue sections in 0.5% Triton X-100/PBS at room temperature for 5 min with gentle agitation.
5. Wash in PBS for 5 min with gentle agitation. Repeat the procedure two times.
6. Outline tissue sections with a hydrophobic pen to contain reagents.
7. Proceed to EdU Click-iT staining following the manufacturer's instructions.
8. Wash slides in PBS for 15 min each with gentle agitation. Repeat the procedure three times.
9. Incubate slides with blocking solution for 1 h in a square bioassay dish lined with a damp paper towel to maintain humidity in the chamber.
10. Aspirate blocking solution. Do not wash slides.
11. Incubate slides with rabbit anti-MKI67 at [1:500] and mouse anti-CDH1 at [1:500] in antibody diluent at 4 °C overnight in a humidity chamber.
12. Wash slides in PBS for 15 min each with gentle agitation (*see* **Note 20**). Repeat the procedure three times.
13. Incubate slides with donkey anti-rabbit Alexa Fluor 488 (binds to anti-MKI67 antibody) and donkey anti-mouse Alexa Fluor 568 (binds to anti-CDH1 antibody) at [1:1000] in antibody diluent at 4 °C overnight in a humidity chamber.

14. Wash slides in PBS for 15 min with gentle agitation while protected from light. Repeat the procedure three times.
15. Incubate slides with 1 μg/mL of DAPI in PBS for 10 min, or other suitable fluorescent counterstain, while protected from light.
16. Wash microscope slides in PBS for 3 × 15 min each, while protected from light.
17. Mount microscope slides with coverslips using 70% glycerol in PBS, or another aqueous mounting medium.
18. Proceed to imaging and cell counting under a fluorescent microscope (Fig. 1a–f).
19. Count by position recording CDH1+/MKI67+ cells as positive and CDH1+/MKI67− as negative. Record with the "position 1" at the base of the crypt working your way upward to "position 20" (*see* **Note 21**).
20. Plot values as a histogram over "position 1" to "position 20"in an appropriate statistical software.
21. Apply an appropriate curve fit to the histogram (i.e., Gaussian) to visualize the zone of highest proliferation (*see* **Note 22**).

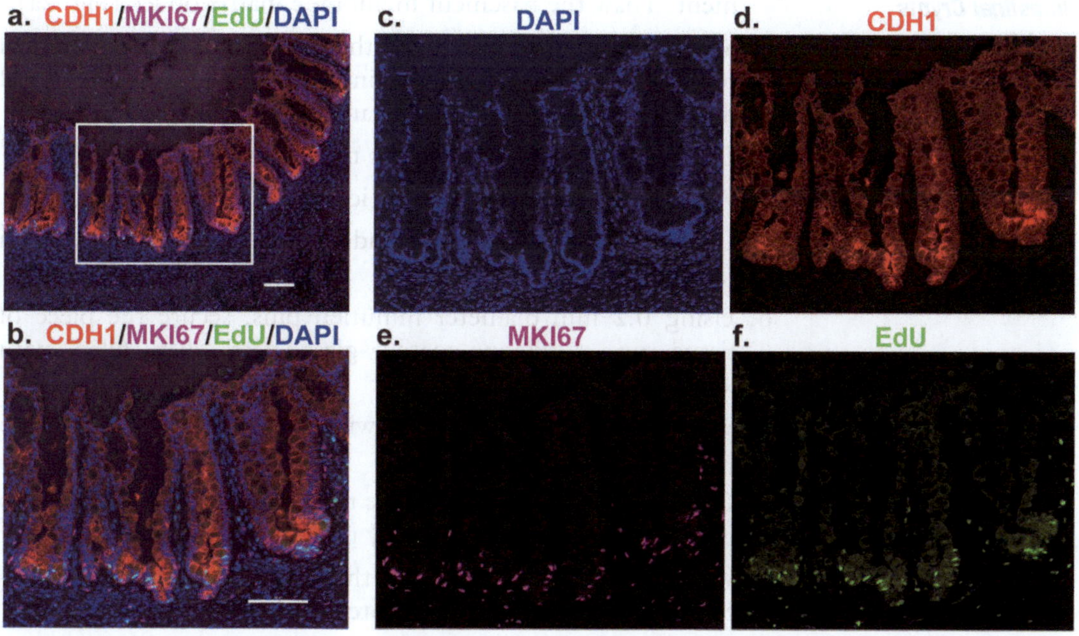

Fig. 1 Intestinal cell proliferation in transplanted human intestinal organoids (HIO). Immunofluorescence staining of tHIO at 12 weeks posttransplantation at 10× (**a**) and 20× (**b**) magnification. DAPI is shown in blue (**c**), CDH1 in red (**d**), MKI67 in purple (**e**), and EdU staining in green (**f**) (All scale bars, 100 μm)

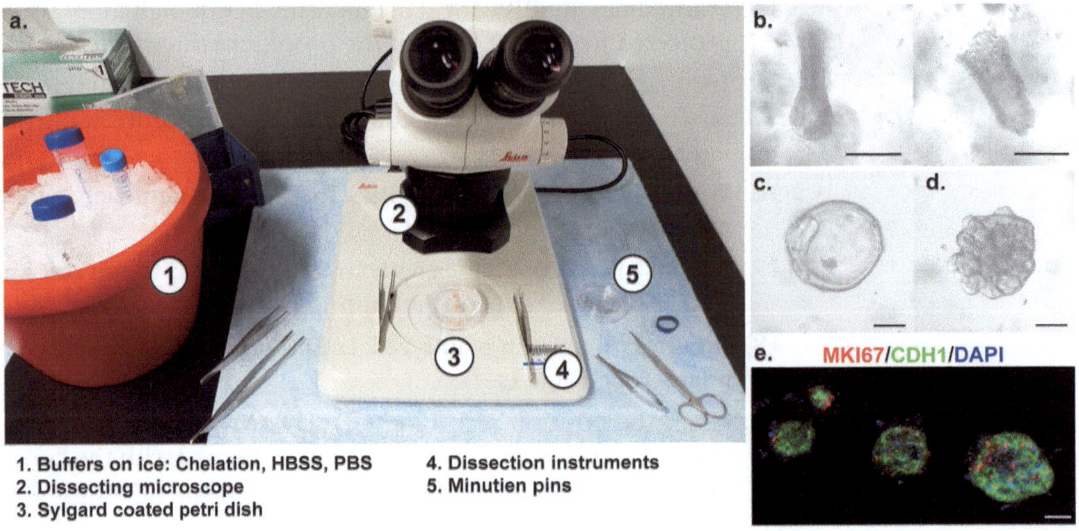

Fig. 2 Enteroids derived from transplanted human intestinal organoids (tHIO). (**a**) Example of an experimental setup to generate tHIO-derived enteroids. (**b**) Close-up picture on representative tHIO crypts picked up before embedding in Matrigel®. (**c**, **d**) tHIO-derived enteroids in Matrigel® after 2 days in culture. (**e**) Immunofluorescence staining of tHIO-derived enteroids. DAPI is shown in blue, CDH1 in green, MKI67 in red (All scale bars, 50 μm)

3.4 Isolation of tHIO Intestinal Crypts and Culturing to Generate Enteroids

1. Prepare all the reagents before the beginning of the experiment. Thaw the basement membrane matrix on ice (Fig. 2a).
2. Sacrifice mice in accordance with the approved IACUC animal protocol in place (isoflurane inhalation followed by cervical dislocation) 8–12 weeks posttransplantation.
3. Dissect out the tHIO from the mouse kidney.
4. Wash the dissected tHIO with ice-cold DPBS.
5. Cut the tHIO using a razor blade to expose the interior lumen (*see* **Note 23**).
6. Using 0.2 mm diameter minutien pins, secure the piece of tissue on a silicone-coated glass petri dish filled with ice-cold DPBS.
7. Stretch and pin the tissue flat with the mucosal side facing up (*see* **Note 24**).
8. Gently scrape the surface of the mucosa with curved forceps to remove villi and debris and any mucus that is present.
9. Wash the tissue 3–4 times with ice-cold chelation buffer to remove villi and debris (*see* **Note 25**).
10. Cover the biopsy with freshly prepared 2 mM EDTA chelation buffer (*see* **Note 26**).
11. Place the petri dish on ice and shake gently for 30 min on a horizontal orbital shaker.

12. Wash the tissue with ice-cold chelation buffer without EDTA. Repeat the procedure four times and leave the tissue in ice-cold chelation buffer.
13. Process the tissue under a dissecting microscope.
14. Gently scrape the mucosal layer to release the intestinal crypts using curved forceps (*see* **Note 27**).
15. Gently remove the crypt suspension from the petri dish using a pipette and transfer it to a 15 mL conical tube (*see* **Note 28**).
16. Filter the crypt suspension through a 150 μm nylon mesh (*see* **Note 29**).
17. Centrifuge the crypt suspension 5 min at 50 × *g*, 4 °C and discard the supernatant.
18. Resuspend the pellet in 1 mL ice-cold chelation buffer.
19. Centrifuge the crypt fraction for 10 min at 150 × *g*, 4 °C and remove the supernatant.
20. Resuspend the crypt pellet in basement membrane matrix using prechilled pipette tips (200–500 crypts/50 μL basement membrane matrix).
21. Apply 50 μL of crypt suspension in basement membrane matrix per well on the prewarmed plate. Slowly eject the basement membrane matrix in the center of the well (Fig. 2b, c).
22. Place the 24-well plate in a 37 °C, 5% CO_2 incubator for 30 min to allow a complete polymerization of the basement membrane matrix.
23. Overlay the basement membrane matrix with 500 μL of Intesticult® Organoid Growth medium.
24. Incubate the plate in a 37 °C, 5% CO_2 incubator.
25. Replace the medium with fresh Intesticult® Organoid Growth medium every 2 days for 8–10 days (*see* **Notes 30** and **31**) (Fig. 2c–e).

4 Notes

1. Divide intestinal growth medium into 10 mL aliquots in 15 mL conical tubes and freeze at −20 °C for up to 3 months. Store thawed aliquots up to 5 days at 4 °C without loss of activity.
2. Males are preferably used for the kidney subcapsular transplantation as their kidneys are bigger and easier to access.
3. The chow diet is supplemented with antibiotics and given to the mice at least 14 days prior any surgeries. The antibiotics decrease inflammation and risk of infection.

4. The anesthesia system delivers an isoflurane and oxygen mixture that can be controlled and monitored to maintain the anesthesia during surgery. The extra anesthetic gas is collected and evacuated into a canister.
5. Aqueous mounting media can also be used to mount coverslips onto microscope slides.
6. Desiccated 5-ethynyl-2′-deoxyuridine (EdU) is resuspended in PBS and heated at 70 °C for 1 min to achieve complete powder dissolution.
7. Intestinal growth medium supplemented with 100 ng/mL recombinant Wnt3a (1:1000 dilution of 100 μg/mL stock), 1 μg/mL R-spondin1 (1:1000 dilution of 1 mg/mL stock), 100 ng/mL Noggin (1:1000 dilution of 100 μg/mL stock), and 100 ng/mL EGF (1:5000 dilution of 500 μg/mL stock) can be used to culture HIO-derived epithelial organoid. Alternatively, recombinant growth factors could be replaced by Wnt3a, R-spondin1, and Noggin conditioned media.
8. Intestinal growth medium aliquots are thawed and can be kept up to 5 days at 4 °C without loss of activity. Add the human recombinant EGF prior media change (1:5000 stock dilution).
9. In our experience, HIOs can be transplanted from 20 to 40 days of culture in vitro.
10. Final anesthetic gas concentration is achieved by delivering 2% isoflurane with 2.5–3 L/min O_2.
11. The left kidney is used for ease of access.
12. Analgesia provisions are most effective at reducing the intensity of painful stimulation when given prior to the surgery. Any animal showing evidence of pain should be provided analgesia. Other opioids like buprenorphine can be used, that is, butorphanol (0.2–2 mg/kg subcutaneous or intraperitoneal) or oxymorphone (0.2–0.5 mg/kg subcutaneous).
13. Keep the animal warm using a 37 °C heating-pad. Adjust anesthetic gas concentration to 1.5–1.75% isoflurane with 2–3 L/min O_2.
14. This technique allows you to hold the kidney outside the abdominal cavity. Do not completely tie the knot to avoid renal vascular ligation and permanent kidney damage. Alternatively, curved forceps can be used to lift the kidney.
15. Slide the closed straight suture tying forceps under the capsule and open the forceps while pulling it back. Repeat the motion until appropriate size of the subcapsular pocket is achieved.
16. Inserted HIOs will not dislodge from under the subcapsular pocket.

17. VICRYL RAPIDE® sutures are synthetic coated absorbable sutures and the animals will not chew them. Alternatively, skin staplers can be used.
18. In our experience 8–12 weeks posttransplanted HIOs provide us with a fully laminated small intestinal tissue that can be further used for downstream applications ranging from physiological to molecular assays. HIOs transplanted beyond a year do not exhibit common intestinal epithelial features probably due to the accumulation of mucus and debris within the lumen.
19. Further characterization of cell cycle kinetics can be achieved by varying the duration of the EdU pulse and/or varying the duration of the chase using additional thymidine analogs, that is, CldU, IdU. For example, dual pulse chase can be achieved by injecting a single dose of EdU, followed 24 h later by a single injection of BrdU (100 mg/kg).
20. A minimum of 15 min is recommended to wash the slides.
21. Count a minimum of ten well oriented crypts per sample. Only count well oriented crypts, and consider either ascending side of the crypt, not both.
22. The means of the curve fits may be directly compared. Other more sophisticated comparisons such as an *F* test to compare models may also be used.
23. Make sure there is mucus present and also observe under the scope for the presence of villi.
24. If the harvested HIO has multiple cystic structures, open them up using small scissors to expose the mucosal layer.
25. Presence of mucus is a good indication of the maturity of the HIO. However, an excess of mucus can make the scrapping step difficult and reduce the efficiency of crypt isolation. If the mucus is too thick, add dithiothreitol (DTT; final concentration 0.5 mM) to the chelation buffer to increase crypt yield.
26. To prepare 2 mM EDTA chelation buffer, add 200 μL of 0.5 M EDTA in 49.8 mL chelation buffer.
27. Unlike adult human tissue, the crypts in these tissues do not have dark Paneth cells, so it may be harder to visualize them under the dissecting scope.
28. Check the tissue to make sure that almost all crypts have been removed from the mucosa.
29. Check the flow-through for crypt enrichment under an inverted microscope.
30. Enteroids can be passaged 7–10 days after seeding and/or frozen for long-term storage [10].
31. Enteroids can be fixed using 2% PFA and stained for immunofluorescence.

References

1. Barker N (2014) Adult intestinal stem cells: critical drivers of epithelial homeostasis and regeneration. Nat Rev Mol Cell Biol 15:19–33
2. Simons BD, Clevers H (2011) Stem cell self-renewal in intestinal crypt. Exp Cell Res 317:2719–2724
3. Noah TK, Donahue B, Shroyer NF (2011) Intestinal development and differentiation. Exp Cell Res 317:2702–2710
4. Date S, Sato T (2015) Mini-gut organoids: reconstitution of the stem cell niche. Annu Rev Cell Dev Biol 31:269–289
5. Spence JR, Mayhew CN, Rankin SA et al (2011) Directed differentiation of human pluripotent stem cells into intestinal tissue in vitro. Nature 470:105–109
6. Watson CL, Mahe MM, Múnera J et al (2014) An in vivo model of human small intestine using pluripotent stem cells. Nat Med 20:1310–1314
7. McCracken KW, Howell JC, Wells JM, Spence JR (2011) Generating human intestinal tissue from pluripotent stem cells in vitro. Nat Protoc 6:1920–1928
8. Múnera JO, Wells JM (2017) Generation of gastrointestinal organoids from human pluripotent stem cells. Methods Mol Biol 1597:167–177
9. Mahe MM, Brown NE, Poling HM, Helmrath MA (2017) In vivo model of small intestine. Methods Mol Biol 1597:229–245
10. Mahe MM, Sundaram N, Watson CL et al (2015) Establishment of human epithelial enteroids and colonoids from whole tissue and biopsy. J Vis Exp (97):e52483. https://doi.org/10.3791/52483

Chapter 13

Generation of Knockout Gene-Edited Human Intestinal Organoids

Chathruckan Rajendra, Tomas Wald, Kevin Barber, Jason R. Spence, Faranak Fattahi, and Ophir D. Klein

Abstract

We discuss a methodology to generate and study knockout gene-edited human intestinal organoids. We describe the generation of knockout human embryonic stem cell lines that we then differentiate into mature human intestinal organoid tissue in Matrigel using several growth factors. We also discuss a pair of assays that can be used to study the integrity of the intestinal epithelial barrier of the human intestinal organoids under inflammatory stress conditions.

Key words Human embryonic stem cells, Human intestinal organoids, Crispr-Cas9, Transfection, Gene editing

1 Introduction

There is great interest in developing methods to model human disease in vitro. We propose a method to functionally analyze genetic risk factors for gastrointestinal disease in a human intestinal organoid (HIO) system. A Crispr-Cas9 gene-editing transfection-based platform [1] is used to knockout specific genes of interest in human embryonic stem cell (hESC) lines and develop monoclonal knockout cell lines. After expansion of these hESC lines, they are differentiated into HIOs as per published protocols [2, 3]. Studying these knockout hESC lines during and after differentiation into HIOs can give us insight into the potential roles of genes of interest during development and in mature human intestinal epithelium and mesenchyme.

2 Materials

2.1 Cell Lines and Culture Conditions

1. Two hESC lines have been used in our laboratory for this methodology, H9 and UCSF4.
2. Human intestinal organoids.

Paloma Ordóñez-Morán (ed.), *Intestinal Stem Cells: Methods and Protocols*, Methods in Molecular Biology, vol. 2171, https://doi.org/10.1007/978-1-0716-0747-3_13, © Springer Science+Business Media, LLC, part of Springer Nature 2020

3. Matrigel or Geltrex.
4. mTeSR media for hESC maintenance.
5. Nunclon delta surface tissue culture plates for HIO culture.

2.2 Transfections

1. Standard Opti-MEM solution.
2. Lipofectamine Stem reagent.

2.3 Genomic DNA Isolation

1. Neutralization solution for blood (trademark solution from Sigma-Aldrich).
2. Lysis solution for blood (trademark solution from Sigma-Aldrich).

2.4 Targeting, Plasmid Design, and Genotyping

1. DNA sequence analysis tool (SnapGene, Clone manager).
2. Sanger sequence data analysis tool (Chromas, SnapGene).

2.5 Staining

Invitrogen's BD permeabilization buffer (specific formulation designed to reduce nonspecific staining of fluorochrome labeled antibodies and increase fluorescent signal-to-noise ratios).

2.6 Reverse Transcriptase PCR

1. Applied Biosystem's High Capacity cDNA Reverse Transcription Kit.
2. DNA gel extraction and RNA extraction kits.

3 Methods

3.1 Maintenance of Human Embryonic Stem Cells (Plate Preparation, Freezing, Thawing, Maintenance in Culture) [4]

3.1.1 Plate Preparation (See Note 1)

1. Aliquot 500 μL of hESC qualified Matrigel or Geltrex at 4 °C into 50 mL tubes. Can store 500 μL aliquots of hESC qualified Matrigel or Geltrex at −20 °C.
2. Dilute hESC qualified Matrigel at a concentration of 1:100 with 1× DMEM/F12 and keep on ice to prevent it from solidifying.
3. Prepare 6-well tissue culture plates by pipetting 1 mL of solution to each well. Gently shake plate to ensure solution evenly covers well.
4. Incubate plates at 37 °C overnight.
5. The plates can be used for hESC culture the next day. Unused plates can be stored in the incubator at 37 °C and remain usable for up to 2 weeks.

3.1.2 Thawing

1. Thaw frozen vial of hESCs in 37 °C water bath or by holding vial lid in gloved hand and gently swirling frozen portion in water (1–2 min).

2. Add cells to 5 mL of hESC cell media (we use mTeSR) in 15 mL conical tube.
3. Spin down tube in centrifuge at 200 × *g* for 1 min.
4. Aspirate media and resuspend with 2 mL of media (mTeSR) with Y-27632 ROCK inhibitor (1:1000 dilution) (10 mM stock solution).
5. Plate the above 2 mL onto 1 well of 6-well tissue culture plate.
6. Aspirate ROCK inhibitor supplemented media after 5–6 h (can leave overnight to increase stem cell yield) from the corner of the well and add 2 mL of fresh mTeSR media.

3.1.3 Feeding

1. Aspirate media from the corner of the well using a sterile pipette tip.
2. Add 2 mL of mTeSR media to the corner of the well.
3. Change media daily.

3.1.4 Passaging (See **Note 2**)

1. Aspirate media from the corner of the well.
2. Add 1 mL PBS to well to wash. Gently rock plate and aspirate PBS.
3. Add 1 mL 1× EDTA (1:1000 in PBS) (0.5 mM final concentration in PBS).
4. For maintenance: Let it sit for 1–2 min and observe cells through microscope. Edges should be coming off colonies, but colonies should not disassociate into single cells.

   For differentiation: Let it sit for 4 min and observe cells through microscope. There should be more disassociation of colonies into single cells versus maintenance passage.
5. Add 1 mL of mTeSR media to well and pipet up and down 5–6 times with serological pipettor to dislodge all the cells from the surface of the well and transfer to tube.
6. Centrifuge for 1 min at 300 × *g*.
7. Aspirate the supernatant.
8. For maintenance: Add 12 mL of mTeSR and resuspend cell pellet in media with serological pipettor. For differentiation: Add 2 mL of mTeSR and resuspend cell pellet in media with serological pipettor.
9. For maintenance: Transfer 2 mL of media with cells into each well of 6-well plate. For differentiation: Transfer 500 μL of media to each well of 24-well plate.
10. Examine wells under microscope to ensure cell clusters floating in well.
11. For maintenance: Passage every 4–6 days. For differentiation: Start after 1–2 days when cells reach >80% confluency.

3.1.5 Freezing

1. Aspirate media from corner of well when hESCs reach 80% or greater confluency.
2. Add 1 mL PBS to well to wash. Gently shake and aspirate.
3. Add 1 mL 1× EDTA (1:1000 in PBS) (0.5 mM final concentration in PBS).
4. Leave it for 2–3 min to allow colonies to detach.
5. Add 1 mL of mTeSR. Pipet up and down 3–4 times to detach colonies from bottom of well and transfer to tube.
6. Spin down at 300 × *g* for 1 min.
7. Aspirate the supernatant.
8. Resuspend the pellet in 1 mL of Stem-CellBanker and transfer to cryogenic vial.
9. Store vials at −80 °C.
10. After 2–3 days, transfer vials to liquid nitrogen for long-term storage.

3.2 CRISPR-Based Genome Editing

3.2.1 Design of Targeting Constructs (Fig. 1)

1. Design sgRNAs using the IDT online tool. The online tool will provide a list of sgRNA targets based on the sequence you are interested in editing. In order to increase the chance of obtaining effectively knocked-out cells, select from the list of candidate sgRNA provided by the online tool with following specifications: both cut within the same exon, high on-target and low off-target scores, and the excised sequence between the two sgRNAs creates a codon frame shift in case the NHEJ does not lead to insertions or deletions. The distance between the PAM sequences should be not dividable by 3.
2. Find all 23 bp genomic sites of the form 5′-N_{20}NGG-3″ near your intended target site (ideally ±50 bp). These may reside on the + or − strand.′
3. Incorporate 19 bp of the selected target sequence as highlighted here: 5′-NNNNN NNNNN NNNNN NNNNN NGG-3′ into the DNA fragment as indicated below:

   TGTACAAAAAAGCAGGCTTTAAAGGAACCAATTCAGTCGACTGGATCCGGTACCAAGGTCGGGCAGGAAGAGGGCCTATTTCCCATGATTCCTTCATATTTGCATATACGATACAAGGCTGTTAGAGAGATAATTAGAATTAATTTGACTGTAAACACAAAGATATTAGTACAAAATACGTGACGTAGAAAGTAATAATTTCTTGGGTAGTTTGCAGTTTTAAAATTATGTTTTAAAATGGACTATCATATGCTTACCGTAACTTGAAAGTATTTCGATTTCTTGGCTTTATATATCTTGTGGAAAGGACGAAACACCGNNNNNNNNNNNNNNNNNNNGTTTTAGAGCTAGAAATAGCAAGTTAAAATAAGGCTAGTCCGTTATCAACTTGAAAAAGTGGCACCGAGTCGGTGCTTTTTTTCTAGACCCAGCTTTCTTGTACAAAGTTGGCATTA.

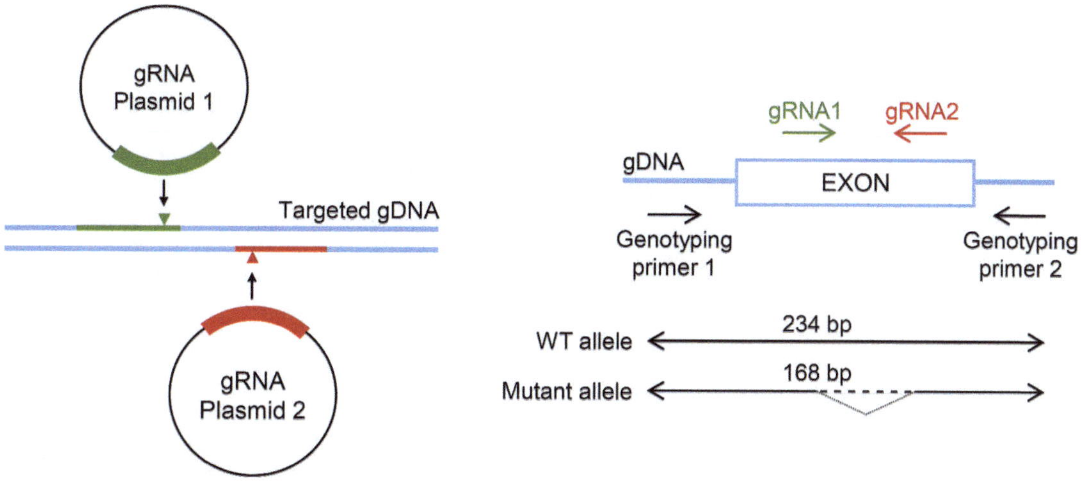

Fig. 1 Schematic showing targeting strategy and design of genotyping primers (adapted from: https:/www.scbt.com/whats-new/crispr-systems)

This 455 bp fragment bears all components necessary for gRNA expression, namely: U6 promoter + target sequence + guide RNA scaffold + termination signal.

4. Synthesize this as a gBlock.
5. Clone the synthesized gBlock into an empty backbone vector such as pCR-Blunt II-TOPO from Invitrogen for transfection and gRNA expression.

3.2.2 Transfection Based Genome Editing (Fig. 2)

1. For the gene of interest, use 2 guide RNAs (on DNA plasmids) targeting 2 separate sites within either exon 1 or exon 2 [Subheading 3.2.1] and a Cas9-GFP expressing plasmid (*see* **Note 3**).
2. For transfection of one well of a six-well plate with cells at >80% confluency, we prepare two separate tubes first. Tube 1: 100 μL of Opti-MEM and 15 μL of Lipofectamine Stem Reagent. Tube 2: 100 μL of Opti-MEM and 1.3 μg of guide RNA 1 plasmid, 1.3 μg of guide RNA 2 plasmid and 1.3 μg of the pSpCas9-p2A-GFP (PX458) plasmid.
3. Pipet up and down to mix contents of each tube and let sit for 5 min at room temperature.
4. Transfer contents of tube 1 to tube 2.
5. Pipet up and down to mix contents of combined tube and let sit for 15 min.
6. Pipet up 200 μL of contents in a 200 μL pipettor.
7. Instead of pushing down, gently rotate top of the pipettor to release droplets slowly over the well. Move pipettor over the entire surface area of the well to increase the number of transfected cells.

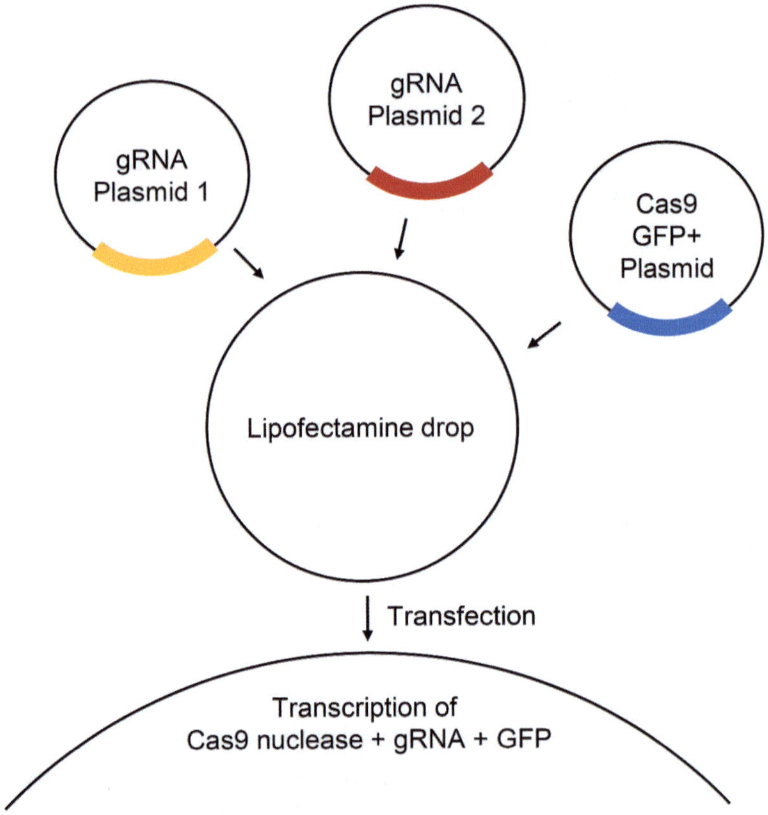

Fig. 2 Schematic showing transfection based CRISPR editing of genome (adapted from: https:/www.scbt.com/whats-new/crispr-systems)

8. Repeat **step 6**. until tube empty.
9. Place transfected cells into 37 °C incubator for 24 h.
10. Prepare 10 cm Matrigel-coated plates (5 mL of DMEM/F12 + Matrigel solution, described in plate preparation) and place them in 37 °C incubator overnight.

3.2.3 Establishment of Monoclonal Colonies

1. Aspirate media from corner of transfected well.
2. Add 1 mL of PBS to well to wash. Gently shake and aspirate.
3. Add 1 mL 1× EDTA (1:1000 in PBS) (0.5 mM final concentration in PBS).
4. Let sit for 4–5 min. Observe under the microscope, and make sure you see single cells.
5. Add 1 mL of mTeSR media to well and pipet up and down 5–6 times with serological pipettor to dislodge all the cells from the bottom of the well and transfer to tube.
6. Centrifuge at 300 × *g* for 1 min.
7. Aspirate media from tube.

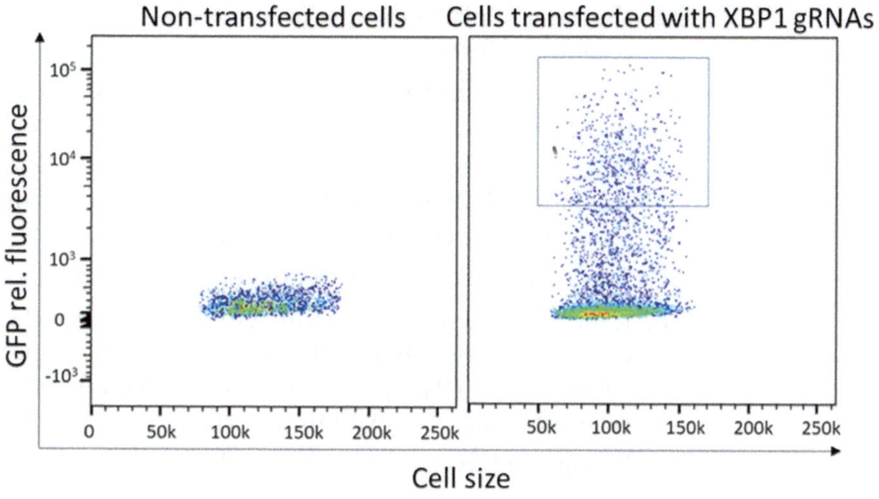

Fig. 3 FACS-Sort with transfected cells being GFP positive

8. Resuspend in 300 μL of mTeSR and transfer solution to flow sorting tube and place on ice.
9. Add DAPI to sample at a concentration on 1:10,000.
10. Add 1 mL of mTeSR media into empty 1.5 mL Eppendorf tube to sort cells into and keep the tube on ice.
11. Set up gating scheme on FACs-Sort machine, to select for GFP-positive and DAPI-negative single cells (Fig. 3).
12. Insert 1.5 cm Eppendorf tube into FACs-sort machine and start sorting, collect as many cells as possible (ideally ~5000 to 10,000 cells).
13. Place the 1.5 cm Eppendorf tube containing sorted cells on Ice immediately after sort.
14. Remove a 10-cm Matrigel plate from incubator.
15. Aspirate DMEM/F12 solution from the corner of the Matrigel plate.
16. Add Y-27632 ROCK inhibitor (1:1000 dilution) (10 mM stock solution) to 10 mL of mTeSR and add solution to the Matrigel plate (*see* **Note 4**).
17. Spin down the 1.5 mL Eppendorf tube with sorted cells.
18. Aspirate supernatant from the tube.
19. Resuspend pellet with 1 mL of mTeSR media from **step 16** and add to the Matrigel plate.
20. Shake plate gently to evenly distribute cells across whole surface of the plate and then, place it in 37 °C incubator overnight.
21. Replace with fresh media (10 mL of mTeSR) the next morning.

22. Continue replacing media every 2 days for 2 weeks to allow colonies to form.
23. Check plate under the microscope to look for colonies as they form.
24. Once colonies are ~1–2 cm in diameter, pick individual clone colonies with 200 μL pipette tip under microscope and transfer to individual wells of 6-well Matrigel-coated plate.
25. Allow colonies 2–3 days to expand, replacing media daily.
26. If there is evidence of differentiation (darkening at the center of the colonies, or abnormal beaded looking cells at periphery of colonies), use 10 μL pipette tip to scrape off differentiated cells. Remove and replace with fresh media.
27. Passage and freeze clones as described in the passaging section above.
28. Isolate 1/3 of 1 well of 6-well plate of cells for genotyping during passaging.
29. Spin down cells for genotyping at 300 × *g* for 1 min in 1.5 mL Eppendorf tube.
30. Aspirate media.
31. Resuspend in 1 mL of PBS.
32. Spin down cells at 300 × *g* for 1 min in 1.5 mL Eppendorf tube.
33. Aspirate PBS.
34. Resuspend in 100 μL of lysis buffer.
35. In thermocycler, heat up lysed mixture to 75 °C for 15 min.
36. Add 900 μL of neutralization buffer to lysed solution and store genomic DNA in −20 °C for respective monoclonal lines.

3.2.4 Genotyping of Monoclonal Colonies

1. Thaw genomic DNA of clone of interest on ice or holding vials in hands.
2. Design and order forward and reverse primers for targeting area of interest from CRISPR design above (Subheading 3.2.1).
3. Dilute genomic DNA (1:20) with H_2O.
4. Prepare PCR reaction as described below to amplify region of interest (total volume of 20 μL): 2 μL of genomic DNA, 1 μL of F Primer (10× dilution), 1 μL of R Primer (10× dilution), 8 μL GoTaq polymerase (2× working concentration), and 8 μL of H_2O.
5. Run PCR reaction as follows: 95 °C for 2 min (1 cycle)—Initial Denaturation, 95 °C for 1 min; 65 °C for 1 min, 72 °C for 10 min (30 cycles)—Denaturation, Annealing, Extension; 72 °C for 5 min (1 cycle)—Final Extension; 4 °C for infinity—Soak.

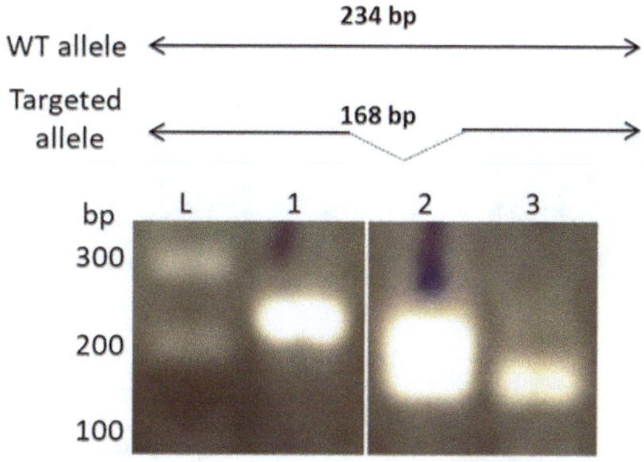

Fig. 4 Genotyping PCR of clone, showing wildtype (1), heterozygous knockout (2) and homozygous knockout (3) for gene of interest

6. Pipet total volume of 20 μL into a well of prepared 2.5% agarose gel.
7. Run gel at 100 V for 20 min or until running band makes it to bottom of the gel.
8. Wearing a face shield, observe the gel under UV light.
9. Cut out band from the gel with a scalpel. If CRISPR was effective, a shorter segment than the expected sequence should be observed on the gel (Fig. 4).
10. Purify PCR product from the gel piece by using a DNA gel extraction kit.
11. Submit purified PCR product for Sanger Sequencing.
12. Analyze DNA sequence and chromatogram of sequence to determine if desired homozygous deletion for gene of interest is achieved.

3.3 Differentiation of Human Embryonic Stem Cells (hESCs) to Human Intestinal Organoids (HIOs) (Fig. 5)

3.3.1 Cell Culture Plate Preparation

1. Prepare 24-well Matrigel-coated plates for differentiation, place 250 μL of DMEM/F12 + Matrigel into each well and leave it in incubator overnight (described in plate preparation section above). Use the eight center wells of each plate for our differentiations.
2. When passaging (refer to passaging section above), plate three wells of >80% confluency to 8 wells of 24-well Matrigel-coated plate.
3. Allow cells to grow for 1–2 days until >80% confluent. Change media every day with 500uL of mTeSR per well.

*3.3.2 hESC to Endoderm Differentiation (See **Note 5**)*

1. Endoderm media (prepare and store at 4 °C):

   Day 1 media—Combine RPMI 1640, L-glutamine (final concentration of 2 mM), Normocin (100 μg/mL), and Activin A (100 ng/mL); Day 2 media—Combine RPMI 1640, 0.2%

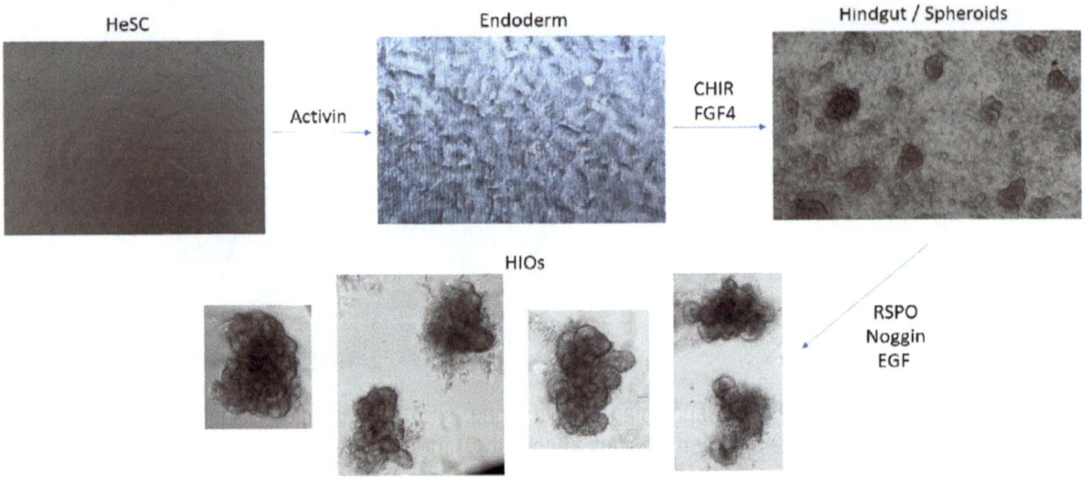

Fig. 5 Differentiation protocol from hESCs to HIOs

dFBS (vol/vol), L-glutamine (final concentration of 2 mM), Normocin (100 μg/mL), and Activin A (100 ng/mL); Day 3 media—Combine RPMI 1640, 2% dFBS (vol/vol), L-glutamine (final concentration of 2 mM), Normocin (100 μg/mL), and Activin A (100 ng/mL).

2. Remove mTeSR from each well.
3. Replace with 500 μL of Day 1 media per well.
4. After 24 h, replace with Day 2 media per well.
5. After 24 h, replace with Day 3 media per well.

3.3.3 Endoderm to Hindgut Differentiation (Prepare and Store at 4 °C)

1. Hindgut media: combine RPMI 1640, 2% dFBS (vol/vol), L-glutamine (final concentration of 2 mM), Normocin (100 μg/mL), FGF4 (500 ng/mL), and CHIR99021 (6 mM).
2. Remove mTeSR from each well.
3. Replace with 500 μL of hindgut media daily from day 4 to 8.
4. On day 7 or day 8 of differentiation, observe plate under microscope and you will see spheroid budding off from layer of hindgut tissue in well.

3.3.4 Hindgut to HIO Maturation (Prepare and Store at 4 °C)

1. Combine advanced DMEM/F12, B27 supplement (1× final dilution), L-glutamine (2 mM final concentration), Normocin (100 μg/mL), HEPES buffer (15 mM final concentration), R-Spondin1 (500 ng/mL), Noggin (100 ng/mL), and EGF (100 ng/mL).
2. If there are budding spheroids on day 7 or day 8 of differentiation, can harvest and plate for HIO maturation.
3. With 1000 μL pipettor, pipet up and down 5–6 times on sides of well to knock off spheroids from basal hindgut layer.

4. Collect media with spheroids in 1.5 cm Eppendorf tube and let them settle to bottom.
5. Spin down at 50 × *g* for 1 min.
6. Aspirate media from Eppendorf tube while avoiding spheroids at the bottom.
7. Thaw 500 μL Matrigel pellet on ice.
8. Resuspend spheroid pellet with 500 μL of Matrigel by gently pipetting up and down to avoid bubbles and ensure spheroids are evenly distributed throughout Matrigel.
9. Work quickly as Matrigel will start to solidify in 5–10 min.
10. Switch to a 200 μL pipette tip to pick up 50 μL of spheroids + Matrigel resuspension.
11. In a new 24-well Nunclon delta surface tissue culture dish, place the resuspended bead of 50 μL into center of one well. Work carefully to avoid bubbles.
12. Repeat for the rest of spheroid-Matrigel suspension.
13. Flip the 24-well plate with beads upside down and incubate in 37 °C incubator for 15 min to allow Matrigel beads to solidify.
14. Add 500 μL human intestinal organoid media to corner of each well.
15. Replace media every 2–3 days.

3.3.5 Passaging HIOs

1. After 10–14 days in culture, HIOs will be ready to passage.
2. Cut tip of 1000 μL pipette tip to avoid disrupting structure of HIOs.
3. Using 1000 μL pipettor with cut tip, pipet up and down in well with HIOs and break Matrigel bead. Place contents of well into 1.5 cm Eppendorf tube.
4. Spin down at 50 × *g* for 1 min.
5. Carefully aspirate media and Matrigel leaving HIOs in bottom of Eppendorf tube.
6. Add 500 μL of fresh media to tube.
7. Using a 100 μL pipettor pipet up and down vigorously for 5–10 min to break and remove mesenchyme/break up HIOs into crypts. Avoid creating bubbles.
8. Spin down solution at 50 × *g* for 1 min.
9. Aspirate media leaving crypt epithelium at the bottom of the Eppendorf tube.
10. Resuspend gently in 200 μL of thawed Matrigel on ice to avoid creating bubbles, and plate as above (Subheading 3.3.1).

3.3.6 Validation of HIO Differentiation by Immunohistochemistry

1. At day 3 and day 8 of differentiation, plan to fix and stain wells for endoderm and hindgut markers as below.
2. Wash one well with cells with 500 μL of 1× PBS.
3. Aspirate PBS and fix cells with 4% PFA, add 500 μL to one well for 20 min at room temperature.
4. Aspirate 4% PFA and wash cells with 0.5 mL of 1× PBS.
5. Aspirate 1× PBS and add 500 μL 1× BD permeabilization/blocking buffer to well.
6. Incubate at room temperature for 20 min.
7. Aspirate 1× BD permeabilization/blocking buffer.
8. Add primary antibodies (endoderm—goat anti-SOX17 1:500 and rabbit anti-FOXA2 1:1000) (hindgut—rabbit anti-CDX2 1:500) to fresh 1× BD buffer and add 500 μL onto each well.
9. Incubate overnight at 4 °C.
10. The next day, aspirate antibody solution and wash three times with 500 μL BD buffer, 5 min per wash.
11. Add secondary antibodies donkey anti-goat Alexa Fluor® 594 1:500 and donkey anti-rabbit Alexa Fluor® 488 1:500 to fresh 1× BD Buffer and pipet 500 μL onto each well. Make sure wells are covered with foil, to protect fluorescent secondary antibodies, from this step forward.
12. Incubate at room temperature for 2 h.
13. Aspirate the antibody solution and wash cells 3 times, for 5 min each wash, with 500 μL BD buffer.
14. Add DAPI (1:10,000 dilution) to 1× PBS and pipet 500 μL onto each well.
15. Incubate for 5 min at room temperature.
16. Aspirate solution and 500 μL of fresh 1× PBS.
17. Image under fluorescent microscope. Definitive endoderm will show positive staining for both SOX17 and FOXA2 and hindgut will show positive staining for CDX2.
18. Store plates in foil to protect from light at 4 °C.

3.3.7 Validation of HIO Differentiation by Quantitative PCR

1. At day 3 and day 8 of differentiation, plan to harvest RNA for qPCR.
2. Aspirate media from well.
3. Use 300 μL of RX buffer and pipet up and down 5–6 times to collect tissue from the bottom of well.
4. Collect solution in 1.5 mL Eppendorf tube.
5. Isolate RNA from RX buffer using RNA isolation kit.
6. Nanodrop sample to check the concentration of RNA eluted.

7. Set up and run RT-PCR reaction as per instructions on cDNA Reverse Transcription Kit.
8. Dilute cDNA fivefolds with H_2O.
9. Set up qPCR for cDNA samples from Day 3 and Day 8 differentiation time points with qPCR primers for endoderm markers (*SOX17* and *FOXA2*) and hindgut markers (*CDX2*).
10. There should be upregulation of *SOX17* and *FOXA2* in endoderm samples and upregulation of *CDX2* in hindgut samples.

3.4 Assays Looking at HIO Integrity and Response After Inflammatory Stimulus

3.4.1 Analysis of Epithelial Cell–Cell Junctional Markers

Immunohistochemistry to analyze E-cadherin expression and ZO-1 expression

1. Cut tip of a 1000 μL pipette tip.
2. Pipet 1 well with HIOs in Matrigel with 1000 μL pipettor up and down to break up Matrigel bead.
3. Place in 1.5 mL Eppendorf tube.
4. Spin down at 50 × *g* for 1 min.
5. Carefully aspirate solution and Matrigel leaving HIOs in tube.
6. Resuspend HIOs in 4% PFA solution and let fix at room temperature for 1 h.
7. Spin down at 50 × *g* for 1 min.
8. Aspirate PFA from tube leaving HIOs at bottom.
9. Add Rabbit ZO-1 Ab (1:250 dilution) or Rabbit E-cadherin Ab (1:350 dilution) to BD buffer.
10. Pipet 500 μL of desired antibody to tube with HIOs.
11. Let incubate at 4 °C overnight.
12. Next day, spin down at 50 × *g* for 1 min.
13. Aspirate primary antibody solution, leaving HIOs at bottom.
14. Wash 3 times with BD buffer, 10 min per wash, spin down and aspirate after last wash.
15. Add donkey anti-rabbit Alexa Fluor® 568 (1:500 dilution) to BD buffer.
16. Add 500 μL of secondary antibody solution to tube.
17. Incubate at room temperature for 2 h, cover with foil from this step forward to protect fluorescent antibody.
18. Spin down at 50 × *g* for 1 min.
19. Aspirate solution and wash 3 times with BD buffer, 10 min per wash, spin down and aspirate after last wash.
20. Add DAPI (1:10,000 dilution) in 1× PBS, leave at room temperature for 5 min.
21. Spin down at 50 × *g* for 1 min.

22. Aspirate solution and resuspend in 500 μL of 1× PBS.
23. Place HIOs on glass bottom petri dish and image under fluorescent microscope.
24. Store in 1× PBS with the dish covered in foil at 4 °C.

3.4.2 Inflammatory Challenge of HIOs, TNF-α Assay

1. Add hTNF-α at a final concentration of 25 nM to HIO media.
2. Pipet 500 μL into 3 wells with HIOs.
3. Cut 1000 μL pipette tip and collect HIOs from separate wells at following time points (0, 2, 6, and 12 h).
4. Collect HIOs in 1.5 mL Eppendorf tubes.
5. Spin down at 50 × *g* for 1 min.
6. Aspirate media from tube.
7. Resuspend in 4% PFA, fix at room temperature for 1 h.
8. Spin down at 50 × *g* for 1 min.
9. Add rabbit cleaved Caspase-3 antibody (1:250 dilution) to BD buffer.
10. Pipet 500 μL of primary antibody solution to tube with HIOs.
11. Let incubate at 4 °C overnight.
12. Next day, spin down at 50 × *g* for 1 min.
13. Aspirate primary antibody solution, leaving HIOs at bottom.
14. Wash with 3 times BD buffer, 10 min per wash, spin down and aspirate after last wash.
15. Add donkey anti-rabbit Alexa Fluor® 568 (1:500 dilution) to BD buffer.
16. Add 500 μL of secondary antibody solution to tube.
17. Incubate at room temperature for 2 h, cover with foil from this step forward to protect fluorescent antibody.
18. Spin down at 50 × *g* for 1 min.
19. Aspirate solution and wash 3 times with BD buffer, 10 min per wash, spin down and aspirate after last wash.
20. Add DAPI (1:10,000 dilution) in 1× PBS, leave at room temperature for 5 min.
21. Spin down at 50 × *g* for 1 min.
22. Aspirate solution and resuspend in 500 μL of 1× PBS.
23. Place HIOs on glass bottom petri dish and image under fluorescent microscope. You should observe increased staining at the greater time points of assay.
24. Store in 1× PBS with the dish covered in foil at 4 °C.

4 Notes

1. When preparing Matrigel-coated plates, care should be made to keep Matrigel at 4 °C or on ice prior to diluting with DMEM/F12. At room temperature, Matrigel will solidify within 5–10 min and will not mix evenly with DMEM/F12. If this occurs, Matrigel will not coat tissue culture plates evenly. This is also important to consider when plating HIOs with Matrigel beads. Since it takes 5–10 min for Matrigel to solidify, plates should be kept upside down after bead plating and/or Matrigel should be warmed up in hands prior to resuspending to allow Matrigel with HIOs to stay in bead form.
2. When working with hESC cultures, careful observation should be made daily in regards to cell morphology. Edges of colonies and centers of colonies may start to differentiate if left in culture without passaging >5–6 days. If this is noted, pipette tips may be used to scrape off the differentiated cells and cells should be passaged immediately. Also of note, spherical structures called embryoid bodies can develop and this will be seen budding off the surface of maintenance stem cell cultures. These should also be scraped off as they can affect efficiency of gene editing and differentiations.
3. We used 2 guide RNA plasmids along with a Cas9-GFP labeled plasmid for our transfections. Another potentially more efficient methodology to consider is to design and use a single plasmid containing both guide sequences and the Cas9-GFP sequence in the same plasmid. This should ensure that all GFP-positive cells sorted during the FACS sort would also contain both target sequences, theoretically, increasing the efficiency of the number of cells being edited.
4. We used a limited dilution method post-FACS sort to obtain clonal colonies for genotyping and expansion. By doing so, we assume that the clonal populations we obtain are from a single cell since we dilute out the cells on a 10-cm plate. It is possible that two cells could form a single polyclonal colony, but this is less likely. Another method to ensure definite monoclonal populations is to seed individual sorted cells into 96-well plates. However, we observed that single hESCs do not survive well when plated in single wells. Hence, we recommend using feeder cells if this methodology is used to help with clonal expansion of monoclonal colonies.
5. Another important consideration for our differentiations is the media in which hESCs are maintained. There are various choices for media such as mTeSR, E8 Flex, and Stemflex. However, not all media choices lead to efficient HIO differentiations. In our experience, using mTeSR for this protocol is

the most successful method since it sets up the hESCs to go down the endodermal pathway most efficiently and to generate HIOs effectively.

Acknowledgments

This work was supportesd in part by the Kenneth Rainin Foundation through a Rainin Innovator Grant.

References

1. Byrne SM, Church GM (2015) Crispr-mediated gene targeting of human induced pluripotent stem cells. Curr Protoc Stem Cell Biol 35:5A 8 1–5A 822
2. McCracken KW, Howell JC, Wells JM, Spence JR (2011) Generating human intestinal tissue from pluripotent stem cells in vitro. Nat Protoc 6(12):1920–1928
3. Spence JR, Mayhew CN, Rankin SA, Kuhar MF, Vallance JE, Tolle K, Hoskins EE, Kalinichenko VV, Wells SI, Zorn AM, Shroyer NF, Wells JM (2011) Directed differentiation of human pluripotent stem cells into intestinal tissue in vitro. Nature 470(7332):105–109
4. Barber K, Studer L, Fattahi F (2019) Derivation of enteric neuron lineages from human pluripotent stem cells. Nat Protoc 14(4):1261–1279

Chapter 14

Direct Lineage Reprogramming of Mouse Fibroblasts to Acquire the Identity of Fetal Intestine-Derived Progenitor Cells

Shizuka Miura and Atsushi Suzuki

Abstract

Intestinal organoids are useful models for studying the characteristics of intestinal diseases and their treatment. However, a major limiting factor in their usability is the need for donor tissue fragments or pluripotent stem cells to generate the organoids. Here, we describe an approach to generate intestinal organoids from fibroblasts, a new source. We used direct reprogramming technology to generate cells with the properties of fetal intestine-derived progenitor cells (FIPCs) from mouse embryonic fibroblasts (MEFs). These induced FIPCs (iFIPCs) can give rise to cells resembling intestinal stem cells (ISCs), henceforth referred to as induced ISCs (iISCs). These iFIPCs and iISCs form spherical and budding organoids, respectively, similar to FIPCs and ISCs. These induced intestinal organoids could be used for studies on intestinal diseases and regenerative therapy.

Key words Intestinal organoid, Direct reprogramming, Induced fetal intestine-derived progenitor cell (iFIPC), Induced intestinal stem cell (iISC), Mouse embryonic fibroblast (MEF), Differentiation, Self-renewal, Infection

1 Introduction

Over the past decade, the use of intestinal organoid cultures has become increasingly frequent [1, 2]. Using specific culture conditions, ISCs obtained from the small intestine of adult mice can be cultured in vitro for a prolonged period, because of their self-renewing cell divisions. During this incubation period, they form epithelial organoids with crypt-villus–like structures [1]. These organoids are called budding organoids (BOs), in which ISCs give rise to four different types of differentiated cells, whilst maintaining the stem cell population. This process resembles the behavior of ISCs residing in the bottom of intestinal crypts. Meanwhile, FIPCs obtained from the developing embryonic mouse intestines form spherical organoids (SOs) in vitro. FIPC-derived SOs can develop into BOs after serial passages without exogenous Wnt

Paloma Ordóñez-Morán (ed.), *Intestinal Stem Cells: Methods and Protocols*, Methods in Molecular Biology, vol. 2171, https://doi.org/10.1007/978-1-0716-0747-3_14,

stimulation. Although there has been a scientific interest in using human-derived intestinal organoids for the construction of models for intestinal disease and transplantation therapy, the collection of intestinal tissue fragments from healthy adult donors and unborn children is difficult. Moreover, although intestinal organoids can be technically developed from pluripotent stem cells (PSCs) [3], the existence of inherent ethical difficulties, the risk of tumor formation, and the complexity of the involved differentiation procedures limit the feasibility of this approach.

Direct reprogramming technology may be able to resolve the above limitations. This technology enables us to induce the conversion of somatic cells into alternative cell types without passing through a pluripotent state. In our previous study, iFIPCs were directly induced from MEFs, using four defined transcription factors (hepatocyte nuclear factor 4α (Hnf4α), forkhead box protein a3 (Foxa3), GATA binding protein 6 (Gata6), and caudal type homeobox 2 (Cdx2)) [4]. Similar to the FIPC-derived SOs, iFIPC-derived SOs can develop into BOs under a specific culture condition (see below) (Fig. 1). These iFIPC-derived BOs contain multipotent iISCs and have the ability to undergo self-renewal. The global gene expression profiles of iFIPC-derived SOs and iISC-derived BOs are similar to those of FIPC-derived SOs and adult mouse crypt-derived BOs, respectively. Moreover, iFIPC-derived SOs and iISC-derived BOs can reconstitute colonic and intestinal epithelial tissues, respectively, after transplantation into an injured mouse colon. Our promising new approach will be a key in the further development of human iFIPCs and iISCs for medical purposes.

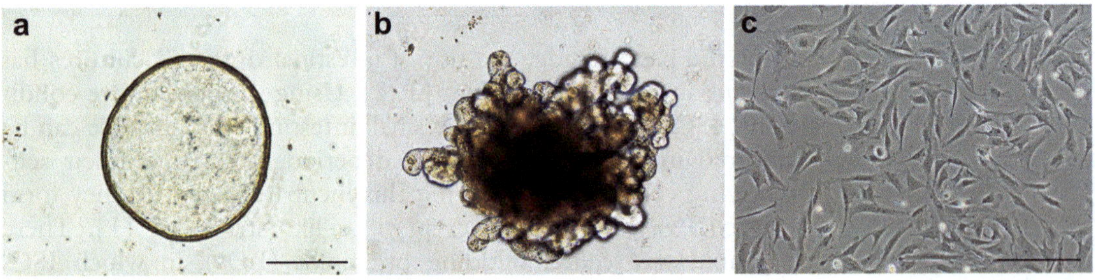

Fig. 1 Representative morphologies of the intestinal epithelial organoids directly induced from MEFs. (**a**) A SO formed from an MEF-derived iFIPC (phase-contrast image). Scale bar, 50 μm. (**b**) A BO formed from an MEF-derived iISC (phase-contrast image). iFIPC-derived SOs can develop into BOs containing iISCs that build crypt-villus–like structures and have multipotency and self-renewal capacity. Scale bar, 50 μm. (**c**) A representative phase-contrast image of MEFs. Scale bar, 50 μm

2 Materials

2.1 Mouse Embryonic Fibroblast (MEF)

1. Embryonic day (E) 13.5 mouse embryos (C57BL/6 mice).
2. Trypsinization solution: 2.5 g/L trypsin and 1 mM ethylenediaminetetraacetic acid (EDTA).
3. 25 μg/mL DNase I.
4. MEF medium: Dulbecco's modified Eagle's medium (DMEM) containing 10% fetal bovine serum (FBS), 2 mM L-glutamine, and 100 units/mL penicillin–100 μg/mL streptomycin mixed solution.
5. The 6-cm tissue culture dish.

2.2 Retrovirus Production and Transduction of Cells

1. Mouse *Hnf4α*, *Foxa3*, *Gata6*, and *Cdx2* cDNAs were obtained by reverse transcription polymerase chain reaction.
2. Retrovirus vector: pGCDNsam (a gift from M. Onodera).
3. Plat-E cells (a gift from T. Kitamura) [5].
4. Plat-E cell medium: DMEM containing 8% FBS, 2 mM L-glutamine, and 100 units/mL penicillin–100 μg/mL streptomycin mixed solution.
5. Linear polyethylenimine (PEI).
6. Poly-L-lysine.
7. Cellulose acetate filters (0.2 μm).
8. Hank's balanced salt solution (1 L) containing 2.38 g HEPES, 0.41 g $CaCl_2$, and 0.35 g $NaHCO_3$.
9. Gelatin.
10. Twelve-well plate.
11. 5 μg/mL protamine sulfate.

2.3 Mouse iFIPC Culture

1. Matrigel.
2. Mouse intestinal basal medium (MIBM): Advanced DMEM/F12, supplemented with 2 mM GlutaMax, 10 mM HEPES, N2 supplement (1×), B27 supplement (1×), 1 mM *N*-acetylcysteine, and penicillin–streptomycin (*see* **Note 1**).
3. WCENR: 100 ng/mL Murine recombinant Wnt3a, 3 μM CHIR99021, 50 ng/mL Human recombinant epidermal growth factor (EGF), 100 ng/mL Murine recombinant Noggin, 500 ng/mL Human recombinant R-spondin1 (*see* **Note 2**).
4. ENR: 50 ng/mL Human recombinant epidermal growth factor (EGF), 100 ng/mL Murine recombinant Noggin, 500 ng/mL Human recombinant R-spondin1 (*see* **Note 2**).

3 Methods

3.1 MEF Culture

1. To prepare the MEFs, carefully remove the heads and visceral tissues from the E13.5 mouse embryos.
2. Mince the remaining tissues using forceps. Next, incubate these fragments on a shaker in the trypsinization solution (20 min, 37 °C).
3. After the trypsinization, add MEF medium and further dissociate the tissue fragments by pipetting. Incubate the suspension on a shaker (20 min, 37 °C).
4. Centrifuge (400 × *g*, 1 min, 4 °C) the suspension and resuspend the triturated cells in MEF medium.
5. Seed the cells on 6 cm tissue culture dishes and incubate the cultures for 3–4 days (37 °C, 5% CO_2) (*see* **Note 3**).

3.2 Retrovirus Production and Transduction of Cells

1. Subclone the cDNAs into pGCDNsam vectors (pGCDNsam-Hnf4α, pGCDNsam-Foxa3, pGCDNsam-Gata6, and pGCDNsam-Cdx2).
2. Three days before transfection, plate Plat-E cells (1.6×10^6) on poly-L-lysine–coated 10 cm dishes (*see* **Note 4**). Three days after having seeded the Plat-E cells, dilute 12 μg of the constructed retroviral plasmid DNA and 36 μL of 1 mg/mL PEI in 1 mL DMEM and incubate (15 min, room temperature). Subsequently add the mixture to the Plat-E cells in a drop-by-drop manner. After incubation (6 h, 37 °C, 5% CO_2), replace the medium with fresh MEF medium.
3. After 24 and 48 h, collect the supernatants from the cultures of transfected cells and filtrate them through 0.2 μm cellulose acetate filters and centrifuge (9000 × *g*, 16 h, 4 °C).
4. Resuspend the viral pellets in Hank's balanced salt solution (1/140 of the initial supernatant volume).
5. One day before viral infection, seed the MEFs ($2–4 \times 10^4$ cells) on gelatin-coated 12-well plates and incubate them.
6. Add the viral supernatants with 5 μg/mL protamine sulfate to the MEFs and incubate the cultures for 6 h. Repeat this viral infection step three times.

3.3 Mouse iFIPC Culture

1. After the last infection step, suspend the retrovirus-infected MEFs in Matrigel and seed them on 24-well plates.
2. The Matrigel will solidify after 10–15 min and form a hemisphere. Next, apply 500 μL MIBM containing Wnt3a, CHIR99021, EGF, Noggin, and R-spondin1 (designated WCENR).

3. Replace the culture medium with fresh MIBM (containing WCENR) every 4 days.
4. After approximately 10 days, intestinal SOs and BOs are formed from MEF-derived iFIPCs.
5. iFIPCs can be passaged approximately 7 days after seeding. Remove the culture medium and mechanically break up the Matrigel that contains the iFIPC-derived SOs and BOs in 1 mL MIBM. Centrifuge the organoid suspensions (200 × *g*, 3 min, room temperature) and remove the supernatant. In turn, add 1 mL fresh MIBM onto the pellets. Gently resuspend the pellets by pipetting and centrifuge the suspension (200 × *g*, 3 min, room temperature). Mix the pellets with Matrigel and seed on 24-well plates.
6. After subsequent passages, remove WCENR from the culture medium and add EGF, Noggin, and R-spondin1 (designated ENR). Consequently, iFIPC-derived SOs develop into BOs after passages without exogenous Wnt stimulation. The resultant BOs contain iISCs.
7. iISCs can be passaged approximately 7 days after seeding, similar to iFIPCs.

4 Notes

1. MIBM can be stored at 4 °C for up to 2 weeks.
2. MIBM containing WCENR or ENR can be stored at 4 °C up to 4 days.
3. For preparation of MEFs, mouse embryos should be quickly processed, and MEFs should be plated as soon as possible. The growth rate and freshness of MEFs will have an impact on the efficiency of cell-fate conversion.
4. Proliferation of Plat-E cells is also important. Check the condition of Plat-E cells before transfection. To maintain Plat-E cells that express transgenes required for the retrovirus production, add 10 μg/mL Blasticidin and 1 μg/mL Puromycin to Plat-E cell medium once 2 weeks.

Acknowledgments

We thank Drs. Toshio Kitamura and Masafumi Onodera for sharing the Plat-E cells and pGCDNsam plasmid, respectively. This work was supported in part by the JSPS KAKENHI (Grant Numbers: 23112002, 25713014, JP16H01850, JP18H06069, FDG6J02459, JP18H05102, JP19H01177, and JP19H05267), the Core Research for Evolutional Science and Technology

(CREST) Program of the Japan Agency for Medical Research and Development (AMED), the Practical Research Project for Rare/ Intractable Diseases of AMED, the Research Center Network for Realization of Regenerative Medicine of AMED, and the Takeda Science Foundation.

References

1. Sato T, Vries RG, Snippert HJ, van de Wetering M, Barker N, Stange DE et al (2009) Single Lgr5 stem cells build crypt-villus structures in vitro without a mesenchymal niche. Nature 459:262–265
2. Fordham RP, Yui S, Hannan NR, Soendergaard C, Madgwick A, Schweiger PJ et al (2013) Transplantation of expanded fetal intestinal progenitors contributes to colon regeneration after injury. Cell Stem Cell 13:734–744
3. Spence JR, Mayhew CN, Rankin SA, Kuhar MF, Vallance JE, Tolle K et al (2011) Directed differentiation of human pluripotent stem cells into intestinal tissue in vitro. Nature 470:105–109
4. Miura S, Suzuki A (2017) Generation of mouse and human organoid-forming intestinal progenitor cells by direct lineage reprogramming. Cell Stem Cell 21:456–471
5. Morita S, Kojima T, Kitamura T (2000) Plat-E: an efficient and stable system for transient packaging of retroviruses. Gene Ther 7:1063–1066

Chapter 15

Single-Molecule RNA FISH in Whole-Mount Organoids

Costanza Borrelli and Andreas E. Moor

Abstract

Single-molecule RNA fluorescent in situ hybridization (smFISH) enables the detection and quantification of single RNA molecules. Three-dimensional organoid cultures have emerged as versatile in vitro primary culture models that recapitulate many physiological features of their tissue of origin. Here we describe a protocol to visualize single RNA molecules in organoid cultures. Our method accommodates both a whole-mount staining workflow which requires spinning disk confocal microscopy, and a cryosectioning workflow which is compatible with widefield microscopy. Organoid smFISH enables to address various biological problems that range from the identification of cell types (e.g., via the intestinal stem cell marker Lgr5) to the quantification of RNA localization in an epithelium.

Key words Intestinal organoids, smFISH, Whole-mount, RNA imaging, Spatial transcriptomics

1 Introduction

Single-molecule RNA fluorescent in situ hybridization (smFISH) has become the standard method for the absolute quantification of gene expression in cultured cells [1, 2] and tissues [3–5]. This method has proven instrumental for the field of spatial transcriptomics, since it enables the detection and quantification of single RNA molecules in their spatial tissue context. SmFISH makes use of libraries of about 50 20-bp DNA oligos that are complementary to different regions of the transcript of interest [6]. These oligos are labeled with single fluorophores, and result in diffraction-limited spots when hybridized to single RNA molecules. Every fluorescent spot corresponds to a single RNA molecule, allowing for precise and absolute quantification of transcripts at the single cell level.

Adult stem cells have the ability to form three-dimensional mini-organs when cultured with growth-factors and a supporting basement membrane [7]. These organoid cultures are relevant in vitro models of their organ of origin that offer many advantages compared to 2D cell culture and animal models [8]. Technology development in the field of organoid culture has been rapidly

Paloma Ordóñez-Morán (ed.), *Intestinal Stem Cells: Methods and Protocols*, Methods in Molecular Biology, vol. 2171, https://doi.org/10.1007/978-1-0716-0747-3_15,

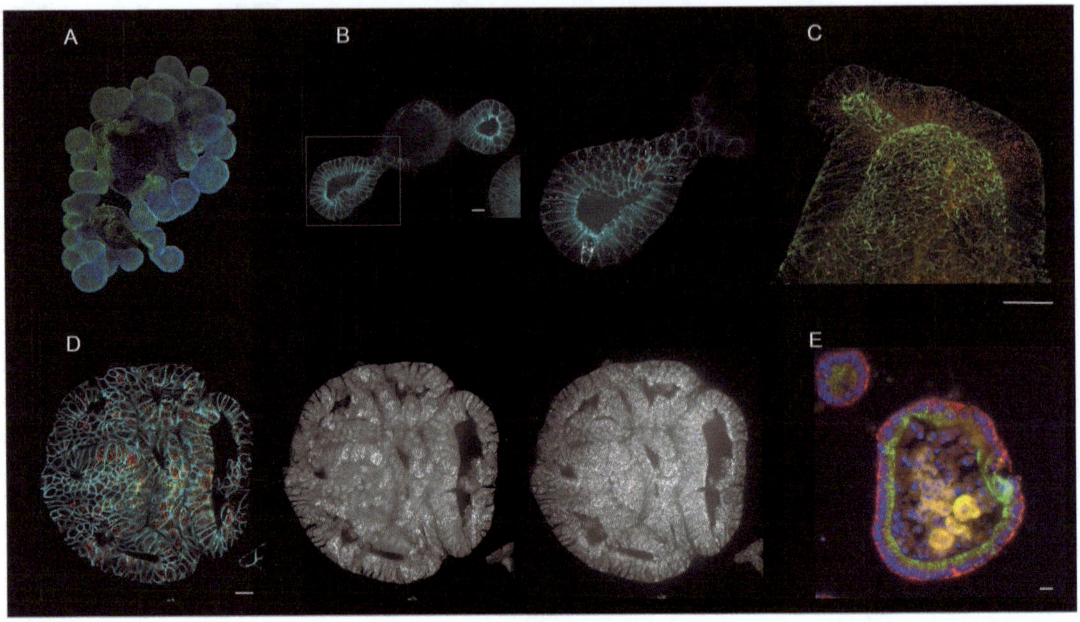

Fig. 1 smFISH in organoid cultures. (**a**) Mouse small intestinal organoid exhibits intact morphology in the whole-mount workflow. E-Cadherin (green) Dapi (blue). (**b**) Overview and zoom of a budding crypt of a small intestinal organoid. The zoom reveals a Paneth cell (Lysozyme1, Lyz1-mRNA, white) and an enteroendocrine cell (ChromograninA, Chga-mRNA, red), E-Cadherin (cyan). (**c**) Projection of a murine small intestinal organoid, Lgr5-mRNA red, E-Cadherin green. (**d**) Cross section of a murine pancreas organoid. Left: E-Cadherin (cyan), Mki67-mRNA (red), Cdh1-mRNA (yellow). Middle: Mki67-mRNA, Right: Cdh1-mRNA. Images (**a–d**) were obtained with the whole-mount workflow (Subheading 3.2). (**e**) smFISH example image of intestinal organoids that were stained with the cryosectioning workflow (Subheading 3.3). Net1-mRNA (red), Apob-mRNA (green), DAPI (blue). Reproduced from Moor et al. [10] with permission from AAAS. All scale bars: 10 μm

progressing and enables in vitro cultures of almost all human or mouse adult stem cell compartments [8].

Here we present a smFISH protocol for use in organoid cultures that we adapted from the excellent tissue smFISH protocol by Lyubimova et al. [9]. We have used this protocol for single molecule transcript imaging in wild-type murine small intestinal and pancreas organoids, and in colon cancer organoids (Fig. 1). Our protocol enables rapid whole-mount staining of entire organoids for imaging with spinning-disk confocal microscopy (after submission of our chapter in 03/2019, a similar method was published by Omerzu et al. [15]). For imaging on a conventional widefield microscope, or in case single molecule resolution in great organoid depth is required, we also provide a lengthier organoid smFISH protocol that is based on embedding and cryosectioning. We regularly image three transcripts, nuclei, and a cell-border antibody staining in parallel on one organoid sample. If the microscope setup is compatible with near-infrared dyes (such as Cy7), one can detect up to four transcripts in a single hybridization.

We have previously used smFISH in intestinal organoids to establish an in vitro model of intestinal intracellular RNA localization [10]. We envision further applications of this technique to study spatial transcriptional dynamics, intestinal stemness and differentiation processes, or to validate single cell RNA sequencing observations in organoid cultures.

2 Materials

2.1 Organoid Cultures

We grow intestinal organoids according to the method of Sato and Clevers [11]. Culture methods for a wide range of adult stem cell compartments differ in their growth factor requirements and have recently been extensively reviewed by Li and Izpisua Belmonte [8]. Small and large intestinal organoids accumulate shed cells in their lumen. These dead cells lead to autofluorescence that can disturb the smFISH microscopy. We therefore use 50–2000 intestinal organoids for smFISH 2–3 days after splitting when they have not yet accumulated large numbers of dead cells in the lumen.

The organoid collection and fixation reagents (*see* **Notes 1** and **2**):

1. Organoid Harvesting Solution.
2. Prelubricated RNase-free 1.7 mL tubes.
3. Paraformaldehyde solution 4% in PBS.

2.2 smFISH Probe Hybridization

Prepare all solutions using RNase-free water and reagents (*see* **Note 1**).

2.2.1 smFISH Hybridization Buffer

For 10 mL Hybridization buffer, dissolve 1 g dextran sulfate in 6 mL water at room temperature with agitation (30 min). Add 1.5 mL formamide (stored at 4 °C, bring to RT before opening), 500 μL *E. coli* tRNA (stock 20 mg/mL), 1 mL 20× SSC, 40 μL BSA (stock 50 mg/mL), 100 μL Vanadyl-ribonucleoside complex (stock 200 mM). Adjust the volume to a total of 10 mL by adding water. We aliquot the hybridization buffer in vials of 900 μL and store them at −20 °C.

2.2.2 smFISH Wash Buffer

For 500 mL Wash buffer, combine 375 mL water, 50 mL 20× SSC and 75 mL formamide. Mix well and store in 50 mL aliquots at RT.

The formamide concentration in both the hybridization and wash buffer can be increased to 20% or 25% if the GC concentration of the transcript of interest is high.

2.3 smFISH Probes

Ready-to-use predesigned or custom smFISH probes can be obtained from Biosearch Technologies (https://www.biosearchtech.com/support/education/stellaris-rna-fish/). Alternatively, uncoupled DNA oligos with 3′ amine modifications can be purchased from Biosearch Technologies in multiwell plates. For a detailed protocol on how to couple oligos to fluorophores and how to select fluorophores and compatible microscope filter cubes, please refer to the sections "Coupling FISH probe libraries to fluorophores" and "Materials" in Lyubimova et al. [9]. When following the published coupling protocol [9], the probe stocks are resuspended in 100 μL TE and probe working solution consists of a 1:100 dilution of probe stock in TE. Ready-to-use smFISH probes have to be diluted according to the instructions of the manufacturer.

2.4 Mounting Reagents

For whole-mount and cryosectioning protocol:

1. Prolong Gold.
2. Gasket (*see* **Note 1**).
3. 22 × 22 mm #1 coverslip.
4. Circular coverslip (12 mm diameter).
5. Microscope slide.

Additionally, for cryosectioning protocol only:

1. OCT.
2. Tissue marking dye (*see* **Note 1**).
3. Cryomold.
4. Liquid blocking pen.

3 Methods

3.1 Organoid Collection and Fixation

1. Remove the culture medium by careful aspiration. Add 500 μL Organoid Harvesting Solution, detach the Matrigel domes and combine all wells in a 15 mL Falcon tube. Rinse the well with an additional 500 μL Organoid Harvesting Solution to collect all organoids and add this to the 15 mL tube.
2. Incubate the tube for 30 min at 4 °C under gentle agitation to completely dissolve the Matrigel.
3. Centrifuge at 200 × *g* for 5 min at 4 °C.
4. Remove the supernatant, add 10 mL cold PBS, resuspend the organoid pellet and centrifuge at 200 × *g* for 5 min at 4 °C.
5. Remove the supernatant, add 1 mL 4% paraformaldehyde (PFA) in PBS and transfer the pellet to a 1.8 mL tube (*see* **Notes 1** and **2**).

6. Fix the organoids for 30 min at 4 °C.
7. Centrifuge at 150 × *g* for 5 min at 4 °C, remove the supernatant and wash the pellet with PBS.
8. Remove the supernatant and resuspend the cells in cold 70% ethanol.

Proceed with Subheading 3.2 for whole-mount staining and image acquisition by spinning-disk confocal microscopy. If the organoids should be embedded for cryosectioning, proceed to Subheading 3.3 (*see* **Note 3**).

3.2 Whole-Mount Staining

1. Permeabilize the cells by incubating them for 3 h in 70% ethanol at 4 °C (*see* **Note 4**).
2. Centrifuge at 150 × *g* for 3 min at 4 °C, remove the supernatant and resuspend the pellet in 2× SSC.
3. Incubate the organoids for 5 min at 4 °C.
4. Centrifuge at 150 × *g* for 3 min at 4 °C, remove the supernatant and resuspend the pellet in smFISH wash buffer.
5. Incubate the organoids for 5 min at RT.
6. Take 5 μL of each smFISH probe working solution and add smFISH hybridization buffer to a total volume of 150 μL hybridization mix, vortex (*see* **Note 5**).
7. Centrifuge at 150 × *g* for 3 min at RT, remove the supernatant and resuspend the pellet in hybridization mix.
8. Protect the samples from light and incubate at 30 °C overnight.
9. Prepare 1 mL of smFISH wash buffer with 1:200 of DAPI 10 μg/mL.
10. Add 1 mL smFISH wash buffer, centrifuge at 150 × *g* for 3 min at RT, remove the supernatant and resuspend the pellet in the prepared smFISH wash buffer with 1:200 DAPI.
11. Protect the samples from light and incubate at 30 °C for 30 min.
12. Centrifuge at 150 × *g* for 3 min at RT, remove the supernatant and resuspend the pellet in 2× SSC.
13. Centrifuge at 150 × *g* for 3 min at RT, remove the supernatant carefully with a P200 tip while leaving the pellet intact.
14. Carefully resuspend the pellet in 8 μL Prolong Gold.
15. Pipet the resuspended organoids on to a 22 × 22 mm #1 coverslip.
16. Cover the drop with a 12 mm circular coverslip (*see* **Note 6**).
17. Mount the coverslip on to a microscope slide equipped with a gasket (Fig. 2). The usage of gaskets protects the whole-mount organoid structure and prevents crushing.

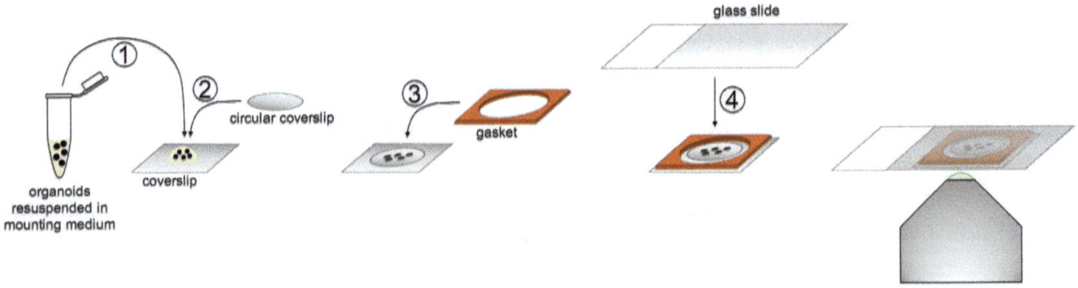

Fig. 2 Mounting of whole-mount smFISH organoids. Carefully resuspend the pellet in 8 μL Prolong Gold (*1*). Pipet the resuspended organoids on to a 22 × 22 mm #1 coverslip (*2*). Cover the drop with a 12 mm circular coverslip (*2*). Attach a gasket to the coverslip (*3*) and mount the coverslip on to a microscope slide (*4*)

18. The sample is now ready for imaging with a spinning disk confocal microscope. The slides can be frozen and stored at −20 °C for several months without visible loss of signal quality (*see* **Notes 7** and **8**).

3.3 Cryosectioning and Staining

Continue here after **step 8** of the "Organoid collection and fixation" protocol (Subheading 3.1) if the samples should undergo cryosectioning instead of whole-mount smFISH staining.

1. Let the organoids sediment by gravity and remove most of the supernatant.
2. Mix 100 μL OCT with 5 μL tissue marking dye by stirring with a pipette tip.
3. Collect the organoid pellet by pipetting with a P20 pipette and reduce the ethanol carry over to a minimum (Fig. 3).
4. Pipet the organoid-ethanol mixture to the middle of an empty cryomold.
5. Use a dissecting microscope to visualize the organoids and arrange them in the center of the cryomold with a pipette tip. Wait until the carried over ethanol has evaporated.
6. Add the stained OCT to the organoids. Using a pipette tip, create a dome of blue OCT containing the organoids in the center of the cryomold.
7. Transfer the cryomold to dry ice and incubate for 1 min.
8. Gently fill the cryomold with unstained OCT and let it solidify on dry ice for 1 h and transfer the block for storage to −80 °C.
9. When proceeding to the cryotome, clean the stage with 70% ethanol and use a new blade for each session.
10. Use the blue dye as landmark for the organoid location during cryosectioning and slowly trim the block until the blue zone is reached.

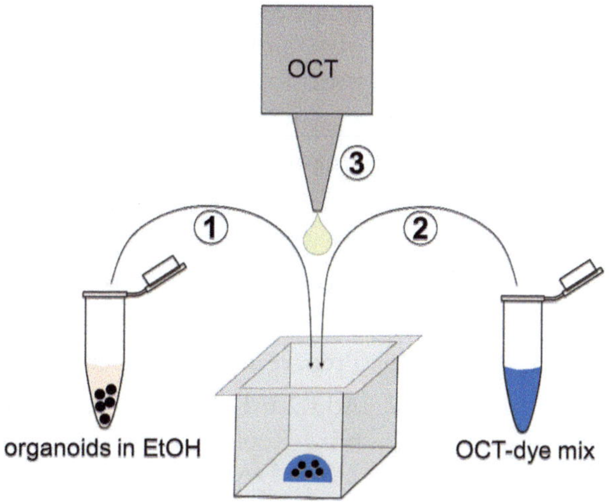

Fig. 3 Embedding of organoids. Collect the organoid pellet by pipetting with a P20 pipette and reduce the ethanol carry over to a minimum (*1*). Pipet the organoid-ethanol mixture to the middle of an empty cryomold. Use a dissecting microscope to arrange the organoids in the center of the cryomold with a pipette tip and wait until the carried-over ethanol has evaporated. Mix the organoids with the stained OCT with a pipette tip with the aim of creating a dome of blue OCT that contains the organoids in the center of the cryomold (*2*). Transfer the cryomold to dry ice and incubate for 1 min. Gently fill the cryomold with unstained OCT (*3*) and let it solidify on dry ice for 1 h and transfer the block for storage to −80 °C

11. Cut 5–8 μm thick sections and capture them with a polylysine-coated 22 × 22 mm coverslip (*see* **Note 9**).
12. Air-dry the section for 5 min at RT. Then place the coverslip into an empty six-well plate and keep it on dry ice until the end of cryosectioning.
13. Use a liquid blocking pen to draw a circle around the organoids.
14. Add 2 mL 4% PFA in PBS and fix the section for 15 min at RT.
15. Remove the fixative and wash the section with 3 mL PBS.
16. Replace the PBS with 3 mL cold 70% ethanol and permeabilize the section for at least 1 h at 4 °C.
19. Replace the ethanol with 3 mL 2× SSC and incubate the section for 5 min at 4 °C.
20. Replace the 2× SSC with 3 mL smFISH wash buffer and incubate the section for 5 min at RT.
21. Take 5 μL of each smFISH probe working solution and add smFISH hybridization buffer to a total volume of 150 μL hybridization mix, vortex (*see* **Note 5**).

22. Remove the smFISH wash buffer completely by tilting the plate and removing any residual liquid with a Kimwipe. Avoid touching the sample area in the center of the coverslip.
23. Gently pipet 150 μL of hybridization mix to the center of the area that is circled with the liquid blocking pen.
24. Pipet RNase-free water into the space around the wells in the six-well plate to prevent drying of the samples.
25. Close the plate with its lid and transfer it to a hybridization oven at 30 °C for overnight incubation.
26. Prepare 2 mL of wash buffer with 1:200 of DAPI 10 μg/mL.
27. Add 2 mL smFISH wash buffer the rinse off the hybridization mix. Replace with the prepared smFISH wash buffer with 1:200 DAPI.
28. Protect the samples from light and incubate at 30 °C for 30 min.
29. Replace the wash buffer with 2 mL 2× SSC.
30. Remove the coverslip with forceps and dry the residual liquid with a Kimwipe.
31. Carefully pipet 8 μL Prolong Gold to the sample area.
32. Cover the sample area with a circular coverslip (12 mm) and gently push on it with a Kimwipe to remove excess mounting buffer. Mount the coverslip sandwich on to a gasket on a microscope slide (Fig. 4).
33. The sample is now ready for imaging with a widefield microscope. The slides can be frozen and stored at −20 °C for months without visible loss of signal quality (*see* **Notes 8** and **10**).

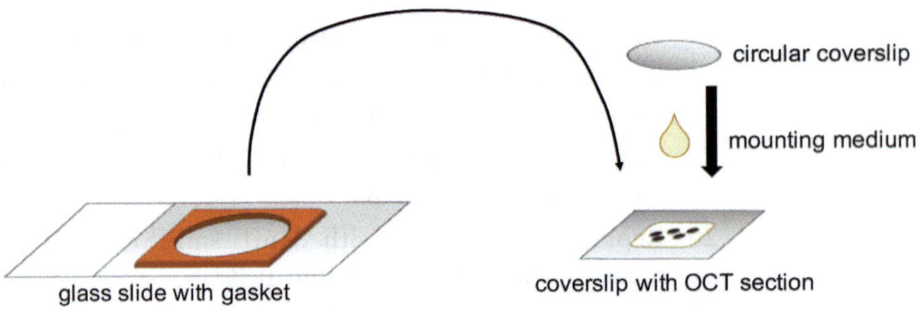

Fig. 4 Mounting of cryosectioned smFISH organoids. Remove the coverslip with forceps from the staining plate and dry the residual liquid with a Kimwipe. Carefully pipet 8 μL Prolong Gold to the sample area. Cover the sample area with a circular coverslip (12 mm) and gently push on it with a Kimwipe to remove excess mounting medium. Mount the coverslip sandwich on to a microscope slide

4 Notes

1. Caution: Formamide and paraformaldehyde are toxic substances. Handle them inside a fume hood, wear protective gloves and a lab coat and consult institutional standard operating procedures. All hybridization and wash buffer reagents are purchased from Ambion unless stated otherwise. Dextran sulfate (#D8906) and *E. coli* tRNA (#R1753) are obtained from Sigma, while vanadyl-ribonucleoside complex (#S1402S) from New England Biolabs. Gaskets are purchased from Grace Biolabs (#JTR20-0.5) and blue tissue marking dye is obtained from Trajan (#YBP-1163-5).
2. Organoids become very sticky after fixation. It is critical to use prelubricated 1.7 mL tubes for the whole protocol (Costar #3207). We additionally coat each tube and all pipette tips that will be in direct contact with organoids by briefly rinsing them with sterile FBS before use.
3. We recommend performing the faster and more efficient whole-mount staining protocol if a spinning-disk confocal microscope with a sensitive camera is available. We have obtained good single molecule signals from depths of ca 50 μm within the organoid with the whole-mount approach. If cells within deeper regions are to be probed one will obtain a better signal quality with the cryosectioning approach.
4. Fixed organoids can be stored in 70% ethanol at 4 °C without detectable decrease of smFISH quality for at least 3 days. We only use a part of the sedimented organoid pellet for each smFISH staining and keep a reserve in ethanol for future smFISH experiments.
5. Organoid cell membranes can be counterstained by adding a FITC-coupled antibody against E-Cadherin (BD Biosciences #612131) in a 1:100 dilution to the hybridization mix.
6. The organoids can be easily crushed. Hence it is important to place the circular coverslip on the organoid drop in a gentle manner and to let it sink by gravity.
7. The samples can be imaged on a spinning-disk confocal microscope by using oil-immersion objectives with a high numerical aperture (1.3 or greater). The faint signal requires a sensitive cooled CMOS or CCD camera and long exposure times (we often use 1–3 s). For confocal smFISH imaging we use a Nikon Ti-E body with Yokogawa W1 Spinning Disk and a Photometrics Prime 95B back illuminated SCMOS camera or a Hamamatsu ImagEM X2-1K EM-CCD camera and the following solid state diode lasers: 405 nm (120 mW), 488 nm (100 mW), 561 nm (100 mW), 640 nm (150 mW).

8. We usually acquire 12–20 z-stack images with a step size of 0.3 μm to account for all RNA molecules in a region of interest. SmFISH yields signal spots that originate from single RNA molecules. The resulting spots can be identified automatically and quantified by several published analysis pipelines in different software packages: Matlab [9, 12], ImageJ [13], or Imaris [14].
9. The protocol to coat coverslips with polylysine for smFISH has been described extensively in Lyubimova et al. [9] in section "equipment setup—Cover glass preparation". We check the coverslips on a stereo microscope after capturing OCT sections to ensure that organoids are present.
10. The samples can be imaged on a widefield inverted microscope by using oil-immersion objectives with a high numerical aperture (1.3 or greater). The faint single molecule signal requires a sensitive cooled CMOS or CCD camera and long exposure times (we often use 1–3 s). For widefield smFISH imaging we use a Nikon Ti2E body with a Photometrics Prime 95B back illuminated SCMOS camera and a Lumencore Spectra X light source.

Acknowledgments

We thank Gerald Schwank, Nina Frey, and Jan Reichmuth for generous donations of pancreatic and colon cancer organoid samples. We thank Efi Massasa and Shalev Itzkovitz for help with creating the organoid cryosectioning protocol. We thank Dario Zimmerli, Nikolaos Doumpas, and Matthias Moor for valuable comments on the manuscript. AEM is funded by the Swiss National Science Foundation grant PCEPP3_181249.

References

1. Femino AM, Fay FS, Fogarty K, Singer RH (1998) Visualization of single RNA transcripts in situ. Science 280:585–590. https://doi.org/10.1126/science.280.5363.585
2. Raj A, van den Bogaard P, Rifkin SA et al (2008) Imaging individual mRNA molecules using multiple singly labeled probes. Nat Methods 5:877–879. https://doi.org/10.1038/nmeth.1253
3. Itzkovitz S, Lyubimova A, Blat IC et al (2012) Single-molecule transcript counting of stem-cell markers in the mouse intestine. Nat Cell Biol 14:106–114. https://doi.org/10.1038/ncb2384
4. Bahar Halpern K, Shenhav R, Matcovitch-Natan O et al (2017) Single-cell spatial reconstruction reveals global division of labour in the mammalian liver. Nature 542:352–356. https://doi.org/10.1038/nature21065
5. Moor AE, Harnik Y, Ben-Moshe S et al (2018) Spatial reconstruction of single enterocytes uncovers broad zonation along the intestinal villus axis. Cell 175(4):1156–1167.e15. https://doi.org/10.1016/j.cell.2018.08.063
6. Moor AE, Itzkovitz S (2017) Spatial transcriptomics: paving the way for tissue-level systems biology. Curr Opin Biotechnol 46:126–133. https://doi.org/10.1016/j.copbio.2017.02.004

7. Clevers H (2016) Modeling development and disease with organoids. Cell 165:1586–1597. https://doi.org/10.1016/j.cell.2016.05.082
8. Li M, Izpisua Belmonte JC (2019) Organoids—preclinical models of human disease. N Engl J Med 380:569–579. https://doi.org/10.1056/NEJMra1806175
9. Lyubimova A, Itzkovitz S, Junker JP et al (2013) Single-molecule mRNA detection and counting in mammalian tissue. Nat Protoc 8:1743–1758. https://doi.org/10.1038/nprot.2013.109
10. Moor AE, Golan M, Massasa EE et al (2017) Global mRNA polarization regulates translation efficiency in the intestinal epithelium. Science 357:1299–1303. https://doi.org/10.1126/science.aan2399
11. Sato T, Clevers H (2013) Primary mouse small intestinal epithelial cell cultures. Methods Mol Biol 945:319–328. https://doi.org/10.1007/978-1-62703-125-7_19
12. Bahar Halpern K, Itzkovitz S (2016) Single molecule approaches for quantifying transcription and degradation rates in intact mammalian tissues. Methods 98:134–142. https://doi.org/10.1016/j.ymeth.2015.11.015
13. Wang S (2018) Single molecule RNA fish (smFISH) in whole-mount mouse embryonic organs. Curr Protoc Cell Biol 83(1):e79. https://doi.org/10.1002/cpcb.79
14. Yang L, Titlow J, Ennis D et al (2017) Single molecule fluorescence in situ hybridisation for quantitating post-transcriptional regulation in Drosophila brains. Methods 126:166–176. https://doi.org/10.1016/j.ymeth.2017.06.025
15. Manja Omerzu, Nicola Fenderico, Buys de Barbanson, Joep Sprangers, Jeroen de Ridder, Madelon M. Maurice, (2019) Three-dimensional analysis of single molecule FISH in human colon organoids. Biology Open 8 (8):bio042812

Chapter 16

Specific Gene Expression in Lgr5$^+$ Stem Cells by Using Cre-Lox Recombination

Pierre Dessen, Joerg Huelsken, and Paloma Ordóñez-Morán

Abstract

Intestinal stem cells are responsible for tissue renewal. The study of stem cell properties has become a major challenge in the field. We describe here a method based on Cre recombinase inducible lentivirus vectors that permits delivery of transgenes, either for overexpression or knockdown, in primary stem cells that can be cultured in an 3D intestinal organoid system. This method is an excellent approach for genetic manipulation and can complement in vivo transgenic experiments.

Key words Intestine, Stem cells, Lgr5, Organoid culture, Cre recombinase, Lentivirus

1 Introduction

In the intestine, the undifferentiated stem cells can self-renew and give rise to more specialized cells. These crypt base columnar stem cells (CBC) can constantly regenerate the epithelium. These cells are not quiescent but divide every 24 h and specifically express the leucine-rich-repeat containing G-protein-coupled receptor 5 (Lgr5) [1]. Interestingly, single sorted Lgr5$^+$ stem cells are sufficient to give rise to organoids in 3D culture. This culture method maintains basic crypt–villus physiology and permits long-term intestinal epithelial expansion sustained by several growth factors regulating mainly Wnt, EGF, and BMP signaling.

The Clevers lab generated a heterozygous Lgr5-EGFP-IRES-creERT2 "knock-in" mouse model. These mice harbor an allele that both abolishes *Lgr5* gene function and expresses EGFP and the CreERT2 fusion protein. When these mice are bred with mice containing a *loxP*-flanked sequence of interest, tamoxifen-inducible, Cre-mediated recombination will result in deletion of the floxed sequences in the *Lgr5*-expressing cells of the offspring [1]. We designed a similar approach for in vitro assays. Our protocol can be performed in a short time frame instead of generating

Paloma Ordóñez-Morán (ed.), *Intestinal Stem Cells: Methods and Protocols*, Methods in Molecular Biology, vol. 2171, https://doi.org/10.1007/978-1-0716-0747-3_16,

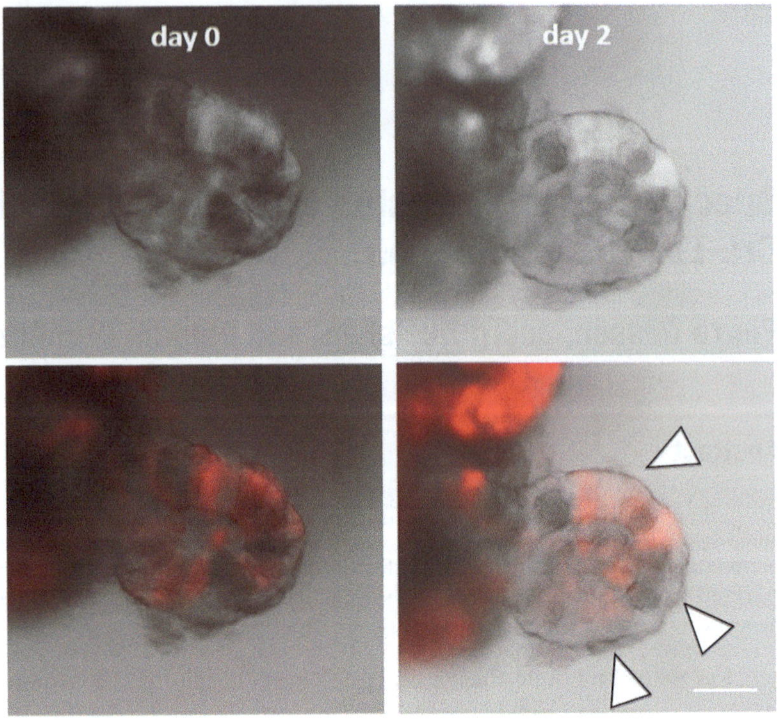

Fig. 1 Treatment with 4-OHT to lentiviral infected mouse intestinal organoids (day 0, left and day 2, right). Upper part, shows phase contrast images. Lower part, shows phase contrast and fluorescence (Turbo-RFP, in red) images. White arrows show the cre activation. Bar, 20 μm

time-consuming tissue-specific mouse models. We used a lentivirus vector expressing RFP, which is flanked by *loxP*-sequences, followed by the gene of interest. Only upon tamoxifen-induced, cre-mediated excision of the RFP cassette does the gene of interest become expressed (Fig. 1). Here, we describe the protocol that allows stem cell-specific gene expression by in vitro technology in organoids which can be used to improve our understanding of stem cell behavior and its potential role in intestinal development, homeostasis, damage, or tumorigenesis in the next years.

2 Materials

2.1 Plasmids

1. Cre-inducible lentiviral vector to generate pSFFV-*loxP*-TurboRFP-*loxP*-(cDNA of gene)-WPRE. The full sequence of this synthetic construct has been uploaded to NCBI (ID: 2282362). The vector contains a gateway cassette, so anyone can insert his cDNA of choice easily by using the Gateway Cloning System [2, 3]. The background vector SIV-GAE-

SFFV was a kind gift from Didier Negre, École NS de Lyon, France.

2. For the production of lentiviral particles:
   (a) Simian immunodeficiency virus (SIV) packaging construct.
   (b) Vesicular stomatitis virus G glycoprotein (VSV-G) envelope vector.

2.2 Lentiviral Production and Virus Concentration

1. HEK293T cells.
2. Cell culture plates: 15 cm and 24-well.
3. 293T medium: DMEM + GlutaMax, 100× penicilin–streptomycin, 10% fetal bovine serum.
4. $CaCl_2$: 2.5 M in bidistilled water.
5. dH_2O.
6. TE 0.1×: Tris 1 mM, EDTA 0.1 mM pH 8.8.
7. HBS 2×: 280 mM NaCl, 100 mM Hepes, 1.5 mM Na_2HPO_4, $7.11 \leq pH \leq 7.13$.
8. 50 mL Falcon tubes.
9. Eppendorfs.
10. Ultracentrifuge.
11. Plastic Tubes for the ultracentrifuge.
12. Adaptors for the ultracentrifuge.
13. Centrifuge.
14. 0.22 μm strainer.
15. 50 mL syringes.

2.3 Organoid Culture, Lentiviral Infection, and Treatment for Cre-Induced Gene Expression

1. 4-hydroxy-Tamoxifen (4-OHT, stock: 50 μM).
2. Fluorescence microscope.
3. Mouse organoids.
4. Matrigel.
5. Trypsin–EDTA 0.05%.
6. 0.1 mM EDTA pH 8.
7. Organoid media: Advanced Media DMEM/F12 supplemented with penicillin, streptomycin, L-glutamine, *N*-acetylcysteine, 100 ng/mL mNoggin-His, 250 ng/mL mR-Spondin1-Fc, 50 ng/mL Epidermal growth factor (EGF), 100× N_2 and 50× B27 supplement. The R-spondin1-Fc and Noggin-His proteins were produced as described in the Chapter 10 part of this same book "Intestinal Stem Cells."

8. Medium from L cells estably expressing Wnt3a (cells that are commercially available).
9. Y-27632 (stock: 10 mM).
10. Optional: CHIR99021 (stock: 20 mM).

3 Methods

3.1 Cloning of Plasmids

1. The full-length cDNA of your gene of interest can be purchased or amplified by PCR.
2. Your cDNA should be cloned into an entry clone (pENTR1a vector: attL1-cDNA-attL2).
3. The SIV-GAE-SFFV vector has been modified to get the final sequence which is uploaded at NCBI ID:2282362 (pRRL or destination vector with attR cassette).
4. Perform the Gateway LR reaction to generate the expression vector by using the LR clonase mix, the pENTR1a and pRRL vector. You will obtain the pSFFV-*loxP*-TurboRFP-*loxP*-gene vector (*see* **Note 1**).

3.2 Lentiviral Production

3.2.1 293T Plating

1. 293T are maintained in 293T medium.
2. Plate five 15 cm plates for one production (2.5×10^6 cells/15 cm plate). Cells are plated the day before transfection.

3.2.2 293T Transfection

1. Ideally cells should be 60–80% confluent.
2. Prepare the following transfection mix in 50 mL falcon tubes for 5 dishes (15 cm plates):
   112.5 μg vector plasmid (pSFFV-*loxP*-TurboRFP-*loxP*-gene).
   73 μg SIV packaging construct.
   39.5 μg envelop plasmid (VSV-G envelope).
3. Then add slowly in this order
   3.3 mL of TE 0.1×.
   1.6 mL dH_2O.
   706 μL $CaCl_2$ 2 M.
4. Mix.
5. Add 5.7 mL of HBS 2×, dropwise under agitation by vortexing.
6. Wait 10 min (no more than 30 min) at RT.
7. Add dropwise 2.25 mL/plate of the precipitate and mix.

3.2.3 Lentiviruses Collection

1. Change the medium the day after.
2. Check for transfection efficiency under microscope as fluorescent reporter is encoded in the vector plasmid (*see* **Note 2**).
3. Collect supernatant for the first time 24 h after medium replacement.
4. Harvest supernatant 2 times, every 12 h. Keep it at 4 °C during the collecting period.
5. Pool the collected supernatants, centrifuge 5 min at 200 × *g* to remove cell debris and filtrate on 0.45 μm membranes (*see* **Note 3**).
6. Concentrate virus particles by ultracentrifugation at 47,000 × *g* for 2 h at 4 °C in a swinging rotor.
7. Discard supernatant and resuspend pellet in 500 μL PBS 1× total if possible (*see* **Note 4**).
8. Aliquots should be stored at −80 °C.
9. Perform lentiviral titration in 293T cells before doing the next steps of the protocol (*see* **Note 5**).

3.3 Organoid Culture and Transduction of Intestinal Cells

Organoid cultures are established from total intestinal crypt preparations (described in several chapters of this book "Intestinal Stem Cells" for example Chapter 11). The lentiviral transduction can be done on organoids already established in culture and alternatively, on fresh crypts isolated the same day (*see* **Note 6**).

3.3.1 Transduction of Organoid Cultures from Lgr5-EGFP-IRES-creERT2 Mice

1. Remove the Matrigel by mechanical dissociation with a 1 mL tip and collect everything with 1 mL pipette in eppendorfs (organoids established 3 weeks before).
2. Centrifuge the samples (200 × *g*, 5 min, RT) and resuspend the pellet in 1 mL of Advanced medium.
3. Centrifuge the samples (200 × *g*, 5 min, RT) and resuspend the pellet in 1 mL of PBS 1×.
4. Centrifuge the samples (200 × *g*, 5 min, RT) and resuspend the pellet in 100 μL of trypsin-EDTA or 15 min PBS/EDTA (2 mM).
5. Incubate for 2 min (Trypsin-EDTA) in a water bath at 37 °C or 10 min (PBS/EDTA) at 37 °C, 5% CO_2.
6. Resuspend in 50 mL Advanced medium + FBS and centrifuge at 200 × *g*, 5 min at RT.
7. Discard the supernatant and resuspend the pellet in L-Wnt3a medium supplemented with 2× growth factors (*see* **Note 7**). Add 10 μM Y-27632 to prevent anoikis.
8. Disperse the cells by pipetting 5 times with a 1 mL pipette.

9. Combine the cells and the viruses (50 μL volume maximum) and resuspend well (*see* **Note 8**).
10. Centrifuge for 1 h at 200 × *g* at RT in a 50 mL Falcon tube.
11. Incubate the Falcon tube at 37 °C, 5% CO_2 for 5 h (*see* **Note 9**).
12. Remove the medium and resuspend the crypt cells in cold Matrigel.
13. Plate the cells in a 24-well and incubate 10 min at 37 °C, 5% CO_2 to solidify Matrigel.
14. Add 400 μL of organoid media into 24-well plates. Organoid media has to be changed every other day.

3.3.2 Cre Activity Induction and Validation

1. Two days after, check the fluorescence reporter in a fluorescence microscope (*see* **Note 10**).
2. For cre activity induction, organoids cultures are treated with 50 nM 4-*hydroxy*-tamoxifen (4-OHT). Perform the treatment with 4-OHT to express (or knockdown) the gene of interest (Fig. 1).
3. Once you express (or knockdown) your cDNA of interest, validate this result by qRT-PCR.

4 Notes

1. We detect a better efficiency of transduction by using the SFFV promoter but CMV and EF-1 promoters also work in these cells.
2. You can generate a lentiviral vector driving TurboRFP expression which enables identification of positive cells by flow cytometry. In case you do not use a fluorescence reporter in your construct, it is better to perform the following control: transfect an EGFP/TurboRFP expressing vector to an extra plate with the same number of 293T cells.
3. The cleared supernatants can be kept at 4 °C for 3–4 days or can be stored at −80 °C for long-term. In this protocol, we do not recommend using supernatants directly to transduce the cells because you will require a high number of particles per volume.
4. For aliquots use tube with screw cap and non-round shape bottom to avoid contamination later.
5. The most frequent cause of a poor production is over- or underconfluence of 293T cells during transfection or using 293T cells with a high number of passages. The titration can also be used to check your vector for cell transduction as

certainly the size of the target gene will have an impact on titer yield. If the vector is big, the transduction will be less efficient. Once the virus has been titered, it would be to the user's benefit to determine the appropriate amount of virus needed to obtain acceptable infections.

6. When the infection is done in freshly isolated crypts, you would need to consider that total number of cells surviving will be lower. During the incubation time, you would need to add CHIR99021 at 2 μM to activate Wnt signaling.
7. We normally use 500 μL, but it depends on the number of cells that you aim to infect and the L-Wnt3a medium preparation.
8. The number of viruses is very variable. We recommend testing your own preparations each time you are going to use a new batch with a different "Multiplicity of Infection" (MOI). Only by doing these experiments you will really be able to determine the final volumes and efficiency in the intestinal cells.
9. This incubation time is also variable. If you are using organoids that have been cultured for more than 2 months or cancer organoids which are more resistant than normal cells, then the cells can be incubated overnight with the lentiviruses. Even if total cell survival is lower, the efficiency of infection will be higher.
10. Alternatively, the vector can be generated with an antibiotic resistance as puromycin instead of a fluorescence reporter, so you can treat the organoids and select the positive clones.

Acknowledgments

We would like to thank Didier Negre (École NS de Lyon, France) for the background vector SIV-GAE-SFFV. P.O-M was supported by EMBO and the University of Nottingham, UK.

References

1. Barker N, van Es JH, Kuipers J, Kujala P, van den Born M, Cozijnsen M et al (2007) Identification of stem cells in small intestine and colon by marker gene Lgr5. Nature 449 (7165):1003–1007
2. Ordóñez-Morán P, Dafflon C, Imajo M, Nishida E, Huelsken J (2015) HOXA5 counteracts stem cell traits by inhibiting Wnt signaling in colorectal cancer. Cancer Cell 28 (6):815–829
3. Chiacchiera F, Rossi A, Jammula S, Piunti A, Scelfo A, Ordóñez-Morán P et al (2016) Polycomb complex PRC1 preserves intestinal stem cell identity by sustaining Wnt/β-catenin transcriptional activity. Cell Stem Cell 18(1):91–103

Chapter 17

Generating and Utilizing Murine Cas9-Expressing Intestinal Organoids for Large-Scale Knockout Genetic Screening

Hossein Kashfi, Nicholas Jinks, and Abdolrahman S. Nateri

Abstract

Organoid culture faithfully reproduces the in vivo characteristics of the intestinal/colon epithelium and elucidates molecular mechanisms underlying the regulation of stem cell compartment that, if altered, may lead tumorigenesis. CRISPR-Cas9 based editing technology has provided promising opportunities for targeted loss-of-function mutations at chosen sites in the genome of eukaryotes. Herein, we demonstrate a CRISPR/Cas9-mediated mutagenesis-based screening method using murine intestinal organoids by investigating the phenotypical morphology of Cas9-expressing murine intestinal organoids. Murine intestinal crypts can be isolated and seeded into Matrigel and grown into stable organoid lines. Organoids subsequently transduced and selected to generate Cas9 expressing organoids. These organoids can be further transduced with the second lentiviruses expressing guide RNA (gRNA) (s) and screened for 8–10 days using bright-field and fluorescent microscopy to determine possible morphological or phenotypical abnormalities. Via phenotypical screening analysis, the candidate knockouts can be selected based on differential abnormal growth pattern vs their untransduced or lenti-GFP transduced controls. Further assessment of these knockout organoids can be done via phalloidin and propidium iodide (PI) staining, proliferation assay and qRT-PCR and also biochemical analysis. This CRISPR/Cas9 organoid mutagenesis-based screening method provides a reliable and rapid approach for investigating large numbers of genes with unknown/poorly identified biological functions. Knockout intestinal organoids can be associated with the key biological function of the gene(s) in development, homeostasis, disease progression, tumorigenesis, and drug screening, thereby reducing and potentially replacing animal models.

Key words Phenotypic screening, Intestinal organoid, CRISPR-Cas9, gRNA, Lentivirus transduction, 3Rs (replacement, reduction and refinement)

1 Introduction

Colorectal cancer (CRC) is the third most common cancer worldwide with approximately 1.1 million new diagnoses (6.1% of all cancers) in 2018 [1]. Regrettably, variations in treatment response and subsequent survival rates may be caused by significant heterogeneity of CRC tumors [2]. This variance makes the modeling of

Hossein Kashfi and Nicholas Jinks contributed equally with all other contributors.

Paloma Ordóñez-Morán (ed.), *Intestinal Stem Cells: Methods and Protocols*, Methods in Molecular Biology, vol. 2171, https://doi.org/10.1007/978-1-0716-0747-3_17, © Springer Science+Business Media, LLC, part of Springer Nature 2020

CRC in the laboratory challenging, with traditional 2D immortalized cell models limited by their inherent simplicity and genetic drift, reducing reproducibility and validity across functional genetics and drug response studies, while full in vivo models are vastly expensive and resource intensive, making them impractical for the majority of research centers. Therefore, studying of CRC and other cancers requires novel methods to be developed that facilitate more accurate investigations without the resource requirements of full in vivo modeling.

One of the most recent and promising modeling systems to be introduced is organoids: a 3D (three-dimensional), multicellular culture that mimics the phenotype of their original tissue and are compatible with many standard laboratory protocols. Stem cells from the organ in question are cultured with appropriate growth factors resulting in a small "organ in a dish" that can be maintained and expanded almost indefinitely. Organoids faithfully reproduce functional genetic or pharmacological changes visible in their originating organ, and have been cultured from the brain [3], retina [4], stomach [5], lung [6], and small intestine [7, 8], among others from both human patients and mice, providing a crucial "middle ground" alternative between immortal cell culture and full in vivo animal models for functional generic analysis, high-throughput drug testing [9], biobanking [10], tumor microenvironment [11], disease modeling [12, 13], and targeted gene editing [14] via CRISPR/Cas9 technology, validating data between experimental models. When generated from both adjacent normal and matched tumor samples of an individual patient, epithelial organoids can provide an excellent tool for researchers to investigate a wide range of cellular and molecular events directly related to individual patient care.

The CRISPR/Cas9 gene editing system provides a robust tool for generating genetically modified in vitro models. With recent systems capable of both "knock-in" and "knock-out" of chosen genes, CRISPR/Cas9 is used for generating in vitro cellular models for study. CRISPR editing has been successfully applied to cultured human organoids [15], demonstrating both an ideal platform and tool for assessing functioning of individual genes in multiple tissue types without the limitations and ethical implications of generating full in vivo models.

Matano et al. introduced CRISPR/Cas9-mediated sequential mutations in organoids derived from the normal human intestinal epithelium [16]. They demonstrated that engineered organoids harboring mutations at driver genes including APC, SMAD4, and TP53 KRAS and/or PIK3CA grew independent of niche factors similar to the tumor counterparts and developed tumors upon transplantation kidney subcapsule in mice. These results highlighted the significance of morphological alterations in organoids. Inconsistent with the previous study, Drost and colleagues also utilized the CRISPR/Cas9 to engineer intestinal organoids mutated for APC, KRAS, TP53, and SMAD4. They revealed that simultaneous inactivation of APC and P53 drive the extensive aneuploidy, the main feature of tumor progression [17]. In another study, CRISPR/Cas9 mediated engineering of intestinal organoids in cystic fibrosis patients was successfully applied to repair the deletion at position 508 of cystic fibrosis transmembrane conductor receptor (CFTR) [18]. These studies have used CRISPR/Cas9 expression in organoids, but protein expression and success rate have varied considerably. Given the fragility of organoid cultures, electroporation and other invasive methods are unlikely to produce sufficient qualities for larger projects.

Herein, we present a method for isolating and culturing mouse intestinal crypts and culturing organoids, establishing Cas9 protein expression via lentiviral transduction, and subsequently using viral transduction of targeted Cas9 plasmids to generate genetic knockout organoid lines suitable for functional exploration via phenotypical or biochemical analysis. We propose that this CRISPR/Cas9-mediated mutagenesis-based screening method can provide a promising approach which facilitates the rapid and reliable identification and selection of new genes play role in development, and disease.

2 Materials

2.1 Mouse Intestinal Organoid Isolation

1. HA-Rspo1-Fc cells: received from Prof. Hans Clevers laboratory.
2. Mice: C57BL/6J mice, approximately 4–6 weeks old were used. Mice were housed and bred in the transgenic animal facility of the Biomedical Service Unit at the University of Nottingham.
3. Fresh organoid medium: basal Advanced DMEM/F-12 medium, 2 mM L-glutamine, 100 U/mL Penicillin/Streptomycin, and 10 mM HEPES was supplemented with N2 supplement (1×), B27 supplement (1×), 1 mM *N*-acetylcysteine to stimulate cell proliferation, 50 ng/mL murine recombinant epidermal growth factor (mEGF) to activate EGF signaling pathway, 1 μg/mL R-spondin 1 (in-house) as Wnt agonist and 100 ng/mL Noggin (in-house) to inhibit BMP pathway.
4. EDTA (UltraPure, 0.5 M).
5. PBS (1×).
6. Matrigel.

2.2 Lentivirus Production, Transduction and Generation of Stable Organoid Lines, Protein and RNA Extraction, and Phalloidin Staining

1. Human HEK-293 (viral production cells).
2. Plasmids: pMD2.G and pCMVR8.74.
3. DMEM.
4. RPMI.
5. FBS.
6. L-Glutamine.
7. Penicillin/Streptomycin (P/S).
8. Opti-MEM.
9. Polyethylenimine (PEI).
10. Polybrene reagent.
11. Fluorescent microscopy system.
12. Phalloidin.
13. DAPI mounting medium.
14. 8 μg/mL polybrene.
15. Paraformaldehyde.
16. Triton X-100.
17. Virus sucrose solution 50 mL in ultrapure water: 10% sucrose, 50 mM Tris–HCl pH 7.4, 100 mM NaCl, 0.5 mM EDTA.
18. RIPA buffer [radioimmunoprecipitation assay buffer (RIPA buffer; 150 mM sodium chloride, 1% NP-40, and 1% sodium dodecyl sulfate) containing 1× protease inhibitor cocktail].
19. TRIzol (TRI reagent).
20. RT-PCR, immunofluorescence (IF), and western blots reagents; see published protocols [19, 20].

3 Methods

The brief step-by-step flowchart and a diagram showing generation of knockout intestinal organoids using the lentivirus-based system for Cas9/gRNA expression and subsequent analysis are provided in Figs. 1 and 2.

3.1 Mouse Intestinal Organoid Isolation

1. Sacrifice the mouse and excise the whole intestine from the abdominal cavity.
2. Clean the specimen carefully with ice-cold PBS to eliminate external connective tissues and internal feces (*see* **Note 1**).
3. Cut a section from duodenum, jejunum, and ileum, and open longitudinally.

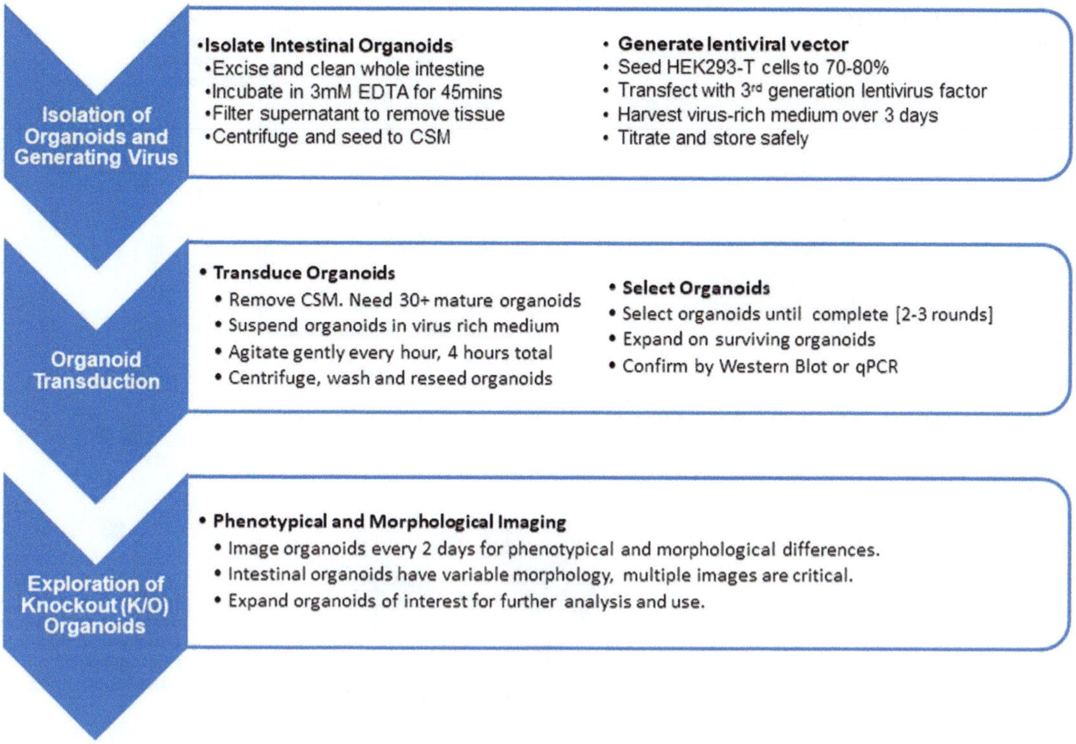

Fig. 1 Brief step-by-step flowchart of generating knockout intestinal organoids using the lentiviral transduction methodology

4. To detach the villi, subdivide every piece into small fragments of approximately 1 mm and wash with cold PBS several times, into a 50 mL falcon tube.
5. Once the PBS become clear, incubate the intestinal pieces with 3 mM EDTA in PBS for 45 min at 4 °C to separate the crypts by chelation.
6. Transfer the fragments to a falcon containing 10% FBS/PBS and agitate frequently and subsequently filter through a 70 μm strainer.
7. Following the filtration, collect the crypt fragments on top of the strainer and transfer to a new falcon containing 10% FBS/PBS.
8. Agitate and pass through the filter (this process repeated 3–4 times).
9. Evaluate the filtered medium by using an inverted microscope in order to choose the best fraction in term of purity and crypt concentration [8, 21].
10. Centrifuge crypts at 300 × *g* for 5 min. Remove the supernatant carefully after the centrifugation step and resuspend crypts in ice-cold Matrigel.

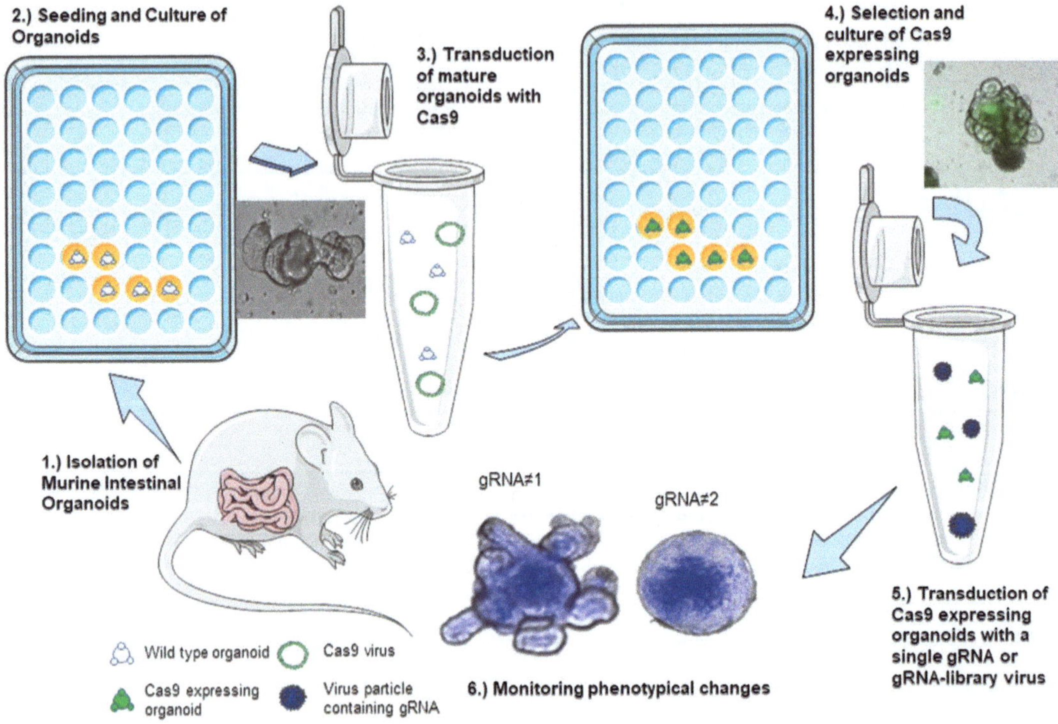

Fig. 2 Summary diagram of generating knockout intestinal organoids using the lentiviral transduction methodology. The art is from the Servier Medical Art Archive (https:/smart.servier.com)

11. Add 25 μL of crypts-containing Matrigel in prewarmed 48-well plate. Next, transfer the plate to a 37 °C, 5% CO_2 incubator for approximately 10 min to allow Matrigel to solidify.
12. After 5–10 min in 37 °C incubator, distribute 300 μL of fresh organoid medium into each well to cover the Matrigel drop. Culture the crypts into 37 °C, 5% CO_2 incubator and replace the organoid medium every 48 h. Monitor regularly organoid growth by inverted BF microscopy (*see* **Note 1**).

3.2 Lentivirus Production

1. Seed HEK293-T cells in a total of 10 mL DMEM per 75 cm^2 flask. Keep the density of cells around 70–80% at the day of transfection (no greater than 90%) (*see* **Note 2**).
2. Agitate plate crossways to spread cells evenly. Incubate overnight at 37 °C.

Day 1

3. Calculate volumes required for plasmid transfection and PEI (final volume per plate 15 μg LV construct, 5 μg pMDG2 packaging, 10 μg pCMV R8.74 packaging plasmid, 1.5 mL Opti-MEM per 75 mL flask) (*see* **Note 3**).

4. Add 1.5 mL Opti-MEM to a 15 mL tube per each flask to be transfected.
5. Add the required amount of packaging and transgene plasmids (no. 4) to the tube per each flask to be transfected.
6. Prepare the PEI concentration (3–4 μL/1 μg plasmid DNA).
7. Incubate for 5 min at room temperature.
8. Add the PEI to each tube containing the plasmid DNA (*see* **Note 4**).
9. Incubate the mixture for 20 min at room temperature.
10. Add the Opti-MEM/plasmid (packaging plasmid and PEI) gently to flask, leave the flasks for 6–9 h. Gently agitate every 2 h for best results.
11. Check cells after 6 h, and then change with complete DMEM medium. Incubate overnight 37 °C.

Day 2–3

12. If gRNA has fluorescent reporter gene, check transduction efficiency. Transfer supernatant from flasks into 50 mL falcon tube and filter through 45 μm filter unit (*see* **Notes 5** and **6**).
13. Replace the medium with complete DMEM medium (3 flasks at a time) and incubate for a further 24 h.
14. Store virus-containing medium at 4° (cover tubes with parafilm. Please note that these samples are highly dangerous and should be securely stored).

Day 4

15. Repeat virus harvest to have 3 days of collection [approx. 60 mL of virus containing medium per plasmid]. Mark samples with day of collection. The titer usually declines with each day. Some detachment of cells may also occur.
16. Pipette into ultracentrifuge tubes containing 1 mL of virus sucrose solution (20%), ~7–8 mL viral supernatant per tube, ensure equal volumes to balance.
17. Ultracentrifuge at 17,664 × *g* (JA-12) for 4 h at 4 °C.
18. Discard supernatant by careful pipetting into beaker of Virkon. Carefully aspirate walls of tube to dry, without touching bottom/pellet (pellet may be invisible, and care must be taken not to disturb).
19. Optional: allow tube to air-dry under hood for 10–15 min to maximize supernatant removed.
20. Add 500 μL of organoid medium or complete ADMEM to each tube. Leave the tubes overnight at 4 °C.

21. Take dry ice. Transfer all virus into one tube, pipette several times before transfer to disaggregate the virus particles. Try to avoid bubbles as they reduce useful volume.
22. Aliquot 50–60 μL into Eppendorfs. Transport on dry ice and store at −80 °C. Try not to freeze-thaw virus aliquots several times as transduction efficiency is reduced with each cycle (*see* **Note 3**).

3.3 Transduction and Generation of Stable Organoid Lines

1. Passage the organoids into sufficient number of wells prior LV transduction [typically 10 confluent organoids per well, 3–4 wells per transduction]. Culture the organoids in Wnt3a supplement to form hyper proliferative structures.
2. Break the basement matrix containing the mature organoids with 1 mL pipette in chilled PBS.
3. Transfer the fragments into a 15 mL tube.
4. Centrifuge for 3 min at 300 × *g*.
5. Aspirate the PBS and the basement matrix residue as much as it is possible and keep the pellet. Pellet may be completely invisible.
6. Repeat **steps 2–6** twice more.
7. Resuspend the pellet with PBS (300 μL) and break down the fragments with 200 mL pipette (20–30×).
8. Centrifuge, 3 min for 300 × *g*.
9. Make up medium containing polybrene to working concentration 16 μg/mL (final 8 μg/mL).
10. Resuspend the pellet with the high-titer LV and polybrene and leave the mixture for 4 h in 37-degree incubator (*see* **Note 5**).
11. Agitate the virus containing the organoid fragments every 30 min, using the 200 μL pipette to maximize the transduction efficacy.
12. After 4 h, centrifuge the transduced organoids for 3 min for 300 × *g*.
13. Discard the supernatant and add required amount of basement matrix and mix the pellet gently using a 200 μL pipette.
14. Seed the organoids (25 μL) in each well of 48-well plate.
15. Incubate the organoids in incubator for 5 min and add the complete organoid medium as described.
16. Start the selection process for the generation of the stable organoid lines from 48 h posttransduction (G418: 200 μg/μ L, puromycin: 1 μg/μL).
17. Continue the selection process until the negative control (untransduced organoids are completely dead).

18. Passage the resistant (survived), stable (transduced) organoids and continue antibiotic selection.
19. After second round of the selection, expand the organoids in free-antibiotic medium (*see* **Notes 7–11**).
20. Freeze organoids. Transport on dry ice and store at −80 °C. The cryopreservation of organoids enables the freezing of knockout organoid lines postscreening (*see* **Note 12**).
21. Validation and exploration the selected organoid line via RT-PCR, WB, and Phalloidin staining (see Subheadings 3.4, 3.5, and 3.6).

3.4 Organoid Protein Extraction

1. Western blot validation requires sufficient number of organoids (at least 6 wells of 20–30 mature organoids/well).
2. When the organoids are grown after 6–9 days, follow the **steps 1–8** of transduction protocol.
3. Resuspend and incubate the organoids pellet in 70–80 μL of RIPA buffer for 20 min in cold room, while rocking.
4. Lyse and homogenize the organoids using a syringe (microlance) occasionally within the incubation time.
5. The lysate can be used directly for the western blot analysis.

3.5 Organoid RNA Extraction

1. Extract total RNA from organoids using TRI reagent according to the manufacturer's instructions.
2. To remove the Matrigel, incubate organoids with Matrigel cell recovery solution for 3 h at 4 °C.
3. Wash with PBS twice.
4. Centrifuge into microcentrifuge tubes.
5. Homogenize organoid pellet manually with 500 μL of TRI reagent and incubate for 5 min at RT. To maximize the homogenizing, the organoids can be agitated using a 200 μL pipette every 1 min.
6. Add 100 μL of chloroform per 500 μL of TRI reagent to the samples and incubate for 3 min at RT. Mix vigorously; then, centrifuge the samples at 12,000 × *g* at 4 °C for 15 min to separate RNA (aqueous phase).
7. Transfer the upper phase containing RNA to a fresh tube and mix with 250 μL of isopropanol per 500 μL of TRIzol. After 10 min at RT, precipitate RNA at 12,000 × *g* for 10 min at 4 °C (*see* **Note 13**).
8. Wash the pellet two times with cold 75% ethanol by centrifugation at 7400 × *g* for 8 min at 4 °C, air-dried and resuspend with 20 μL of DNase/RNase free water.
9. Measure RNA concentration by NanoDrop and store the samples at −80 °C.

3.6 Phalloidin Staining of Organoid Culture

1. Select 5–6 wells of confluent organoids and add 500 μL of 4% PFA for fixation. Incubate the organoids in 4 °C for 1 h.
2. Discard the PFA gently, collect the organoids in a tube (15 mL falcon) and wash the organoids with PBS. Then centrifuge for 3 min, 300 × *g*.
3. Repeat this step twice.
4. Discard the PBS and permeabilize the organoids with 500 μL of 0.5% Triton X-100 for 30 min at RT.
5. Wash twice with PBS, 5 min.
6. Add 800 μL in each tube with Phalloidin 1:500 diluted, keep in darkness and incubate for 40 min at RT.
7. Wash twice with PBS, 10 min.
8. Discard the PBS and lay out the organoids in the microscopy slide.
9. Mounting with DAPI.

4 Notes

1. Intestinal organoids provide an excellent in vitro model to investigate the molecular mechanisms of CRC. Our experience indicates that initial isolation of can produce between 8 and 12 wells of 10–20 organoids per well. Initial isolation will also produce a significant amount of extraneous debris that can be cleared during the first and second postisolation subcultures. Not all organoids will successfully proliferate/differentiate and will die during this initial stage. Each well of 10–12 organoids can generally be subcultured 1:3 with minimal loss. Overseeding organoids will cause starvation and loss during the growth period, while underseeding can lead to reduced subculture efficiency.
2. The HEK293T cells should be healthy, of low passage number and in the exponential phase of growth. Make sure they are passaged regularly and do not allow to reach confluence.
3. Do not add antibiotics to virus harvesting medium and try not to freeze-thaw virus aliquots several times as transduction efficiency is reduced with each cycle.
4. For the transfection, polyfectamine and other related reagents can be used instead.
5. We found viral transduction efficiency was very poor when initial plasmid transfection was lower than 50%. Reducing viral dilution posttitration may improve efficiency if gRNA has low transfection rate.

6. Please ensure that your Cas9 plasmid selection marker is different to that in your prospective plasmids.
7. Organoid morphology is not exact and large number of images is necessary to quantify any potential changes based on knockout.
8. In the present study, monitoring the phenotypical alterations of knockout organoids vs. controls enabled the analysis and selection of candidate genes mediated by the abnormal morphology. Of note, not all the proteins of interest are expressed in intestinal organoids and some may not show a significant phenotypical change due to nonessential functions or functional compensation by their paralogues.
9. As far as we are aware, this is the first study to utilize organoids for large-scale screening of a targeted group of genes using stable and abundant Cas9 expressing organoids. While several studies have been published on the uses of organoids in cancer studies [22, 23] and drug resistance [24], the use of engineered organoids has been limited up to this point.
10. Organoids represent a new opportunity and hope for both biomedical research and personalized, regenerative medicine, particularly for the generation and transplant of patient-derived tissue that will prevent rejection and immune responses. However, current organoid methodology is still limited in both size and practicality, with the lack of vasculature limiting the maximum size of organoid models before necrosis occurs in the central cells due to lack of growth factors, and extracellular matrix materials such as Matrigel will break down if samples expand greatly. One recent paper by Grebenyuk and Ranga [25] discussed combining multiple organoid types in the same dish to potentially produce hybrid tissue samples that could grow too much greater sizes due to vascularization, possibly removing that limitation. While organoids have been demonstrated to self-assemble into ad hoc structures roughly analogous to their source anatomy [26], they are currently unsuitable to such delicate physical engineering.
11. With CRISPR/Cas9 producing accurate mutations and organoids faithfully representing their origin tissues, it is now possible to model how individual genes may affect multiple tissue types or explore how a specific tissue type is affected by knockout without the cost and ethical considerations of in vivo alternatives. This method demonstrates generating CRISPR/Cas9 organoids and subsequent knockout organoid lines, representing a significant step forward in reducing and replacing the use of animal models in research. While organoids will not completely replace animal models in research, they provide both a middle ground option and triage method for future research.

12. This provides the opportunity to rapidly expand organoid use across research groups, collaborations and bioscience by allowing collaborative researchers to share engineered organoids, and the production of an "organoid bank" in a similar way to standard cell lines.
13. Incubation of the organoid lysate in isopropanol step overnight can be done to achieve a higher yield of RNA.

Acknowledgments

This work was supported by the National Centre for the Replacement, Refinement & Reduction of Animals in Research [grant number NC/P001793/1] via a NC3Rs training grant to A.S.N.; and the University of Nottingham. We also appreciate the fantastic fundraising efforts of Alison Sims and her family in memory of Daz Sims to support the work in our laboratory.

References

1. Bray F, Ferlay J, Soerjomataram I, Siegel RL, Torre LA, Jemal A (2018) Global cancer statistics 2018: GLOBOCAN estimates of incidence and mortality worldwide for 36 cancers in 185 countries. CA Cancer J Clin 68 (6):394–424. https://doi.org/10.3322/caac.21492
2. Munro MJ, Wickremesekera SK, Peng L, Tan ST, Itinteang T (2018) Cancer stem cells in colorectal cancer: a review. J Clin Pathol 71 (2):110–116. https://doi.org/10.1136/jclinpath-2017-204739
3. Di Lullo E, Kriegstein AR (2017) The use of brain organoids to investigate neural development and disease. Nat Rev Neurosci 18:573. https://doi.org/10.1038/nrn.2017.107
4. Fligor CM, Langer KB, Sridhar A, Ren Y, Shields PK, Edler MC, Ohlemacher SK, Sluch VM, Zack DJ, Zhang C, Suter DM, Meyer JS (2018) Three-dimensional retinal organoids facilitate the investigation of retinal ganglion cell development, organization and neurite outgrowth from human pluripotent stem cells. Sci Rep 8(1):14520. https://doi.org/10.1038/s41598-018-32871-8
5. Seidlitz T, Merker SR, Rothe A, Zakrzewski F, von Neubeck C, Grützmann K, Sommer U, Schweitzer C, Schölch S, Uhlemann H, Gaebler A-M, Werner K, Krause M, Baretton GB, Welsch T, Koo B-K, Aust DE, Klink B, Weitz J, Stange DE (2019) Human gastric cancer modelling using organoids. Gut 68(2):207–217. https://doi.org/10.1136/gutjnl-2017-314549
6. Barkauskas CE, Chung M-I, Fioret B, Gao X, Katsura H, Hogan BLM (2017) Lung organoids: current uses and future promise. Development 144(6):986–997. https://doi.org/10.1242/dev.140103
7. Dow Lukas E, O'Rourke Kevin P, Simon J, Tschaharganeh Darjus F, van Es JH, Clevers H, Lowe Scott W (2015) Apc restoration promotes cellular differentiation and reestablishes crypt homeostasis in colorectal cancer. Cell 161(7):1539–1552. https://doi.org/10.1016/j.cell.2015.05.033
8. Sato T, Clevers H (2013) Growing self-organizing mini-guts from a single intestinal stem cell: mechanism and applications. Science 340(6137):1190–1194. https://doi.org/10.1126/science.1234852
9. Skardal A, Shupe T, Atala A (2016) Organoid-on-a-chip and body-on-a-chip systems for drug screening and disease modeling. Drug Discov Today 21(9):1399–1411. https://doi.org/10.1016/j.drudis.2016.07.003
10. Kashfi SMH, Almozyan S, Jinks N, Koo B-K, Nateri AS (2018) Morphological alterations of cultured human colorectal matched tumour and healthy organoids. Oncotarget 9 (12):10572–10584. https://doi.org/10.18632/oncotarget.24279
11. Neal JT, Li X, Zhu J, Giangarra V, Grzeskowiak CL, Ju J, Liu IH, Chiou S-H, Salahudeen AA, Smith AR, Deutsch BC, Liao L, Zemek AJ, Zhao F, Karlsson K, Schultz LM, Metzner TJ, Nadauld LD, Tseng Y-Y, Alkhairy S, Oh C, Keskula P, Mendoza-Villanueva D, De La

Vega FM, Kunz PL, Liao JC, Leppert JT, Sunwoo JB, Sabatti C, Boehm JS, Hahn WC, Zheng GXY, Davis MM, Kuo CJ (2018) Organoid modeling of the tumor immune microenvironment. Cell 175(7):1972–1988.e1916. https://doi.org/10.1016/j.cell.2018.11.021

12. Huang HL, Jiang Y, Wang YH, Chen T, He HJ, Liu T, Yang T, Yang LW, Chen J, Song ZQ, Yao W, Wu B, Liu G (2015) FBXO31 promotes cell proliferation, metastasis and invasion in lung cancer. Am J Cancer Res 5 (5):1814–1822

13. Crespo M, Vilar E, Tsai S-Y, Chang K, Amin S, Srinivasan T, Zhang T, Pipalia NH, Chen HJ, Witherspoon M, Gordillo M, Xiang JZ, Maxfield FR, Lipkin S, Evans T, Chen S (2017) Colonic organoids derived from human induced pluripotent stem cells for modeling colorectal cancer and drug testing. Nat Med 23:878. https://doi.org/10.1038/nm.4355. https://www.nature.com/articles/nm.4355#supplementary-information

14. Schwank G, Clevers H (2016) CRISPR/Cas9-mediated genome editing of mouse small intestinal organoids. Methods Mol Biol (Clifton, NJ) 1422:3–11. https://doi.org/10.1007/978-1-4939-3603-8_1

15. Fujii M, Clevers H, Sato T (2019) Modeling human digestive diseases with CRISPR-Cas9--modified organoids. Gastroenterology 156 (3):562–576. https://doi.org/10.1053/j.gastro.2018.11.048

16. Matano M, Date S, Shimokawa M, Takano A, Fujii M, Ohta Y, Watanabe T, Kanai T, Sato T (2015) Modeling colorectal cancer using CRISPR-Cas9-mediated engineering of human intestinal organoids. Nat Med 21 (3):256–262. https://doi.org/10.1038/nm.3802

17. Drost J, van Jaarsveld RH, Ponsioen B, Zimberlin C, van Boxtel R, Buijs A, Sachs N, Overmeer RM, Offerhaus GJ, Begthel H, Korving J, van de Wetering M, Schwank G, Logtenberg M, Cuppen E, Snippert HJ, Medema JP, Kops GJ, Clevers H (2015) Sequential cancer mutations in cultured human intestinal stem cells. Nature 521 (7550):43–47. https://doi.org/10.1038/nature14415

18. Schwank G, Koo BK, Sasselli V, Dekkers JF, Heo I, Demircan T, Sasaki N, Boymans S, Cuppen E, van der Ent CK, Nieuwenhuis EE, Beekman JM, Clevers H (2013) Functional repair of CFTR by CRISPR/Cas9 in intestinal stem cell organoids of cystic fibrosis patients. Cell Stem Cell 13(6):653–658. https://doi.org/10.1016/j.stem.2013.11.002

19. Muhammad BA, Almozyan S, Babaei-Jadidi R, Onyido EK, Saadeddin A, Kashfi SH, Spencer-Dene B, Ilyas M, Lourdusamy A, Behrens A, Nateri AS (2018) FLYWCH1, a novel suppressor of nuclear β-catenin, regulates migration and morphology in colorectal cancer. Mol Cancer Res 16(12):1977–1990. https://doi.org/10.1158/1541-7786.mcr-18-0262

20. Li N, Babaei-Jadidi R, Lorenzi F, Spencer-Dene B, Clarke P, Domingo E, Tulchinsky E, Vries RGJ, Kerr D, Pan Y, He Y, Bates DO, Tomlinson I, Clevers H, Nateri AS (2019) An FBXW7-ZEB2 axis links EMT and tumour microenvironment to promote colorectal cancer stem cells and chemoresistance. Oncogenesis 8(3):13. https://doi.org/10.1038/s41389-019-0125-3

21. Lorenzi F, Babaei-Jadidi R, Sheard J, Spencer-Dene B, Nateri AS (2016) Fbxw7-associated drug resistance is reversed by induction of terminal differentiation in murine intestinal organoid culture. Mol Ther Methods Clin Dev 3:16024. https://doi.org/10.1038/mtm.2016.24

22. Buczacki SJA, Popova S, Biggs E, Koukorava C, Buzzelli J, Vermeulen L, Hazelwood L, Francies H, Garnett MJ, Winton DJ (2018) Itraconazole targets cell cycle heterogeneity in colorectal cancer. J Exp Med 215(7):1891–1912. https://doi.org/10.1084/jem.20171385

23. Xu H, Jiao Y, Qin S, Zhao W, Chu Q, Wu K (2018) Organoid technology in disease modelling, drug development, personalized treatment and regeneration medicine. Exp Hematol Oncol 7:30. https://doi.org/10.1186/s40164-018-0122-9

24. Jabs J, Zickgraf FM, Park J, Wagner S, Jiang X, Jechow K, Kleinheinz K, Toprak UH, Schneider MA, Meister M, Spaich S, Sutterlin M, Schlesner M, Trumpp A, Sprick M, Eils R, Conrad C (2017) Screening drug effects in patient-derived cancer cells links organoid responses to genome alterations. Mol Syst Biol 13(11):955. https://doi.org/10.15252/msb.20177697

25. Grebenyuk S, Ranga A (2019) Engineering organoid vascularization. Front Bioeng Biotechnol 7:39. https://doi.org/10.3389/fbioe.2019.00039

26. Sachs N, Tsukamoto Y, Kujala P, Peters PJ, Clevers H (2017) Intestinal epithelial organoids fuse to form self-organizing tubes in floating collagen gels. Development 144 (6):1107–1112. https://doi.org/10.1242/dev.143933

Part IV

In Vivo Models

Chapter 18

Mouse Model for Sporadic Mutation of Target Alleles to Understand Tumor Initiation and Progression and Stem Cell Dynamics

Theresa N. Nguyen, Elise C. Manalo, Taryn E. Kawashima, and Jared M. Fischer

Abstract

Recent evidence has shown that many different tissues accumulate mutations even though the tissue is phenotypically normal. Therefore, generating mouse models for visualizing the tissue level effects that happen after oncogenic mutation in a single, isolated cell are critical for understanding tumor initiation and the role of competition in stem cell dynamics. Most mouse models have oncogenic mutations at the level of the entire mouse, the entire tissue, or all cells of a specific type in a tissue. However, these mouse models do not mimic the microenvironmental interactions that occur after an isolated cell acquires an oncogenic mutation because of the large number of mutant cells. We developed a mouse model for sporadic and isolated mutation of target alleles to better address the questions of sporadic cancer and stem cell competition. The following chapter describes methods for utilizing this mouse model and a few examples of the novel findings of using such a model.

Key words Apc, Intestine, Stem cell, Sporadic, Cancer, Mice

1 Introduction

The intestinal crypt contains a small number of stem cells at its base, which produce differentiated progeny that typically progress up the crypt, migrate up villi, and are ultimately sloughed from the end of the villus [1]. Recent studies have revealed that multiple stem cells occur per crypt, have a defined set of markers ($Lgr5^{high}$), are highly proliferative and undergo neutral drift [2–6]. In addition, studies indicate that the stem cell population is the cell of origin for intestinal cancer in mice [7]. Therefore, understanding stem cell dynamics before and after driver mutations is important for understanding the early stages of tumor initiation.

Colorectal cancer (CRC) affects ~150,000 people annually in the USA, both men and women equally, leading to ~50,000 deaths

Paloma Ordóñez-Morán (ed.), *Intestinal Stem Cells: Methods and Protocols*, Methods in Molecular Biology, vol. 2171, https://doi.org/10.1007/978-1-0716-0747-3_18,

per year (www.cancer.gov). Phenotypic changes in the colon are well-characterized in CRC: starting with aberrant crypt formation to benign adenomas to adenocarcinomas and finally to metastasis [8]. CRC is divided into sporadic and familial forms. Familial adenomatous polyposis (FAP) is an inherited CRC syndrome involving germline *Apc* mutation, accounting for less than 1% of all CRCs [9]. FAP is characterized by having hundreds to thousands of adenomas, with a few eventually progressing to adenocarcinomas. FAP patients inherit one mutated *Apc* allele. Somatic loss of the normal *Apc* allele is generally considered to be sufficient for adenoma formation; however, there is evidence that loss of *Apc* alone is not enough for sporadic tumorigenesis [10, 11]. In addition, there is evidence from the skin, esophagus, and blood that cancer causing mutations can occur in phenotypically normal cells [12–14].

To better understand the fates of normal and mutant intestinal stem cells, we developed the *Pms2*^{cre} mouse system that features an out-of-frame *cre* allele to stochastically recombine floxed target genes and a marker gene (e.g., β-gal or tdTomato) [15]. This system not only enables the monitoring of tumor formation following stochastic genetic manipulations but also facilitates the tracking of normal or mutant stem cell fates following a targeted, genetic manipulation. Replication slippage into frame is a function of cell division (DNA replication) and therefore Cre activation can occur any time after conception in many tissues (Fig. 1). In the

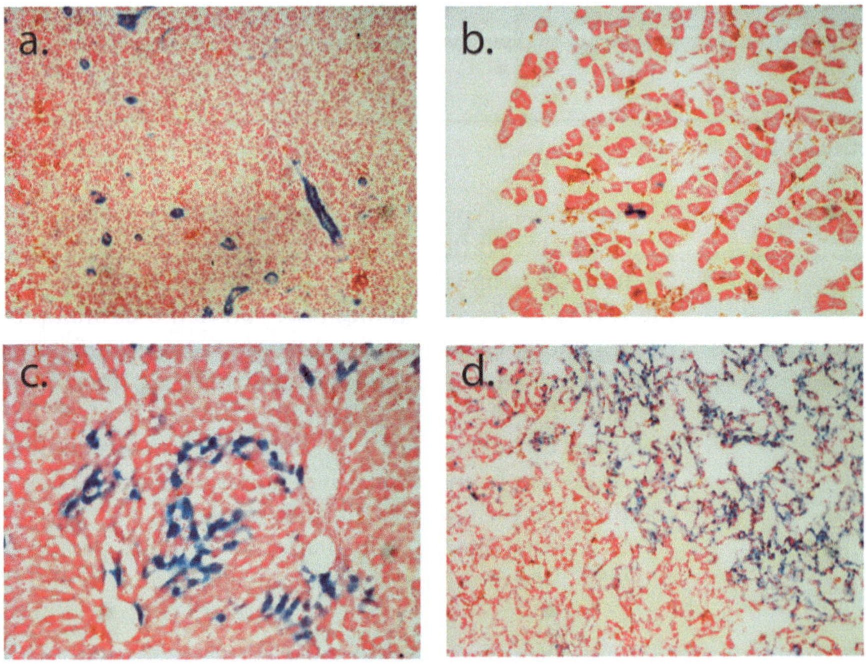

Fig. 1 The Pms2^{cre} mouse model has Cre activity in many different tissues as shown by the β-gal+ cells from the R26R reporter. (**a**) Kidney. (**b**) Pancreas. (**c**) Liver. (**d**) Lung

intestine, *Cre* reversion can happen in any dividing cell, but only the products of events that happen in intestinal stem cells (no matter the location of the stem cell) will persist, such as after TgfβR2 loss [16]. Importantly, by utilizing a reporter gene, we can follow genetic mutations in the absence of phenotypic change. This is significant as we better understand that oncogenic mutations occur in phenotypically normal tissues [12–14].

2 Materials

2.1 Mice

1. Pms2^{cre} mice are generated in house and are currently located at Oregon Health and Science University (OHSU). We are in the process of distributing these mice to the Jackson Laboratory, but they can be obtained from OHSU. Pms2$^{cre/+}$ mice are MMR-proficient. Pms2$^{cre/cre}$ mice are MMR-deficient.
2. R26R LacZ mice.
3. tdTomato reporter mice (Ai14).
4. R26R Confetti reporter mice.
5. Apc conditional mice (Apc^{CKO}) are acquired from Dr. Kucherlapati's group [17].
6. Kras^{G12D} mice (LSL-K-ras G12D).
7. c-myc^{T58A} mice are acquired from Dr. Sears's group [18].
8. Smad4 conditional mice (Smad4^{fx}) are acquired from Dr. Deng's group [19].
9. TgfβR2 conditional mice (TgfβR2^{fx}) are acquired from Dr. Moses's group [20].

2.2 Primers

1. Pms2A (TTCGGTGACAGATTTGTAAATG), Pms2W (TCACCATAAAAATAGTTTCCCG).
2. CreF (AACATTCTCCCACCGTCAGT), CreR (CATTTGGGCCAGCTAAACAT).
3. ApcF2.5 (TAGTACTTTTCAGACGTCATG), ApcR2 (AGTGCTGTTTCTATGAGTCAAC).
4. LacZ mut (GCGAAGAGTTTGTCCTCAACC), LacZ common (AAAGTCGCTCTGAGTTGTTAT), LacZ wt (GGAGCGGGAGAAATGGATATG).
5. Kras1 (GTCTTTCCCCAGCACAGTGC), Kras2 (CTCTTGCCTACGCCACCAGCTC).
   Kras3 (AGCTAGCCACCATGGCTTGAGTAAGTCTGCA).
6. Tom Fwd (CTGTTCCTGTACGGCATGG), Tom Rev. (GGCATTAAAGCAGCGTATCC).

7. Cmyc F3 (TGTACCTCGTCCGATTCCACG), Cmyc R3 (GATGGAGATGAGCCCGACTCCG).
8. Smad4 9 (GGGCAGCGTAGCATATAAGA), Smad4 10 (GACCCAAACGTCACCTTCAG).
9. Tgf Genl (GCAGGCATCAGGACCTCAGTTTGATCC), Tgf Gen2 (AGAGTGAAGCCGTGGTAGGTGAGCTTG).

2.3 Reagents

1. Lysis buffer (50 mM KCl, 10 mM Tris pH 8.3, 2 mM $MgCl_2$, 0.1 mg/mL gelatin, 0.45% NP-40, 0.45% Tween 20).
2. PLP fixative: 2% paraformaldehyde, 75 mM lysine, 75 mM Na_2HPO_4, 10 mM $NaIO_4$.
3. Phosphate Buffered Saline (PBS): 137 mM NaCl, 2.7 mM KCl, 8 mM Na_2HPO_4, 1.5 mM KH_2PO_4, pH 7.4.
4. DTT solution: 20 mM DTT, 20% ethanol, 15 mM Tris pH 8.0.
5. β-gal staining solution: 2 mM X-gal, 4 mM $K_4Fe(CN)_6$, 4 mM $K_3Fe(CN)_6$, 2 mM $MgCl_2$ in PBS.
6. Fixative for fluorescence: 2–4% Formaldehyde.
7. PCR buffer: 10 mM Tris–HCl, 50 mM KCl, 1.5 mM $MgCl_2$, 0.5 μM each primer, 0.2 mM dNTPs, 0.69 mM Cresol Red, 10% sucrose.
8. Cryomold.

3 Methods

3.1 Stochastic Genetically Engineered Mouse Model (GEMM)

1. We have utilized the mismatch repair (MMR) status of *Pms2*^{cre} mice to alter the frequency of *Cre* reversion and thus the timing. Compromised MMR will result in both increased *Cre* reversion and a higher mutation rate throughout the genome. However, while an increased mutation rate could complicate our observations, we offer several reasons why our data are not hindered by Pms2-deficiency. First, we compare Pms2-deficient experimental mice with Cre target genes to Pms2-deficient control mice that lack Cre target genes. Second, we analyze hundreds of reversion events per mouse, thus the stochastic nature of mutation makes modifying a consequential gene in a significant fraction of β-gal$^+$ foci highly unlikely. Finally, *Pms2* null mice are not prone to either intestinal adenomas or carcinomas [21]. Therefore, while we cannot rule out some undefined synergistic effect between the Cre targeted mutations and Pms2 deficiency, we believe such an effect is highly unlikely to play a significant role in phenotypes of Cre-expressing cells in *Pms2*$^{cre/cre}$ mice.
2. Because *Cre* reversion rates are low (~1 in 700 cell divisions in *Pms2*$^{cre/cre}$ mouse embryonic fibroblasts, and ~50 times less in *Pms2*$^{cre/+}$ mice), Cre activation overwhelmingly occurs in

isolated cells. In *Pms2*^{cre} mice carrying a *lox-stop-lox β-gal* allele (*R26R*), clonal patches composed of different numbers of β-gal^{+} crypts/villi are observed. Each β-gal^{+} patch should represent a single *Cre* reversion and is classified by size as either small (1–3 villi), medium (4–9 villi), or large (10+ villi per patch). The different sized patches likely reflect the timing of *Cre* reversion, with reversion earlier in life more likely to form larger patches due to the normal process of crypt fission that takes place during growth and development of the intestine. Consistent with this interpretation, numerous and larger clonal patches are observed in *Pms2*$^{cre/cre}$ mice, while fewer and smaller patches are observed in *Pms2*$^{cre/+}$ mice, in which the mutation rate is ~50-fold lower [15, 22]. Therefore, the pattern of β-gal^{+} crypts in either *Pms2*$^{cre/cre}$; *R26R* or *Pms2*$^{cre/+}$; *R26R* mice provides a baseline of how individual stem cells normally form clonal patches throughout life. Depending on the constellation of the Cre-target alleles, we can measure whether mutant intestinal stem cells have altered crypt succession (number of β-gal^{+} patches) and/or crypt fission (size of β-gal^{+} patches). Here, we concentrate on stochastic *Apc* loss and/or *Kras* activation because these genes are frequent mutational targets in human intestinal cancer; however, any combination of conditional alleles can be used. In addition, since the Pms2 promoter is ubiquitous, most tissue types can be studied (*see* **Note 1**).

3.2 Stochastic GEMM of Intestinal Cancer

1. *Pms2*^{cre}; *Apc*$^{CKO/CKO}$; *Kras*^{G12D}; *R26R* mice are generated by interbreeding *Pms2*$^{cre/+}$; *R26R* mice with *Apc*^{CKO} and *Kras*^{G12D} mice.
2. Pms2^{cre}; Smad4$^{fx/fx}$; c-myc^{T58A}; R26R are generated by interbreeding Pms2$^{cre/+}$; R26R mice with Smad4^{fx} and c-myc^{T58A} mice.
3. Pms2^{cre}; TgfβR2$^{fx/fx}$; Apc$^{CKO/CKO}$; R26R are generated by interbreeding Pms2$^{cre/+}$; R26R mice with TgfβR2^{fx} and Apc^{CKO} mice.
4. *Pms2*^{cre} mice are backcrossed at least three times to C57Bl/6J and subsequently maintained via interbreeding. The *R26R* (*Rosa26 Reporter* mice with *lox-stop-lox LacZ* or with *tdTomato*) allele is used (*see* **Note 2**).
5. Mice are housed in a specific pathogen free HEPA filtered room and are fed a diet of Purina PicoLab Rodent Diet 20. Pms2$^{cre/cre}$ mice are fertile [23], but we do not use these mice as breeders (*see* **Note 3**).
6. A small amount of tissue from each mouse is used for genotyping. DNA is isolated in 200 μL of Lysis buffer with 150 μg of Proteinase K incubated at 55 °C overnight, then inactivated at

95 °C for 10 min. DNA is genotyped in a solution containing PCR buffer and 0.5 U Taq. PCR conditions are 93 °C for 5 min, then 38 cycles of 93 °C for 20 s, 60 °C for 20 s, 72 °C for 40 s, then a final 72 °C for 3 min. Mice are genotyped for the wild-type *Pms2* allele using the Pms2A and Pms2W primers and for the *cre* allele using the CreF and CreR primers. Mice are genotyped for the Apc *CKO* allele with ApcF2.5 and ApcR2, for the c-myc^{T58A} allele with cmyc F3 and cmyc R3, and for the Smad4^{fx} allele with smad4–9 and smad4–10. *R26R* mice are genotyped for *LacZ* with mut, common and wt or for Tomato with Tom fwd and rev. *Kras*^{G12D} mice are genotyped with Kras1, Kras2, and Kras3 at an annealing temperature of 65 °C. TgfβR2^{fx} mice are genotyped with Tgf-Gen1 and Tgf-Gen2 at an annealing temperature of 64 °C.

3.3 Stochastic GEMM of Intestinal Cancer Prevention

1. Sulindac is administered in the drinking water starting at the different time points and continued until time of sacrifice ad libitum. The sulindac solution is 180 mg/L of sulindac and 4 mM sodium phosphate dibasic in distilled water (pH ~7.4). The solution is made fresh every 2 weeks.

3.4 Isolation of Small Intestine

1. Intestines are isolated and flushed with cold PLP fixative (*see* **Note 4**). Intestines are then cut longitudinally and washed in PBS. Intestines are cut into sixths and pinned in a dish with villi facing up. Pinned intestines are incubated in cold PLP fix for 1 h shaking at room temp. Next, pinned intestines are washed in PBS and incubated in DTT solution for 45 min shaking at room temp. Next, pinned intestines are washed in PBS and incubated in β-gal staining solution overnight shaking in the dark at 4 °C. Lastly, pinned intestines are washed in PBS and used for whole-mount analysis or further processing for sectioning.
2. After staining the intestines for β-gal, the number of positive villi are scored under a Leica MZ6 dissecting microscope at 20× power, a 25 mm^2 field of view. The field of view represented 1/28 of each strip of small intestine, thus 1/168 of the entire small intestine (each strip is 1/6 of the entire small intestine). The 25 mm^2 field of view represents 1200 villi, which extrapolates to 201,600 villi in the entire small intestine. The number of β-gal$^+$ villi are counted in each field of view and at least 20 β-gal$^+$ foci are counted for each third of the small intestine (*see* **Note 5**). Nearby β-gal$^+$ foci are considered independent if not arising from the same crypt and surrounded by nonstaining crypts. Adenomas, which involved multiple villi, are scored by whole mount and in cross sections. Microadenomas, which involved a single villus, are determined by scoring cross sections.

3. The process for using a fluorescent reporter is similar to the process stated above, but with the following changes. Intestines are isolated and flushed with cold fixative. Intestines are then cut longitudinally and cleaned in PBS. Intestines are cut into sixths and pinned with villi facing up. Pinned intestines are incubated in cold fixative for 1 h shaking at room temperature in the dark to protect tdTomato fluorescence from quenching. Finally, pinned intestines are washed in PBS, adenomas are counted and processing continued for sectioning.
4. Pinned intestines are incubated in 30% sucrose in PBS overnight at 4 °C in the dark. Intestines are then layered into a standard cryomold with each mold containing 1/3 of the small intestine or both the colon and cecum. OCT is placed between each section in the mold, typically 6 sections per mold. Cryomolds with tissue are frozen and stored at −80 °C.
5. Intestines are sectioned in a cryostat at −20 °C at 12 μm. Sections are placed on SuperFrost PlusGOLD slides. For β-gal stained intestines, slides are washed in PBS, counter stained in Nuclear Fast Red, washed in water, dehydrated in 90% ethanol, cleared with CitriSolv, mounted with Permount and finally covered with Fisherbrand Microscope Cover Glass. For fluorescence, slides are washed in PBS and then mounted with VectaShield Antifade mounting media. Antibody staining can be performed before the counterstaining and mounting process.

3.5 Application to Cancer initiation and Progression

1. The following is an example of the usefulness of this mouse model and method to understand tumor initiation. A popular view is that *Apc* loss in intestinal crypt stem cells is sufficient for adenoma formation [7]. However, our previous studies demonstrated a form of phenotypic plasticity following isolated *Apc* loss in an otherwise *Apc* wild-type mouse, in that whereas adenoma formation could ensue, the majority of *Apc*-deficient intestinal crypts retained a normal phenotype [10]. In both cases (adenoma formation or field formation) *Apc* loss functions as a gatekeeper mutation with net increases in mutant cells [24]. Interestingly, a significant fraction of the normal, *Apc*-deficient crypts exhibited a growth advantage resulting in clonal expansion and a field of mutant crypts, thus raising the possibility that crypt fission leading to an occult, horizontal spread of mutations is an important intermediate during tumorigenesis. Our work has shown that the timing of *Apc* loss can affect field formation by strongly influencing the crypt fission advantage of *Apc*-deficient crypts, with later *Apc* loss resulting in decreased mutant field size and adenoma formation [11]. In addition, the low frequency cre reversion mouse combined with Apc loss ($Pms2^{cre/+}$; $Apc^{CKO/CKO}$) results in a

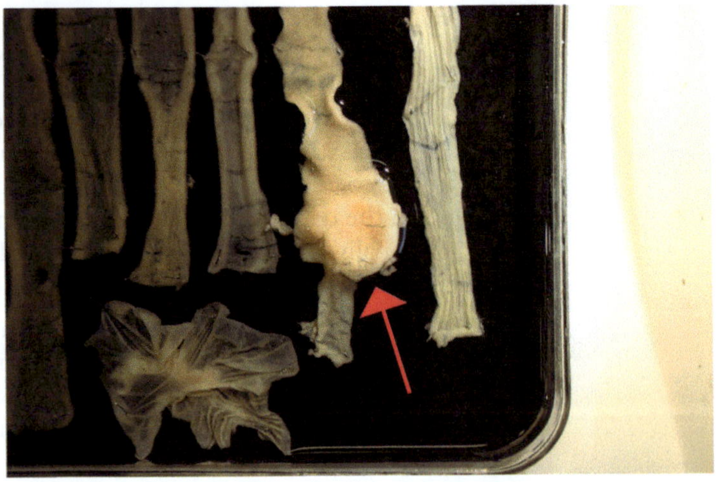

Fig. 2 Whole-mount image of the small intestine showing a single tumor from a Pms2$^{cre/+}$; Apc$^{CKO/CKO}$ mouse

single tumor per mouse, which is highly useful for studying sporadic cancer since humans tend to get a single tumor (Fig. 2). Finally, we showed that the NSAID, sulindac, can act as a chemopreventive by inhibiting the *Apc*-deficient field expansion that appears to precede adenoma formation.

2. The Pms2^{cre} mouse model system allows for the study of cells following loss or gain of multiple genes. Depending on the combination of conditional alleles, the Pms2^{cre} system can reveal data about cancer initiation and progression. The following are a couple of examples of the useful of this mouse model. First, Apc loss and Kras activation resulted in a third of tumors in the *Pms2*$^{cre/+}$; *Apc*$^{CKO/CKO}$; *Kras*^{G12D} (33% (5/15)) progressing to carcinoma compared to only 4% (1/24) of tumors in *Pms2*$^{cre/+}$; *Apc*$^{CKO/CKO}$ mice.
3. Apc loss and TgfβR2 loss (Pms2$^{cre/+}$; Apc$^{CKO/CKO}$; TgfβR2$^{fx/fx}$) resulted in progression from adenoma to carcinoma and metastasis to the lung (Fig. 3).
4. Finally, depending on the combination of conditional alleles, the tumor cell-type of origin can change. For example, Apc and Smad4 conditional alleles (Pms2$^{cre/cre}$; Apc$^{CKO/CKO}$; Smad4$^{fx/fx}$; R26R) result in a tumor of epithelial origin, but c-myc^{T58A} and Smad4 conditional alleles (Pms2$^{cre/cre}$; c-myc^{T58A}; Smad4$^{fx/fx}$; R26R) result in a tumor of stromal origin (Fig. 4).
5. These experiments reveal the usefulness of this mouse model in understanding cancer initiation, progression and metastasis. Beyond the scope of intestinal cancer and stem cell dynamics, this mouse model allows for following isolated, phenotypically

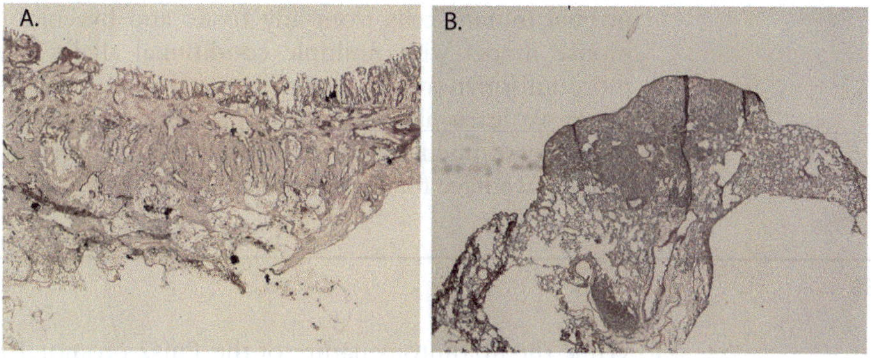

Fig. 3 Different combination of target alleles ($Pms2^{cre/+}$; $Apc^{CKO/CKO}$; $Tgf\beta R2^{fx/fx}$) can result in cancer progression and metastasis. (**a**) Small intestinal adenocarcinoma. (**b**) Lung metastasis

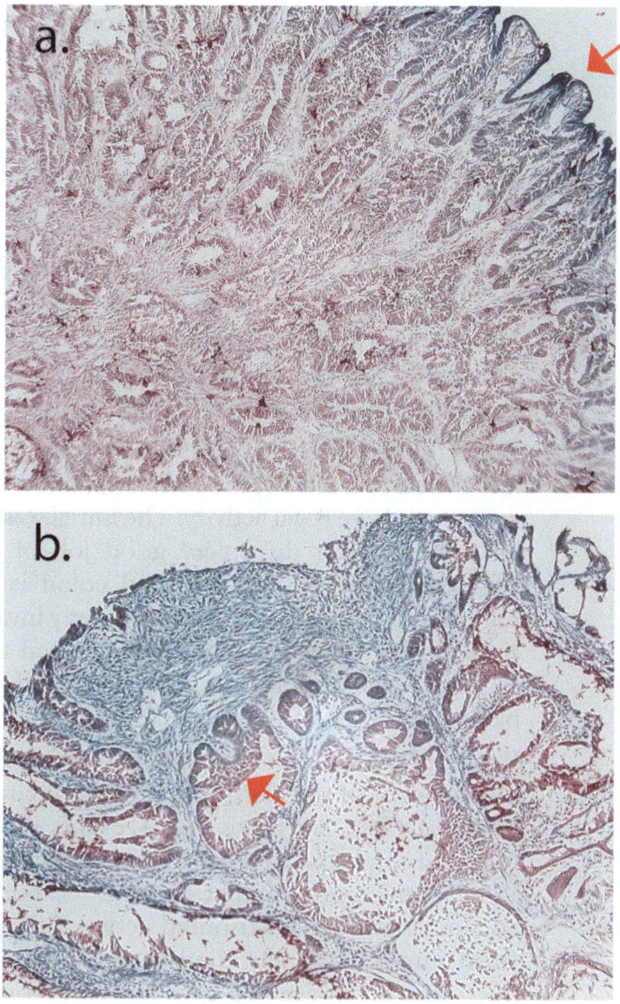

Fig. 4 Different combinations of target alleles result in different cell-type of origin for the adenoma. (**a**) Epithelial tumor from $Pms2^{cre/cre}$; $Apc^{CKO/CKO}$; $Smad4^{fx/fx}$ mouse colon. (**b**) Stromal tumor from $Pms2^{cre/cre}$; c-myc^{T58A}; $Smad4^{fx/fx}$ mouse colon

normal mutant cells from any tissue and by combining this mouse model with multiple conditional alleles can provide more information on cancer progression and metastasis. The more we learn about the interplay between mutations, the microenvironment, and cancer formation, the better the chances we have for understanding and preventing cancer.

4 Notes

1. While the ubiquitous nature of the Pms2 promoter allows for examining the effects of mutations on different tissues, it also has its drawbacks. For example, when trying to study the role of oncogenic Kras mutations in the gut, the mice develop lung tumors and need to be sacrificed around 1 year of age. Thus, preventing the intestinal tumors from fully progressing.
2. While any cre reporter can work with the $Pms2^{cre}$ system, we have found that the Rosa Confetti reporter does not show expression with the $Pms2^{cre}$ mouse. Most likely this is due to a combination of the levels of cre expressed from Pms2 and the difficult nature of recombining out the Stop cassette in the Confetti reporter.
3. While $Pms2^{cre/cre}$ mice are fertile, these mice do not have mismatch repair and therefore will accumulate a large number of mutations in their germline. Thus, we do not use these mice as breeders; however, these mice could be useful if you want to generate a large number of random mutations.
4. When using X-gal to stain for β-galactosidase activity in the gut, it is critical to wash away the bacteria, since many bacteria in the mouse gut have β-gal activity. The initial washes and shaking on an orbital shaker do a very good job of getting rid of the bacteria. However, the proximal colon is the most difficult region to get rid of bacteria due to many invaginations. Positive signal from bacteria can be distinguished from real signal by location outside of intestinal cells and punctate staining pattern.
5. Reporter expression within the intestine can follow a gradient, independent of the distribution of the cre reversion event. For example, there are more β-gal+ events in the proximal small intestine compared to the distal small intestine. However, when $Pms2^{cre}$ mice are crossed to Apc^{CKO} mice, the number of adenomas is evenly distributed between the proximal and distal small intestine. When combining Apc^{CKO} mice with R26R mice, the percentage of β-gal+ tumors follows a gradient similar to the β-gal+ cell distribution. This result suggests that $Pms2^{cre}$ reversion is even throughout the small intestine, but the stop cassette is not excised from the reporter.

Acknowledgments

This work was supported by NIH grant R00CA181679.

References

1. Noah TK, Donahue B, Shroyer NF (2011) Intestinal development and differentiation. Exp Cell Res 317(19):2702–2710. https://doi.org/10.1016/j.yexcr.2011.09.006
2. Barker N, van Es JH, Kuipers J, Kujala P, van den Born M, Cozijnsen M, Haegebarth A, Korving J, Begthel H, Peters PJ, Clevers H (2007) Identification of stem cells in small intestine and colon by marker gene Lgr5. Nature 449(7165):1003–1007. https://doi.org/10.1038/nature06196
3. Snippert HJ, van der Flier LG, Sato T, van Es JH, van den Born M, Kroon-Veenboer C, Barker N, Klein AM, van Rheenen J, Simons BD, Clevers H (2010) Intestinal crypt homeostasis results from neutral competition between symmetrically dividing Lgr5 stem cells. Cell 143(1):134–144. https://doi.org/10.1016/j.cell.2010.09.016
4. Sato T, van Es JH, Snippert HJ, Stange DE, Vries RG, van den Born M, Barker N, Shroyer NF, van de Wetering M, Clevers H (2011) Paneth cells constitute the niche for Lgr5 stem cells in intestinal crypts. Nature 469 (7330):415–418. https://doi.org/10.1038/nature09637
5. Sato T, Vries RG, Snippert HJ, van de Wetering M, Barker N, Stange DE, van Es JH, Abo A, Kujala P, Peters PJ, Clevers H (2009) Single Lgr5 stem cells build crypt-villus structures in vitro without a mesenchymal niche. Nature 459(7244):262–265. https://doi.org/10.1038/nature07935
6. Lopez-Garcia C, Klein AM, Simons BD, Winton DJ (2010) Intestinal stem cell replacement follows a pattern of neutral drift. Science 330 (6005):822–825. https://doi.org/10.1126/science.1196236
7. Barker N, Ridgway RA, van Es JH, van de Wetering M, Begthel H, van den Born M, Danenberg E, Clarke AR, Sansom OJ, Clevers H (2009) Crypt stem cells as the cells-of-origin of intestinal cancer. Nature 457 (7229):608–611. https://doi.org/10.1038/nature07602
8. Khare S, Chaudhary K, Bissonnette M, Carroll R (2009) Aberrant crypt foci in colon cancer epidemiology. Methods Mol Biol 472:373–386. https://doi.org/10.1007/978-1-60327-492-0_17
9. Kinzler KW, Vogelstein B (1996) Lessons from hereditary colorectal cancer. Cell 87 (2):159–170
10. Fischer JM, Miller AJ, Shibata D, Liskay RM (2012) Different phenotypic consequences of simultaneous versus stepwise Apc loss. Oncogene 31(16):2028–2038. https://doi.org/10.1038/onc.2011.385
11. Fischer JM, Schepers AG, Clevers H, Shibata D, Liskay RM (2014) Occult progression by Apc-deficient intestinal crypts as a target for chemoprevention. Carcinogenesis 35 (1):237–246. https://doi.org/10.1093/carcin/bgt296
12. Martincorena I, Roshan A, Gerstung M, Ellis P, Van Loo P, McLaren S, Wedge DC, Fullam A, Alexandrov LB, Tubio JM, Stebbings L, Menzies A, Widaa S, Stratton MR, Jones PH, Campbell PJ (2015) Tumor evolution. High burden and pervasive positive selection of somatic mutations in normal human skin. Science 348(6237):880–886. https://doi.org/10.1126/science.aaa6806
13. Lee-Six H, Obro NF, Shepherd MS, Grossmann S, Dawson K, Belmonte M, Osborne RJ, Huntly BJP, Martincorena I, Anderson E, O'Neill L, Stratton MR, Laurenti E, Green AR, Kent DG, Campbell PJ (2018) Population dynamics of normal human blood inferred from somatic mutations. Nature 561(7724):473–478. https://doi.org/10.1038/s41586-018-0497-0
14. Martincorena I, Fowler JC, Wabik A, Lawson ARJ, Abascal F, Hall MWJ, Cagan A, Murai K, Mahbubani K, Stratton MR, Fitzgerald RC, Handford PA, Campbell PJ, Saeb-Parsy K, Jones PH (2018) Somatic mutant clones colonize the human esophagus with age. Science 362(6417):911–917. https://doi.org/10.1126/science.aau3879
15. Miller AJ, Dudley SD, Tsao JL, Shibata D, Liskay RM (2008) Tractable Cre-lox system for stochastic alteration of genes in mice. Nat Methods 5(3):227–229. https://doi.org/10.1038/nmeth.1183
16. Fischer JM, Calabrese PP, Miller AJ, Munoz NM, Grady WM, Shibata D, Liskay RM (2016) Single cell lineage tracing reveals a role for TgfbetaR2 in intestinal stem cell dynamics and differentiation. Proc Natl Acad Sci U S A

113(43):12192–12197. https://doi.org/10.1073/pnas.1611980113

17. Kuraguchi M, Wang XP, Bronson RT, Rothenberg R, Ohene-Baah NY, Lund JJ, Kucherlapati M, Maas RL, Kucherlapati R (2006) Adenomatous polyposis coli (APC) is required for normal development of skin and thymus. PLoS Genet 2(9):e146. https://doi.org/10.1371/journal.pgen.0020146
18. Wang X, Cunningham M, Zhang X, Tokarz S, Laraway B, Troxell M, Sears RC (2011) Phosphorylation regulates c-Myc's oncogenic activity in the mammary gland. Cancer Res 71 (3):925–936. https://doi.org/10.1158/0008-5472.CAN-10-1032
19. Yang X, Li C, Herrera PL, Deng CX (2002) Generation of Smad4/Dpc4 conditional knockout mice. Genesis 32(2):80–81
20. Chytil A, Magnuson MA, Wright CV, Moses HL (2002) Conditional inactivation of the TGF-beta type II receptor using Cre:lox. Genesis 32(2):73–75
21. Prolla TA, Baker SM, Harris AC, Tsao JL, Yao X, Bronner CE, Zheng B, Gordon M, Reneker J, Arnheim N, Shibata D, Bradley A, Liskay RM (1998) Tumour susceptibility and spontaneous mutation in mice deficient in Mlh1, Pms1 and Pms2 DNA mismatch repair. Nat Genet 18(3):276–279. https://doi.org/10.1038/ng0398-276
22. Larson JS, Stringer SL, Stringer JR (2004) Impact of mismatch repair deficiency on genomic stability in the maternal germline and during early embryonic development. Mutat Res 556(1–2):45–53
23. Fischer JM, Dudley S, Miller AJ, Liskay RM (2016) An intact Pms2 ATPase domain is not essential for male fertility. DNA Repair (Amst) 39:46–51. https://doi.org/10.1016/j.dnarep.2015.12.011
24. Kwong LN, Dove WF (2009) APC and its modifiers in colon cancer. Adv Exp Med Biol 656:85–106

Chapter 19

Hemagglutinating Virus of Japan Envelope (HVJ-E)-Guided Gene Transfer to the Intestinal Epithelium

Masamichi Imajo

Abstract

The rapidly self-renewing epithelium of the small intestine represents an exquisite model for the stem cell-driven tissue renewal and tumorigenesis. Intestinal stem cells (ISCs) are located in the crypt base, where they produce rapidly dividing progenitors that undergo cell-cycle arrest and terminal differentiation upon several rounds of cell division. So far, genetic studies in mice have played a central role in analyzing function of genes during the stem cell-driven renewal of the intestinal epithelium. However, generation and maintenance of genetically engineered mice are a time-consuming endeavor, which limits the progress in intestinal biology. Recently, we have established a novel method that serves as an alternative to mouse genetics in intestinal biology. The method, termed intestine-specific gene transfer (iGT), enables rapid and efficient delivery of small molecules, such as siRNAs and plasmids, into the intestinal epithelium of living mice by utilizing the hemagglutinating virus of Japan envelope (HVJ-E). Here, we describe a detailed protocol for iGT and discuss how this method can accelerate progress in intestinal biology and elucidate the mechanisms of intestinal epithelium self-renewal.

Key words Intestinal epithelium, Hemagglutinating virus of Japan envelope (HVJ-E), Intestine-specific gene transfer (iGT), Self-renewal

1 Introduction

In multicellular organisms, many adult tissues require stem cell-driven renewal, abnormalities of which are strongly associated with a number of diseases including cancer. The intestinal epithelium is representative of such tissues, and organized into the proliferative crypt compartments and the differentiated villus compartments [1, 2]. Intestinal stem cells (ISCs) reside at the bottom of crypts and generate rapidly proliferating progenitor cells continuously to fuel self-renewal of the tissue [3]. Upon several rounds of cell division, these progenitors undergo lineage commitment, cell-cycle arrest, and terminal differentiation into the functional intestinal cell types, such as enterocytes, goblet, enteroendocrine, Paneth, and tuft cells. To date, the stem cell-driven renewal of the intestinal

Paloma Ordóñez-Morán (ed.), *Intestinal Stem Cells: Methods and Protocols*, Methods in Molecular Biology, vol. 2171,
https://doi.org/10.1007/978-1-0716-0747-3_19,

epithelium has been studied mainly by genetic studies in mice, leading to the elucidation of significant roles of several signaling pathways, including the Wnt, Notch, BMP, and Hedgehog signaling pathways, in intestinal homeostasis and tumorigenesis [4–7]. Major limitations of mouse genetics, however, are its laborious and time-consuming nature and inability to analyze function of many genes simultaneously. To overcome these difficulties, we have recently developed a novel in vivo gene transfer method for the intestinal epithelium [8]. This method, termed intestine-specific gene transfer (iGT), utilizes a hemagglutinating virus of Japan envelope (HVJ-E)-based transfection reagent, which can incorporate small molecules including siRNAs and plasmids and transfer them into target cells through fusion of HVJ-E with the plasma membrane [9], thereby enabling efficient transfection to the mouse intestinal epithelium. Since the intestinal epithelium undergoes rapid self-renewal, transient transfection by this method can be a robust platform to analyze function of genes in the homeostatic renewal of the tissue [8, 10, 11]. In this chapter, we describe a detailed protocol for iGT, technical notes for the procedures, and a protocol to evaluate transfection efficiency of iGT.

2 Materials

2.1 Reagents and Solutions

1. Isoflurane.
2. Phosphate buffered saline (PBS).
3. GenomeONE-Si (for transfer of siRNA).
4. GenomeONE-Neo (for transfer of plasmids).
5. Cy3-labeled siRNA.
6. Mucus-removing solution: 20 mM dithiothreitol, 0.05% Tween 20 in PBS.
7. Nylon string.
8. 1 mL syringe and 29-gauge needle.
9. 4% paraformaldehyde in PBS.
10. 12, 15, and 18% sucrose in PBS.
11. O.C.T. compound.
12. Cryomold.
13. Hoechst 33342.
14. Microscope slide.
15. Microcentrifuge tube.
16. 70% ethanol.
17. Mowiol.

2.2 Equipment

1. Small animal anesthetizer.
2. Surgical instruments (dissecting scissors, forceps, and surgical suture (4-0 nylon)).
3. Illuminated magnifier or surgical loupe.
4. Refrigerated microcentrifuge (for microtubes).
5. Cryostat microtome.
6. Epifluorescence microscope equipped with filters and illuminators suitable for observation of Cy3 fluorescence.

3 Methods

3.1 Preparation of Transfection Solution

The following protocol is for transection of siRNAs into the intestinal epithelium by using GenomeONE-Si transfection reagent that contains buffer solution, reagents D and E, and freeze-dried HVJ-E.

3.1.1 Preparation of siRNA Transfection Solution

1. Add 260 μL of buffer solution to freeze-dried HVJ-E, mix gently but thoroughly by pipetting up and down, and then aliquot 120 μL of the HVJ-E solution into a 1.5 mL microcentrifuge tube.
2. Add 24 μL of the reagent D to the HVJ-E solution and mix by gently tapping the tube.
3. Centrifuge the tube at 10,000 × *g* for 10 min at 4 °C. After centrifugation, aspirate the supernatant and resuspend the pellet in 45 μL of 20 μM Cy3-labeled siRNA solution. Place the tube on ice until just before the injection into the mouse intestine.
4. Add 175 μL of buffer solution and 80 μL of the reagent E and mix by tapping the tube (*see* **Note 1**). Inject the solution into the mouse intestinal lumen according to the procedures described below (Subheading 3.2).

3.1.2 Preparation of Plasmid Transfection Solution (Optional)

For the transfection of plasmids into the intestinal epithelium, we used a GenomeONE-Neo transfection reagent containing reagents A, B, and C, buffer solution, and HVJ-E solution.

1. Aliquot 120 μL of the HVJ-E solution into a 1.5 mL microcentrifuge tube. Add 30 μL of the reagent A to the solution, mix by tapping the tube, and incubate on ice for 5 min.
2. After incubation, add 30 μL of plasmids (1–2 μg/μL) and 18 μL of the reagent B to the HVJ-E solution, and mix well by tapping the tube.
3. Centrifuge the tube at 10,000 × *g* for 10 min at 4 °C. After centrifugation, aspirate the supernatant, resuspend the pellet in 260 μL of buffer solution, and mix gently but thoroughly by

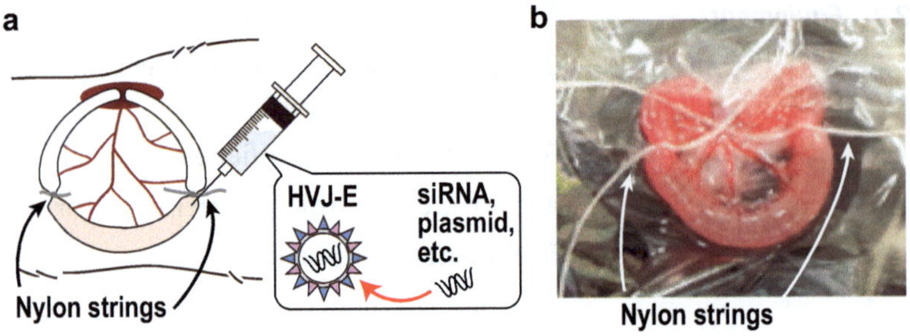

Fig. 1 Surgical procedures for intestine-specific gene transfer. (**a**) A schematic representation of the surgical procedures for iGT. (**b**) A picture showing a mouse whose intestine was bound with nylon strings and injected with the transfection solution during iGT

pipetting up and down. Place the tube on ice until just before the injection into the mouse intestine.

4. Add 40 μL of the reagent C and mix by tapping the tube (*see* **Note 1**). Inject the solution into the mouse intestinal lumen.

3.2 Surgical Procedures for In Vivo Gene Transfer to the Mouse Intestinal Epithelium

1. Anesthetize mice with isoflurane. Mice should be fasted over night to empty the proximal half of the small intestine. The abdomen is shaved and wiped with 70% ethanol. All surgical instruments are also disinfected with 70% ethanol.
2. Make an abdominal midline incision and blunt-dissect the skin from the peritoneum. Cut the peritoneum along the midline to address the small intestine.
3. Exteriorize a portion (about 5 cm long) of the small intestine, and gently suture (bind) its proximal and distal sides with nylon strings to seclude from the outer portion of the intestinal tract (Fig. 1). Use illuminated magnifier during suturing to avoid large blood vessels, as suturing can easily damage the blood vessels and cause bleeding (*see* **Note 2**). We usually wrap the mouse abdomen with plastic films to avoid direct contact of the intestinal tract with the mouse skin and hairs.
4. Fully distend the intestine by injecting 250–400 μL of mucus-removing solution into the secluded region of the intestine with a 29-gauge needle syringe. Keep the distended state for 15 min.
5. Aspirate and inject the solution several times by a 29-gauge needle syringe to remove the mucus from the intestinal lumen, and then remove the solution as completely as possible.
6. Repeat **steps 4** and **5**.
7. Wash the same region of the intestine by injecting and aspirating PBS several times. Repeat this wash step at least three times.

8. Inject 250–400 μL of the siRNA transfection solution or plasmid transfection solution (*see* Subheading 3.1) into the intestinal lumen. The tissue should be fully distended at this time (*see* **Note 3**). Leave the mouse in this state for 1 h.
9. Unfasten the nylon strings at the proximal and distal ends of the injected region. Place the intestine back in the peritoneal cavity, close the skin with wound clips, and recover the mouse from anesthesia.

3.3 Dissection, Fixation, and Observation of the Transfected Tissue

1. 3 h after in vivo gene transfer (*see* **Note 4**), dissect the mouse and remove the injected region of the small intestine. Flush the intestinal lumen with cold PBS several times, and then fix the tissue for 1 h in 4% paraformaldehyde/PBS at 4 °C with gentle agitation. Avoid the exposure to light throughout the following procedures to prevent quenching of Cy3 fluorescence.
2. After fixation, discard the fixative and wash the tissue with cold PBS three times for 10 min each.
3. Discard PBS and incubate the tissue in 12% sucrose/PBS for 2 h at 4 °C with gentle agitation. After incubation, discard the solution and incubate the tissue sequentially in 15 and 18% sucrose/PBS for 2 h each.
4. After incubation in 18% sucrose/PBS, embed and freeze the tissue in O.C.T Compound.
5. Cut 4–8 μm sections of the frozen intestinal specimen by using a cryostat microtome and mount the sections onto a microscope slide. Allow sections to air dry for a few minutes.
6. Wash sections in PBS for 5 min to remove O.C.T. Compound. To stain the nuclei of cells, incubate the sections with Hoechst 33342 diluted in PBS (1/500–1000) for 10 min at room temperature.
7. Aspirate Hoechst 33342 and wash sections in PBS three times for 5 min each. Rinse the slide briefly in DDW, and then seal the sections with the Mowiol mounting medium and coverslips.
8. Observe Cy3 and Hoechst 33342 fluorescence on the tissue sections with an epifluorescence microscope (*see* **Note 5**).

4 Notes

1. After addition of the reagents C and E, HVJ-E particles are sticky and can easily form large aggregates in the solution. For the efficient transfection, the reagent C and E should be added immediately before the injection of transfection solution.
2. Intestinal injury can cause the formation of fibrous bands (adhesions) in the abdomen, which induces severe life-

threatening intestinal obstruction. To reduce the risk for intestinal obstruction, it is critical to minimize injury and bleeding during the iGT procedures. Especially, care should be taken to avoid damaging visible blood vessels when binding the intestinal tract with nylon strings and injecting solution into the lumen.

3. Insufficient distention of the intestinal tract during iGT results in inefficient transfection into the crypt compartment. Even in such cases, the villus compartments are usually transfected with high efficiency. Since crypts are hidden between and below the villi, distention of the tract is required to expose them to the transfection solution.

4. HVJ-E-mediated transfer of the incorporated molecules into the intestinal epithelium is achieved within a few hours. For the observation of fluorescently labeled siRNAs, it is better to observe the transfected tissue within 3–5 h, as fluorescence would be gradually decreased thereafter. However, it should be important to keep in mind that the knockdown of target genes by siRNAs and expression of transgenes from plasmids would take more time (typically 2–3 days) and last for up to 5–7 days after transfection. Thus, the effects of transfection on the tissue phenotypes should be analyzed during this period.

5. The typical results of transfection of Cy3-labeled siRNAs are shown in Fig. 2. Cy3 fluorescence could be observed from

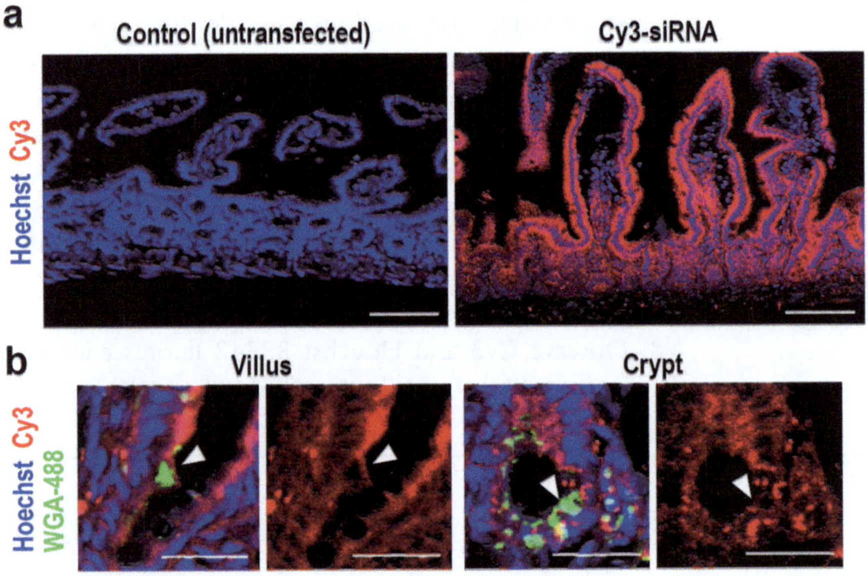

Fig. 2 Pictures showing sections of the small intestine transfected with Cy3-siRNAs. (**a**) The entire epithelium, but not submucosal or muscular layers, was efficiently transfected with Cy3-siRNAs by iGT. Scale bars, 100 μm. (**b**) Granules in goblet (left) and Paneth cells (right) were stained with wheat germ agglutinin-Alexa Fluor 488 (WGA-488) (white arrowheads). The cells harboring these granules were also efficiently transfected with Cy3-siRNAs. Scale bars, 25 μm

immediately after transfection in the entire epithelium including both villus and crypt compartments (Fig. 2a). By contrast, the submucosal layer or smooth muscle layer does not emit fluorescence, indicating that iGT transfers incorporated molecules specifically into the epithelium. By costaining granules in goblet and Paneth cells with Alexa Fluor 488-labeled lectin (wheat germ agglutinin (WGA)), one could confirm that the two types of cells are efficiently transfected with Cy3-labeled siRNAs (Fig. 2b). The other types of epithelial cells, including intestinal stem cells, progenitor cells, enterocytes, and enteroendocrine cells, could be also transfected by iGT [8].

Acknowledgments

This work was supported by the Takeda Science Foundation and Grant-in-Aid for Scientific Research (KAKENHI) on Innovative Areas, "Integrated analysis and regulation of cellular diversity" (18H05100) and for Scientific Research (C) (18K06929).

References

1. van der Flier LG, Clevers H (2009) Stem cells, self-renewal, and differentiation in the intestinal epithelium. Annu Rev Physiol 71:241–260
2. Fre S, Vignjevic D, Schoumacher M et al (2008) Epithelial morphogenesis and intestinal cancer: new insights in signaling mechanisms. Adv Cancer Res 100:85–111
3. Barker N, van Es JH, Kuipers J et al (2007) Identification of stem cells in small intestine and colon by marker gene Lgr5. Nature 449:1003–1007
4. Kretzschmar K, Clevers H (2017) Wnt/b-catenin signaling in adult mammalian epithelial stem cells. Dev Biol 428:273–282
5. van den Brink GR, Bleuming SA, Hardwick JC et al (2004) Indian Hedgehog is an antagonist of Wnt signaling in colonic epithelial differentiation. Nat Genet 36:277–282
6. van Es JH, van Gijn ME, Riccio O et al (2005) Notch/γ -secretase inhibition turns proliferative cells in intestinal crypts and adenomas into goblet cells. Nature 435:959–963
7. He XC et al (2004) BMP signaling inhibits intestinal stem cell self-renewal through suppression of Wnt–β-catenin signaling. Nat Genet 36:1117–1121
8. Imajo M, Ebisuya M, Nishida E (2015) Dual role of YAP and TAZ in renewal of the intestinal epithelium. Nat Cell Biol 17:7–19
9. Kaneda Y, Nakajima T, Nishikawa T et al (2002) Hemagglutinating virus of Japan (HVJ) envelope vector as a versatile gene delivery system. Mol Ther 6:219–226
10. Ordóñez-Morán P, Dafflon C, Imajo M et al (2015) HOXA5 counteracts stem cell traits by inhibiting Wnt signaling in colorectal cancer. Cancer Cell 28:815–829
11. Kon S, Ishibashi K, Katoh H et al (2017) Cell competition with normal epithelial cells promotes apical extrusion of transformed cells through metabolic changes. Nat Cell Biol 19:530–541

Acknowledgments

References

Chapter 20

An Intrasplenic Injection Model for the Study of Cancer Stem Cell Seeding Capacity

Caroline Dafflon, Albert Santamaría-Martínez, and Paloma Ordóñez-Morán

Abstract

In many tumor types, only a minor pool of cancer cells—the so-called cancer stem cells—is able to colonize distant organs and give rise to secondary tumors. In humans, the liver is one of the main target organs for many metastatic tumor types, including colorectal cancer. However, mouse tumour models only rarely spontaneously metastasize to the liver. Therefore, reliable in vivo experimental metastasis assays are crucial to study cell seeding capacity and the mechanisms controlling these metastatic stem cell properties. Here, we describe an intrasplenic injection model that mimics the process of liver metastasis occurring in cancer patients.

Key words Portal vein, Liver metastases, Stem cells, Cell lines, Nude mice, Cancer

1 Introduction

In solid tumors, only a small percentage of cells can extravasate and seed secondary sites during the metastatic process. These metastatic stem cells (MetSCs) are a subset of the so-called cancer stem cells (CSCs) [1, 2]. Besides self-renewal and differentiation capacities, MetSCs are able to spread from one organ to another and induce the formation of a suitable niche in the secondary site. These cells have been identified in different tumor types such as pancreatic, intestinal, and breast tumors [3–5]. MetSCs have specific properties that allow them to survive and grow in new, rather hostile, microenvironments. For instance, CSCs produce and/or induce the expression of extracellular matrix proteins such as tenascin C and periostin in the secondary site, which are essential for metastatic colonization [3, 6].

To study the ability of cancer cells to seed secondary tumors, different methods are used depending on the target organ under study. Here, we describe in detail the model of cancer cell injection

Paloma Ordóñez-Morán (ed.), *Intestinal Stem Cells: Methods and Protocols*, Methods in Molecular Biology, vol. 2171, https://doi.org/10.1007/978-1-0716-0747-3_20,

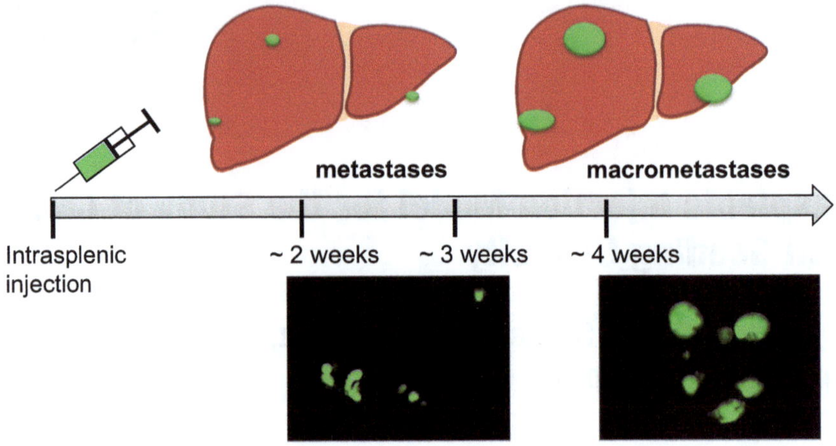

Fig. 1 Scheme of an intrasplenic injection protocol of GFP-labeled tumor cells

into the spleen to generate liver metastases. The results obtained from experiments using colorectal cancer (CRC) or pancreatic cancer cells, for example, can potentially be extrapolated to patients' data, as the liver is the main organ where these tumors metastasize. In this assay, the cells are injected into the portal vein, which connects directly the spleen to the liver. The liver metastasis assay can be performed both syngeneically—injecting murine cells into immunocompetent mice—or xenogeneically, injecting human cells into immunocompromised mice. In this chapter, we describe injection of HCT116 human colon cancer cell line into BALB/c nude mice. This cell line has a high metastatic capacity; indeed, 90% of these cells are positive for the CSC marker AC133 [7–10]. To facilitate the detection of the cells that have successfully seeded the liver, we stably express the green fluorescent protein (GFP) by using lentiviral vectors.

This method provides several advantages. First, cell seeding can be easily controlled by the number of cells injected. Second, the mice can be analyzed at different time points where the disease is still at an early stage and does not cause significant health issues (from 24 h to 30 days after injection depending on the cells used for each approach and the type of metastases of interest, micro- or macrometastases). Third, it is highly reproducible due to the well-defined steps described in this protocol (Fig. 1).

2 Materials

2.1 Plasmids

1. pRRL SIN.cPPT.hPGK-EGFP.WPRE vector.
2. For the production of lentiviral particles (third generation lentiviral vectors):

(a) pMD2.G plasmid containing the vesicular stomatitis virus G glycoprotein (VSV-G) envelope.

(b) Packaging plasmid pMDLg/pRRE containing Gag (group of antigens) and Pol (enzymes).

(c) Packaging plasmid pRSV-Rev containing Rev (gene regulatory proteins).

2.2 Lentiviral Production, Virus Concentration, and Transduction

1. HEK293T cells.
2. Tissue culture dishes (15 cm) and multidishes (24 well).
3. 293T medium: DMEM + GlutaMAX, 1% Penicillin/Streptomycin 100×, 10% Fetal bovine serum.
4. $CaCl_2$ 2.5 M dissolve 27.5 g of $CaCl_2{\cdot}6H_2O$ in 50 mL of bidistilled water. Filter solution through a 0.22 μm filter. Store at room temperature (RT).
5. dH_2O.
6. TE 0.1×: Tris 1 mM, EDTA 0.1 mM pH 8.8. Filter solution through a 0.22 μm filter. Store at 4 °C.
7. HBS 2×: 280 mM NaCl, 100 mM HEPES, 1.5 mM Na_2HPO_4, $7.11 \leq pH \leq 7.13$. Filter solution through a 0.22-μm filter. Store 50 mL aliquots at −20 °C.
8. Polypropylene conical tubes (50 mL).
9. Microcentrifuge tubes (1.7 mL).
10. Chloroquine diphosphate (stock 25 mM). Dissolve 20 mg of chloroquine diphosphate in 1 mL of dH_2O. Filter solution through a 0.22 μm filter. Store the filtrate protected from light at −20 °C.
11. Ultracentrifuge (with rotor, polyallomer tubes, and adaptors).
12. Shaker.
13. 0.22 μm syringe filters.
14. 50 mL syringes.
15. Cell culture centrifuge machine.
16. Tissue culture hood.
17. CO_2 incubator set to 5% CO_2 and 37 °C.
18. Serological disposable pipettes (5, 10 and 25 mL).
19. Pipettes (2, 200, 1000 μL).
20. Pipette controller.
21. Filtered pipette tips (10, 200, 1000 μL).
22. Inverted fluorescence microscope.

2.3 Generation of GFP⁺ Cells

1. Immortalized cell line: HCT116 (ATCC® CCL-247™) human colon cancer cell line.

2. Tissue culture dishes (15 cm).
3. Medium: DMEM + GlutaMAX, 1% Penicillin/Streptomycin 100×, 10% Fetal bovine serum.
4. Phosphate buffer solution 1×, pH 7.4.
5. Polypropylene conical tubes (50 mL).
6. Microcentrifuge tubes (1.7 mL).
7. Microcentrifuge machine.

2.4 Cell Dissociation

1. Tissue culture dishes (15 cm).
2. Medium: DMEM + GlutaMAX, 1% Penicillin/Streptomycin 100×, 10% Fetal bovine serum.
3. Phosphate buffer solution 1×, pH 7.4.
4. Polypropylene conical tubes.
5. Microcentrifuge tubes (1.7 mL).
6. Microcentrifuge machine.
7. 0.45 μm strainer.
8. Trypsin–EDTA (0.05%).

2.5 Surgery

1. Forceps.
2. Scissors.
3. Tweezers.
4. 4-0 needle/string for sewing.
5. 1 mL syringes.
6. 26G3/8 needles.
7. Stiches.
8. Ice.
9. Heat pad.
10. Analgesic (Paracetamol).
11. Antiseptic solution (Povidone–iodine).
12. Ophthalmic gel (Lacryvisc).
13. Freshly trypsinized cells ready to inject.
14. NaCl 0.9% solution.
15. Ethanol 70%.

2.6 Monitoring Cell Dissemination and Liver Metastases

1. Stereomicroscope.
2. Tweezers.
3. Scissors.
4. Phosphate buffer solution 1×, pH 7.4.
5. Tissue culture dishes (10 cm).

6. Digital microscope camera.
7. Optimal cutting temperature compound (OCT).

3 Methods

3.1 Generation of GFP+ Cells

3.1.1 Transfection of 293T Cells (Days 0 and 1)

1. One day before transfection (day 0), plate 293T (2.5×10^6 cells/15 cm plate) and grow them in their culture medium. Normally, five 15 cm plates are used for one lentivirus production.
2. Ideally, next day (day 1) cells should reach 60–80% confluency.
3. Two hours before transfection, replace the medium with 22.5 mL of fresh DMEM medium preheated at 37 °C and supplemented with chloroquine (CQ) at a final concentration of 6 μM.
4. For five 15 cm tissue culture dishes, prepare the following transfection mix in 50 mL polypropylene conical tubes under the tissue culture hood:
   112.5 μg vector plasmid.
   39.5 μg pMD2.G plasmid containing VSV-G envelope.
   37 μg pMDLg/pRRE plasmid containing Gag and Pol.
   73 μg pRSV-Rev plasmid containing Rev.
5. Then add the following solutions in this order and mix gently:
   3.3 mL of TE 0.1×.
   1.6 mL dH_2O.
   706 μL $CaCl_2$ 2 M.
6. Add 5.7 mL of HBS 2× dropwise under agitation by vortexing.
7. Let it rest for 10–30 min (but not more than 30 min) at RT.
8. Add dropwise 2.25 mL/plate of the transfection solution and mix (use a pipette controller and a 5 mL serological disposable pipette).
9. Keep plates in a P2 (biosafety level 2) CO_2 incubator set to 5% CO_2 and 37 °C.

3.1.2 Medium Replacement and Supernatant Collecting (Day 2)

1. Change the medium the day after transfection (day 2).
2. Check for transfection efficiency under an inverted fluorescence microscope.
3. On day 3, collect supernatant for the first time (24 h after medium replacement, day 3).
4. Supernatant can be harvested 2 or 3 times, every 12 h. Store it at 4 °C during the collecting period (*see* **Note 1**).

3.1.3 Concentration and Titration (Day 3 to Day 6)

1. Pool the collected supernatants, centrifuge for 5 min at 200 × *g* to remove cell debris and filter through 0.45 μm membranes.
2. Virus particles are concentrated by ultracentrifugation at 55,000 × *g* for 2 h at 4 °C in a swinging rotor.
3. Discard supernatant and resuspend pellet in 100–200 μL PBS 1× total (make a 100 or 1000-fold concentration) and freeze the particles at −80 °C (*see* **Note 1**).
4. Perform lentiviral titration using 293T cells and calculate the multiplicity of infection (MOI). A MOI of 1 means that you have equal number of cells and virus particles. The cells should be plated the day before (day 2).
5. Infect the cancer cell lines.
6. Change the medium the day after.
7. Expand the cells for a week.
8. Sort the cells by fluorescence-activated cell sorting (FACS) for GFP^+ expression.

3.2 Preparation of Cells Before Surgery

1. Plate the GFP^+ cells that are going to be injected several days before the surgery and keep them in the incubator at 37 °C and 5% CO_2. The cells should be 60–70% confluent before collecting them.
2. Remove medium.
3. Wash with PBS 1×.
4. Remove PBS 1× and trypsinize the cells (10 min at 37 °C).
5. Add 20 mL of medium to resuspend the cells (collect them in a 50 mL polypropylene conical tube).
6. Centrifuge (5 min, 4 °C, 200 × *g*) and discard the supernatant.
7. Resuspend the pellet in 40 mL of cold PBS 1×.
8. Centrifuge (5 min, 4 °C, 200 × *g*) and discard the supernatant.
9. Resuspend the pellet in 40 mL of cold PBS 1×.
10. Centrifuge (5 min, 4 °C, 200 × *g*) and discard the supernatant.
11. Resuspend the pellet in 1 mL cold PBS 1×.
12. Count the cells.
13. Resuspend the cells in PBS 1× to have 25×10^6 cells/mL.
14. Keep the cells on ice.

3.3 Surgery

1. Prepare the heating pad (disinfect with EtOH) and the tools inside the hood (Fig. 2a, b).
2. Prepare the anesthesia machine (isoflurane) (Fig. 2c) (*see* **Note 2**).

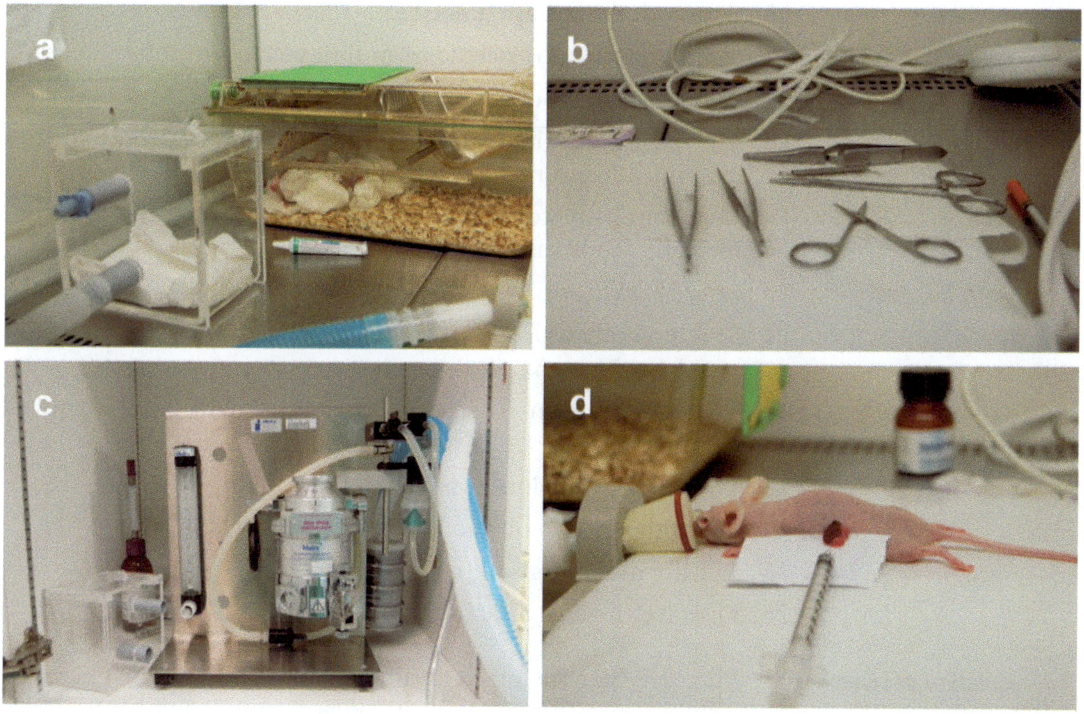

Fig. 2 Tools required for the surgery (a-d)

3. Anesthetize the mouse in the mouse induction chamber.
4. Once the mouse is asleep, place the mouse on the heating pad and apply ophthalmic gel to both eyes to prevent dryness.
5. Introduce the nose of the mouse into the nose cone of the isoflurane support.
6. Flush ethanol 70% on the left flank of the mouse.
7. Make a small incision and cut carefully the skin (1.5 cm).
8. Cut the connective tissue between the skin and the body wall.
9. Search for the spleen and cut the body wall (~5 mm; avoid capillaries).
10. Gently take out the spleen and place a sterile dressing between the skin and spleen (Fig. 2d).
11. Resuspend the cells with the syringe and take 0.1 mL from the mixture (it should contain 2.5×10^6 HCT116 cells) (*see* **Note 3**).
12. Leave the cells back on ice.
13. Inject the cells very slowly into the spleen. The needle should be placed parallel to the spleen.
14. Leave the needle-syringe in the spleen for 4 min (cells are circulating via the portal vein to the liver).

15. Ear mark the mouse if needed.
16. Slowly remove the needle from the spleen.
17. Apply cotton wool for a few seconds.
18. Tie a string knot in the upper part of the spleen, cut the string's extremities, and remove the spleen.
19. Apply pressure to the wound with sterile cotton wool until bleeding stops (usually a few seconds).
20. Inject 500 μL of NaCl 0.9% (*see* **Note 4**).
21. Sew the body wall with the sewing needle and the string (tight knot) (*see* **Note 5**).
22. The skin incision is closed with staples, which are removed within 10 days.
23. Use antiseptic solution to clean the wound.
24. Leave the mouse on the heat pad until awake.
25. Put the mouse back into the cage.
26. Dissolve the analgesic in the drinking water (*see* **Note 6**).

3.4 Isolation of Liver and Visualization of GFP-Labeled Cells

1. Mice can be euthanized at different time points (for micro- or macrometastases observation) (Figs. 1 and 3) (*see* **Note 7**).
2. After removal of the liver, examine the extent of tumor development by eye and use a stereomicroscope to select the area of interest and the magnification required.
3. Use the camera's zoom function to increase the size of the liver metastases as required and take several images (Fig. 3).
4. Record the number of GFP^+ tumor metastases visible on the hepatic surface.
5. Isolate the fluorescent micro- or macrometastases with surrounding tissue and embed them directly in OCT.
6. Store them at −80 °C for histology.

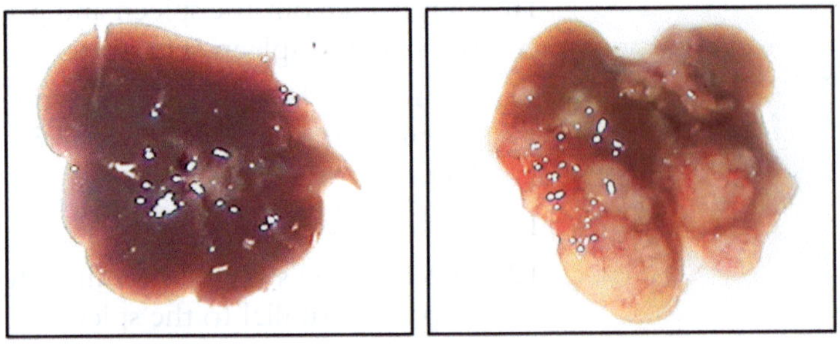

Fig. 3 Liver metastases. Left, 30 days after injecting placebo. Right, 30 days after injecting HCT116 cancer cell lines

4 Notes

1. The lentiviral particles can be kept at 4 °C for 4–5 days or stored at −80 °C in small aliquots.
2. Mice are anaesthetized using isoflurane mixed with oxygen. Anesthesia induction is performed at 4% isoflurane in O_2 in an adapted plexiglass box connected to a vaporizer. The mouse reaction can be visually controlled. The box is linked to a gas evacuation system coupled to an active charcoal filter to absorb isoflurane. Always monitor mouse breathing/heart rate and perform regular pedal reflex tests to establish anesthetic delivery.
3. This number is variable. It will depend on the cell type. You will need to test several concentrations before performing the final experiment.
4. After the operation, the loss of blood volume is compensated by i.p. injection of 0.5 mL saline (at 37 °C) and the muscle incision is sutured with absorbable thread.
5. Make sure not to include internal organs, and clean connective tissue between the body wall and the skin before sewing and closing the knot.
6. Mice need to be treated for pain relief after surgery. We recommend: Temgesic: twice a day, on the day of surgery and during the next 2 days (3 days in total) and Paracetamol: 1 day prior to the surgery and during the next 3 days.
7. The collection time point depends on the cell lines that are used in the experiment. For micrometastases, livers are collected after approximately 2 weeks. For macrometastases, livers are collected after 1 month or more (Table 1 and Fig. 1).

Table 1
Time course for micro- and macrometastases formation in the liver after intrasplenic injection

Cell line (tissue type)	Micrometastases (days)	Macrometastases (days)
PANC-1 (pancreas)	16	36
CAPAN-1 (pancreas)	12	28
LoVo (colon)	14	36
Ls174T (colon)	14	36
HCT116 (colon)	12	30

Acknowledgments

A.S.-M. was supported by an SNSF Ambizione career award (PZ00P3_154751), and P.O-M was supported by the University of Nottingham. This protocol was developed in the Huelsken's lab, so we would like to thank past and current members of this lab for their help in developing this method, specially to Prof. Joerg Huelsken.

References

1. Fico F, Bousquenaud M, Ruegg C, Santamaria-Martinez A (2019) Breast cancer stem cells with tumor—versus metastasis-initiating capacities are modulated by TGFBR1 inhibition. Stem Cell Reports 13 (1):1–9. https://doi.org/10.1016/j.stemcr.2019.05.026
2. Oskarsson T, Batlle E, Massague J (2014) Metastatic stem cells: sources, niches, and vital pathways. Cell Stem Cell 14(3):306–321. https://doi.org/10.1016/j.stem.2014.02.002
3. Malanchi I, Santamaria-Martinez A, Susanto E, Peng H, Lehr HA, Delaloye JF, Huelsken J (2012) Interactions between cancer stem cells and their niche govern metastatic colonization. Nature 481(7379):85–89. https://doi.org/10.1038/nature10694
4. Pang R, Law WL, Chu AC, Poon JT, Lam CS, Chow AK, Ng L, Cheung LW, Lan XR, Lan HY, Tan VP, Yau TC, Poon RT, Wong BC (2010) A subpopulation of CD26+ cancer stem cells with metastatic capacity in human colorectal cancer. Cell Stem Cell 6 (6):603–615. https://doi.org/10.1016/j.stem.2010.04.001
5. Hermann PC, Huber SL, Herrler T, Aicher A, Ellwart JW, Guba M, Bruns CJ, Heeschen C (2007) Distinct populations of cancer stem cells determine tumor growth and metastatic activity in human pancreatic cancer. Cell Stem Cell 1(3):313–323. https://doi.org/10.1016/j.stem.2007.06.002
6. Oskarsson T, Acharyya S, Zhang XH, Vanharanta S, Tavazoie SF, Morris PG, Downey RJ, Manova-Todorova K, Brogi E, Massague J (2011) Breast cancer cells produce tenascin C as a metastatic niche component to colonize the lungs. Nat Med 17(7):867–874. https://doi.org/10.1038/nm.2379nm.2379
7. Ordonez-Moran P, Dafflon C, Imajo M, Nishida E, Huelsken J (2015) HOXA5 counteracts stem cell traits by inhibiting Wnt signaling in colorectal cancer. Cancer Cell 28 (6):815–829. https://doi.org/10.1016/j.ccell.2015.11.001
8. Chowdhury S, Ongchin M, Sharratt E, Dominguez I, Wang J, Brattain MG, Rajput A (2013) Intra-tumoral heterogeneity in metastatic potential and survival signaling between iso-clonal HCT116 and HCT116b human colon carcinoma cell lines. PLoS One 8(4): e60299. https://doi.org/10.1371/journal.pone.0060299
9. Ricci-Vitiani L, Lombardi DG, Pilozzi E, Biffoni M, Todaro M, Peschle C, De Maria R (2007) Identification and expansion of human colon-cancer-initiating cells. Nature 445 (7123):111–115. https://doi.org/10.1038/nature05384
10. O'Brien CA, Pollett A, Gallinger S, Dick JE (2007) A human colon cancer cell capable of initiating tumour growth in immunodeficient mice. Nature 445(7123):106–110. https://doi.org/10.1038/nature05372

Chapter 21

Organoid Derivation and Orthotopic Xenotransplantation for Studying Human Intestinal Stem Cell Dynamics

Shinya Sugimoto, Masayuki Fujii, and Toshiro Sato

Abstract

Intestinal stem cells continuously self-renew throughout life to maintain gut homeostasis. With the advent of the organoid culture system, we are now able to indefinitely expand healthy and diseased tissue-derived human intestinal stem cells in vitro and use them for various applications. Nonetheless, investigating the behavior of human intestinal stem cells in vivo still remains challenging. We recently developed an orthotopic xenotransplantation system that realizes in vivo reconstruction of human intestinal epithelial tissue with preserved stem cell hierarchy by engrafting human normal colon organoids onto the mouse colon surface. We also introduced new growth factors, namely, insulin-like growth factor-1 (IGF-1) and fibroblast growth factor-2 (FGF-2), into the culture condition for human intestinal organoids that significantly increase scalability and transfectability of the organoids. By integrating these recent advances, we organized a tissue-oriented platform encompassing derivation of patient-derived intestinal organoids and their orthotopic xenotransplantation. The research platform based on orthotopic xenotransplantation of human intestinal organoids provides a powerful tool for studying human intestinal stem cell biology in native tissue-relevant contexts as well as for establishing novel disease modeling systems.

Key words Intestinal stem cells, LGR5, Crypt isolation, Organoid culture, Mini-gut, Colon, Human, Electroporation, Orthotopic transplantation, Endoscopy

1 Introduction

The intestinal epithelium is one of the most rapidly self-renewing tissues in our body which turns over every 4–5 days. This dynamic self-renewal is fueled by intestinal stem cells (ISCs) that self-replicate themselves and give rise to all types of intestinal epithelial cells. Over the past decades, researchers using genetically engineered mouse models have exhaustively addressed fundamental questions regarding the location and regulation of ISCs, as exemplified by the identification of self-renewing Lgr5-expressing stem cells which contributed to the dramatic advance of the field. Such mouse models produced genetic insights into the signaling pathways that regulate ISC self-renewal and led to the development of the culture platform that enables isolated single mouse ISCs to

Paloma Ordóñez-Morán (ed.), *Intestinal Stem Cells: Methods and Protocols*, Methods in Molecular Biology, vol. 2171, https://doi.org/10.1007/978-1-0716-0747-3_21,

indefinitely propagate ex vivo. In this culture system consisting of defined soluble growth factors (EGF, Noggin and R-spondin) and laminin-rich basement membrane matrix, single mouse ISCs self-organize to form crypt-like structures, namely, organoids [1]. The organoid system provides a sufficient environment for Lgr5$^+$ ISCs to constantly self-renew and produce all types of differentiated intestinal cells and therefore can be interpreted as the ex vivo reconstitution of the ISC niche. In other words, as long as niche factors are properly supplied, ISCs can execute their functions even after relocation to foreign environments.

A slight modification to the mouse intestinal organoid culture condition subsequently realized organoid culture of human ISCs [2]. The successful derivation of human intestinal organoids not only demonstrated the existence of human ISCs but also opened up a new avenue for human disease biology research by allowing direct expansion and intervention of living disease tissues as patient-derived organoids. Despite the utility, the initial version of human intestinal organoid culture system carried several limitations. While mouse small intestinal organoids can produce all differentiated intestinal cell types in a unified culture condition, human intestinal organoids require medium switching to generate differentiated cells. Human ISCs also require highly active Wnt3a, typified by the serum-stabilized Wnt3a-conditioned medium, which may interfere with some serum-sensitive experiments. We recently resolved these issues by adopting serum-free Afamin-stabilized Wnt3a [3] and the modified culture condition using IGF-1 and FGF-2 [4]. With this refined culture protocol, human intestinal organoids retain multilineage differentiation capacity, including Paneth cell differentiation which has been difficult to maintain in the original condition. This condition also permits a long-term culture and passaging of the cells. The refined condition also increases CRISPR-mediated gene-editing efficiency of the organoids and facilitates organoid-based prospective genetic analysis. In this review, we include the details of this refined culture condition for human intestinal organoids.

The human intestinal organoid culture system provides ISCs with minimal but highly potent and uniform niche factors to drive their self-renewal. Human intestinal organoids in the expansion phase may therefore represent regenerative rather homeostatic states. Indeed, we recently showed that the cycling status of human LGR5$^+$ ISCs in organoids differ from that in human tissues [5]. Besides the niche condition, factors that do not exist in the culture system, such as the growth factor gradient, gut flora, and nonepithelial cells, may also contribute to the distinctions between ISCs in organoids and human tissues. To overcome this issue, we recently developed an orthotopic xenograft system for human intestinal organoids by modifying the previous orthotopic transplantation method for mouse intestinal organoids [5, 6]. This system allows xenotransplanted organoids to recreate crypt structures that mimic the tissue size and cell cycle status in human tissues, offering a pragmatic strategy for understanding the biological behavior of human ISCs in near-native tissue environments. Orthotopic xenotransplantation of CRISPR-Cas9-mediated LGR5 reporter knock-in organoids further visualized lineage formation from single human colonic stem cells without damaging the tissue architecture and demonstrated the stem cell function of human colonic stem cells in the colon environment [5]. In this review, we provide step-by-step protocols for this xenografting experiment with troubleshooting tips. This protocol can be combined with the CRISPR-Cas9 technology and provides a unique opportunity for investigating human intestinal stem cell biology as well as intestinal diseases in native tissue-relevant contexts.

2 Materials

2.1 Human Intestinal Crypt Isolation

1. Dulbecco's phosphate-buffered saline without Ca^{2+} and Mg^{2+} (DPBS).
2. 2.5 mM EDTA in DPBS.
3. Basal culture medium: advanced DMEM/F12 supplemented with 10 mM HEPES, 2 mM GlutaMAX, 100 U/mL penicillin, and 100 μg/mL streptomycin (*see* **Note 1**).
4. 50 mL tube coated with 10% (w/v) bovine serum albumin in PBS (BSA/PBS) (*see* **Note 2**).
5. 10 mL disposable serological pipette coated with 10% BSA/PBS (*see* **Note 2**).
6. Human intestinal samples (*see* **Note 3**).

2.2 Human Intestinal Organoid Culture

1. Matrigel, basement membrane matrix, growth factor reduced (GFR), phenol red-free (*see* **Note 4**).
2. B27 supplement (50×).
3. N-acetyl-L-cysteine [500× stock; 81.5 mg/mL in distilled water (500 mM)].
4. 0.1% BSA/PBS: 0.1% (w/v) BSA in PBS, filter sterilized with a 0.22-μm filter.
5. Afamin-Wnt3a serum-free conditioned medium prepared from a cell line (*see* **Note 5**).
6. Recombinant mouse EGF (10,000× stock; 500 μg/mL in 0.1% BSA/PBS).
7. Recombinant mouse Noggin (1000× stock; 100 μg/mL in 0.1% BSA/PBS) (*see* **Note 6**).

8. Recombinant human R-spondin1 (100× stock; 100 μg/mL in 0.1% BSA/PBS) (*see* **Note 6**).
9. [Leu^{15}]-Gastrin I (10,000× stock; 100 μM 0.1% BSA/PBS).
10. Dimethyl sulfoxide (DMSO).
11. A83-01 (1000× stock; 500 μM in DMSO).
12. SB202190 (3000× stock; 30 mM in DMSO).
13. Y-27632 (1000× stock; 10 mM in PBS).
14. Recombinant human IGF-1 (1000× stock; 100 μg/mL in 0.1% BSA/PBS).
15. Recombinant human FGF-2 (FGF-basic) (1000× stock; 50 μg/mL in 0.1% BSA/PBS).
16. WENRAS medium: 20% (v/v) Afamin-Wnt3a serum-free conditioned medium, 50 ng/mL recombinant mouse EGF, 100 ng/mL recombinant mouse Noggin, 1 μg/mL recombinant human R-spondin1, 500 nM A83-01, 10 μM SB202190, 1× B27 supplement, 1 mM *N*-Acetyl-L-cysteine, and 10 nM [Leu^{15}]-Gastrin I in basal culture medium (*see* Table 1, **Note 7**).

Table 1
Culture media components of human intestinal organoids

Human intestinal organoids:		**Culture condition**		
Reagents	**Final concentration**	**WENRAS** Organoid establishment and expansion	**WENRAIF** Organoid establishment and expansion, and genome editing	**WNRAIF** Generating differentiated cells
Afamin-Wnt3a	20% (v/v)	+	+	+
EGF	50 ng/mL	+	+	−
Noggin	100 ng/mL	+	+	+
R-spondin1	1 μg/mL	+	+	+
A83-01	500 nM	+	+	+
SB202190	10 μM	+	−	−
IGF-1	100 ng/mL	−	+	+
FGF-2	50 ng/mL	−	+	+
Gastrin	10 nM	+	+	+
B27 supplement	1×	+	+	+
N-Acetyl-L-cysteine	1 mM	+	+	+

17. WENRAIF medium: 20% (v/v) Afamin-Wnt3a serum-free conditioned medium, 50 ng/mL recombinant mouse EGF, 100 ng/mL recombinant mouse Noggin, 1 μg/mL recombinant human R-spondin1, 500 nM A83-01, 100 ng/mL recombinant human IGF-1, 50 ng/mL recombinant human FGF-2, 1× B27 supplement, 1 mM *N*-Acetyl-L-cysteine, and 10 nM [Leu^{15}]-Gastrin I in basal culture medium (*see* Table 1, **Note 7**).
18. Primosin.
19. Tissue culture plates, 48-well flat bottom.
20. Ultra-low attachment plates, 6-well flat bottom.
21. TrypLE Select Enzyme (10×), no phenol red.

2.3 Cryopreservation of Organoids

1. Freezing solution: commercial Recovery Cell Culture Freezing Medium.
2. CoolCell Cell Freezing Container.
3. 1 or 2 mL cryo tubes.

2.4 Thawing Cryopreserved Organoids

1. Matrigel.
2. WENRAIF or WENRAS medium.
3. Tissue culture plates, 48-well flat bottom.

2.5 GFP Labeling

1. Matrigel.
2. WENRAIF medium (*see* Table 1, **Note 8**).
3. Y-27632.
4. DMSO.
5. TrypLE Select Enzyme (10×), no phenol red.
6. Opti-MEM I Reduced Serum Medium.
7. BTXpress buffer.
8. PiggyBac-CMV-MCS-EF1a-GFP-T2A-Puro vector (PB513B-1).
9. Super PiggyBac transposase expression vector (PB210PA-1).
10. Nepa Electroporation Cuvettes 2 mm gap w/pipettes.
11. Puromycin.

2.6 Cell Expansion and Cell Preparation for Transplantation

1. Ultralow attachment plates, 6-well flat bottom.
2. Cell Recovery Solution.
3. Matrigel.

2.7 Orthotopic Xenotransplantation

1. NOD.Cg-*Prkdc*^{scid} *Il2rg*^{tm1Sug}/Jic (NOG) mice, age 7–12 weeks.
2. Isoflurane.

3. 250 mM EDTA in DPBS.
4. Flexible animal feeding needle.
5. Handmade balloon catheter (*see* **Note 9**).

2.8 Endoscopic Observation

1. Isoflurane.
2. Flexible animal feeding needle.

2.9 Tissue Processing

1. 4% paraformaldehyde.
2. Sucrose.
3. OCT compound.

3 Methods

3.1 Human Intestinal Crypt Isolation

1. Prewarm 48-well cell culture plates in a 37 °C CO_2 incubator overnight. Prior to crypt isolation procedures, thaw Matrigel aliquots on ice and keep them cold.
2. Wash human intestinal samples with ice-cold DPBS on a petri dish to remove visible luminal contents (*see* **Note 3**).
3. Remove the stroma using fine scissors on a petri dish, and further shred the remaining epithelium into 1-mm^3 pieces.
4. Transfer the epithelial fragments into a 15-mL centrifuge tube, and add ice-cold DPBS up to 10 mL. Thoroughly wash the fragments by pipetting at least ten times with a 10-mL disposable pipette coated with 10% BSA/PBS. Stand the tube still for approximately a minute to allow the fragments to settle down by gravity, and discard the supernatant. Repeat this step at least five times until the supernatant is free of debris.
5. Add 10 mL of 2.5 mM EDTA in DPBS and secure the tube on a rocking shaker. Rock the tube gently at 4 °C for 60 min to release the crypts.
6. Discard the supernatant after allowing the fragments to settle.
7. Add 10 mL of ice-cold DPBS and pipette up and down vigorously at least ten times with a 10 mL disposable pipette coated with 10% BSA/PBS. Allow the fragments to settle and examine one drop of the supernatant under a microscope to check whether the crypts are sufficiently released. Filter the supernatant through a 70 μm cell strainer to removed debris and collect the crypts into a BSA-coated 50 mL tube. Repeat this procedure several times until sufficient amount of crypts is obtained.
8. Centrifuge the collected supernatants at 200 × *g* for 3 min at 4 °C. Discard the supernatant carefully without disturbing the pellet.

9. Resuspend the pellet in 1 mL of ice-cold DPBS and transfer into a new 15 mL centrifuge tube or a 1.5 mL protein LoBind microcentrifuge tube.
10. Transfer 10–20 μL of the crypt suspension onto a petri dish and count the number of crypts under a microscope. Estimate the total number of crypts in the remaining sample.

3.2 Human Intestinal Organoid Culture

1. Centrifuge the crypt suspension at 400 × *g* for 3 min at 4 °C. Discard the supernatant carefully without disturbing the cell pellet (*see* **Note 10**).
2. Using a 200 μL pipette, suspend the crypt pellet in Matrigel (25 μL of Matrigel for 50–200 crypts). Take care not to aspirate air bubbles into the tip. Perform the procedure on ice to keep Matrigel from solidifying.
3. Apply 25 μL of the crypt–Matrigel suspension onto the center of each well of a prewarmed 48-well plate.
4. Place the plate in a CO_2 incubator (5% CO_2, 37 °C) for 10 min to let Matrigel polymerize.
5. Add 250 μL of the culture medium to each well and incubate at 37 °C. Use the WENRAS or WENRAIF medium for human intestine (*see* Table 1, **Note 7**). To avoid anoikis, supplement the culture medium with 10 μM Rho-associated kinase inhibitor, Y-27632, for the first 2–3 days. To avoid contamination of bacteria, supplement the culture medium with 100 μg/mL Primosin for the first 7 days.
6. Replace the medium with 250 μL of the medium every 2–3 days.
7. Examine the cultures daily and passage the organoids upon outgrowth. Human intestinal organoids generally require passaging every 7 days with a 1:5–6 split ratio (Fig. 1).
8. Before passaging, thaw aliquots of Matrigel on ice and keep them cold. Prewarm a 48-well plate in a 37 °C CO_2 incubator.
9. Add 500 μL of TrypLE Select/DPBS to each well using a 1000 μL pipette (*see* **Note 11**).
10. Scrape and suspend the organoids in TrypLE Select/DPBS and transfer them into a 15 mL centrifuge tube.
11. Incubate the tube at 37 °C in a water bath for 10 min. Every 5 min disrupt the organoids by gently pipetting up and down ten times using a 1000 μL pipette (*see* **Note 12**).
12. Add 10 mL of the basal culture medium and centrifuge at 400 × *g* for 3 min at 4 °C.
13. Discard the supernatant carefully without disturbing the cell pellet.

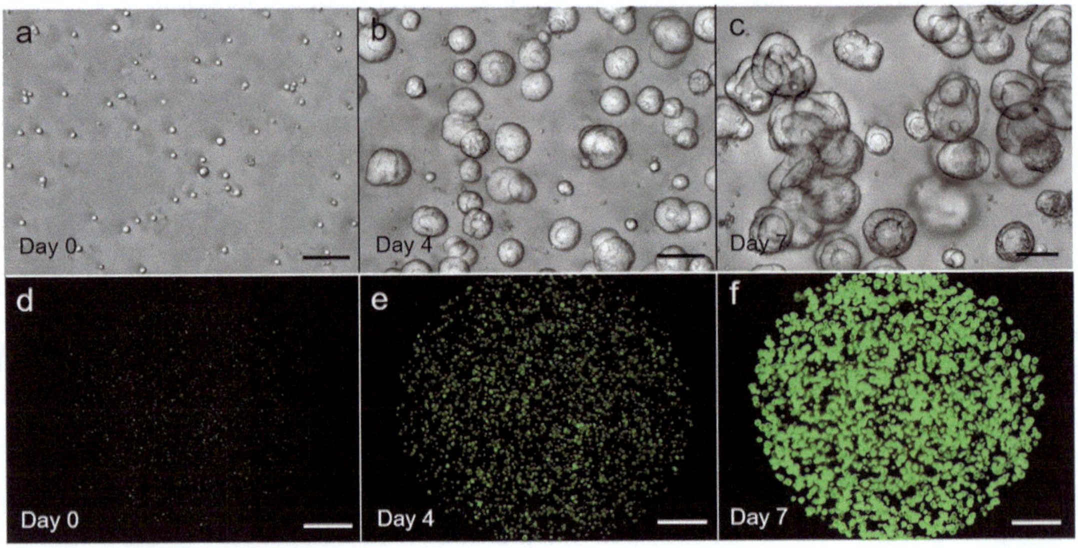

Fig. 1 Human colonic organoids. Single-cell passage day 0 (**a**, **d**), day 4 (**b**, **e**), and day 7 (**c**, **f**). Observation of GFP (**d**–**f**). Scale bars, 100 μm (**a**–**c**) and 1 mm (**d**–**f**)

14. Suspend the cell pellet in Matrigel. Empirically, adequate cell density for passaging human intestinal organoids is around 1000–5000 cells per 25 μL Matrigel.
15. Repeat **steps 3–5** on a new 48-well plate for plating.
16. Alternatively, floating culture on an ultralow attachment plate can be performed as described in Subheading 3.6.

3.3 Cryopreservation of Organoids

1. Use organoids cultured for 2–5 days after passaging for cryopreservation (*see* **Note 13**).
2. Remove the culture medium and add 500 μL of the freezing medium (Subheading 2.3) to each well with a 1000 μL pipette.
3. Scrape the organoids off from the plate and pipet briefly using a 1000-mL pipette to disrupt Matrigel. Transfer the organoid suspension into a 1 or 2 mL cryotube.
4. Place the cryotubes in a cell freezing container and store at −80 °C. After overnight freezing, transfer the cryotubes into liquid N_2 or −150 °C freezer. The organoids can be cryopreserved for at least several years.

3.4 Thawing Cryopreserved Organoids

1. Place a 48-well cell culture plate in a 37 °C CO_2 incubator overnight. Before thawing the cryopreserved organoids, thaw aliquots of Matrigel on ice and keep them cold.
2. Quickly thaw the cryotube containing cryopreserved organoids at 37 °C in a water bath.

3. Transfer the suspension of organoids into a 15 mL centrifuge tube, add 10 mL of the basal culture medium to the tube, and centrifuge at 400 × *g* for 3 min at 4 °C.
4. Discard the supernatant carefully without disturbing the cell pellet (*see* **Note 10**).
5. Resuspend the organoid pellet in Matrigel. One cryovial of organoids is typically plated onto 5–6 wells of a 48-well plate. Maintain organoids as described in Subheading 3.2.

3.5 GFP Labeling by Electroporation

1. Single-cell passage the organoids with TrypLE Select/DPBS 3–4 days before electroporation (Fig. 1a) (*see* **Notes 11** and **12**).
2. Add 250 μL of WENRAIF medium supplemented with 10 μM Y-27632 to each well and incubate at 37 °C. Grow approximately 6–24 wells of intestinal organoids in 48-well plate (*see* **Note 8**).
3. Replace the medium with 250 μL of WENRAIF medium supplemented with 10 μM Y-27632 and 1.25% (v/v) DMSO to each well 1 day before electroporation.
4. Before electroporation, thaw aliquots of Matrigel on ice and keep them cold. Prewarm a 48-well plate in a 37 °C CO_2 incubator.
5. Scrape and suspend the crypt cultures in TrypLE Select/DPBS and transfer into a 15 mL centrifuge tube with a 1000 μL pipette.
6. Incubate the tube at 37 °C in a water bath for 20 min and dissociate the colonies into single cells by pipetting every 5 min with a 1000 μL pipette.
7. Set up and configure the NEPA21 electroporator at the following setting [7]: Poring pulse; Voltage 175 V, Pulse length 5 ms, Pulse interval 50 ms, Number of pulse 2, Delay rate 10%, Polarity +, Transfer pulse; Voltage 20 V, Pulse length 50 ms, Pulse interval 50 ms, Number of pulse 5, Delay rate 40%, Polarity ±.
8. After enzymatic single-cell dissociation, add 10 mL of basal culture medium to the tube. Centrifuge the cells at 400 × *g* for 3 min at 4 °C.
9. Discard the supernatant carefully without disturbing the cell pellet.
10. Add 1 mL of Opti-MEM, pipet well, and filter them through a 20 μm cell strainer.
11. After cell counting or visual estimation of the number in the pellet, transfer 5×10^5 cells (range from 1×10^5 to 1×10^6 cells per each condition) into a 1.5 mL protein LoBind tube.

12. Centrifuge at 400 × *g* for 3 min and completely discard the supernatant carefully without disturbing the cell pellet.
13. Add 90 μL of BTXpress buffer to the cell pellet.
14. Add 7.2 μL of 1 μg/μL PiggyBac vector and 2.8 μL of 1 μg/μL transposase vector into 90 μL of the cell suspension and mix well.
15. Transfer 100 μL of the cell-plasmid mixture into a 2-mm electroporation cuvette.
16. Electroporate using NEPA21 electroporator according to the configured setting.
17. Add 400 μL of BTXpress buffer supplemented with 10 μM Y-27632 to the cuvette. Pipet well and transfer 500 μL of mixture into a new 1.5 mL protein LoBind tube using the plastic pipette provided with the cuvette.
18. Place the tube on a tube rack at room temperature (20–25 °C) for 30 min.
19. Centrifuge at 400 × *g* for 3 min.
20. Discard the supernatant carefully without disturbing the cell pellet and resuspend with 250 μL of Matrigel (1000–5000 cells/well).
21. Plate the organoids in Matrigel and culture in a 48-well plate.
22. Add 250 μL of WENRAIF medium supplemented with 10 μM Y-27632 and 1.25% (v/v) DMSO to each well.
23. Incubate the plate in a 30 °C incubator for cold shock effect [8].
24. One day after electroporation, replace the medium with 250 μL of WENRAIF medium supplemented with 10 μM Y-27632.
25. 2–3 days after electroporation, replace the medium with 250 μL of WENRAIF medium.
26. Three days after electroporation, place the plate to 37 °C incubator.
27. Over at least 3 days after electroporation when the organoids grow as usual, add 2 μg/mL puromycin to the medium until the control organoids die (usually for 2–3 days) (*see* **Note 14**).
28. Replace the medium and grow the surviving organoids as above.
29. Check the GFP at the fluorescence microscope (Fig. 1d–f).

3.6 Cell Expansion and Cell Preparation for Transplantation

1. For large-scale cell expansion, we recommend using the floating culture method: dissociate organoids into single cells using TrypLE Select/DPBS and suspend the cell pellet in the WENRAS or WENRAIF culture medium supplemented with 2.5%

Matrigel. Empirically, adequate cell density for passaging human intestinal organoids is around 10^5–5×10^5 cells per 3 mL of culture medium with 2.5% Matrigel.

2. Dispense 3 mL of the cell suspension to each well of a 6-well ultralow attachment plate and incubate at a 37 °C CO_2 incubator. To avoid anoikis, supplement the culture medium with 10 μM Y-27632.
3. Examine the organoids daily and pipet the culture medium in a well with a 1000 μL pipette when the organoids are aggregated.
4. Passage the organoids upon outgrowth. Organoids typically require passaging once a week. Transfer the organoids in culture medium with 2.5% Matrigel into a 15 mL centrifuge tube, and centrifuge at 400 × *g* for 3 min at 4 °C.
5. Discard the supernatant carefully without disturbing the cell pellet. Repeat **steps 1–3** and replate the dissociated organoids onto a 6-well ultralow attachment plate.
6. Grow GFP-labeled intestinal organoids (equivalent to at least 10^6 cells per mouse) for 3–4 days on a 6-well ultralow attachment plate by floating culture.
7. Before transplantation (*see* Subheading 3.7, **step 15**), transfer the organoids in culture medium with 2.5% Matrigel into a 15 mL centrifuge tube, and centrifuge at 400 × *g* for 3 min at 4 °C.
8. Centrifuge at 400 × *g* for 3 min and discard the supernatant carefully without disturbing the cell pellet.
9. Transfer organoid pellet with approximately 1000 μL of medium into a 1.5 mL Protein LoBind tube. Pipet up and down 50 times with a 200 μL pipette and centrifuge at 800 × *g* for 1 min.
10. Suspend the organoid pellet in 70 μL per mouse of advanced DMEM/F12 with 10% Matrigel and keep them on ice.

3.7 Orthotopic Xenotransplantation

1. Prewarm DPBS and 250 mM EDTA/DPBS in a water bath (50 °C) in mouse room.
2. Anesthetize the NOG mice by inhalation of 2–3% isoflurane in a plastic anesthetizing box. After confirming adequate levels of anesthesia, place the mouse in a supine position putting on a 40 °C heating pad to keep them warm (*see* **Note 15**).
3. To empty the luminal contents within 3 cm from the anus, promote defecation by rubbing the mouse abdomen. Flush the mouse colon with minimal amount (around 200 μL) of DPBS using a thin catheter (Fig. 2a) (*see* **Note 16**).

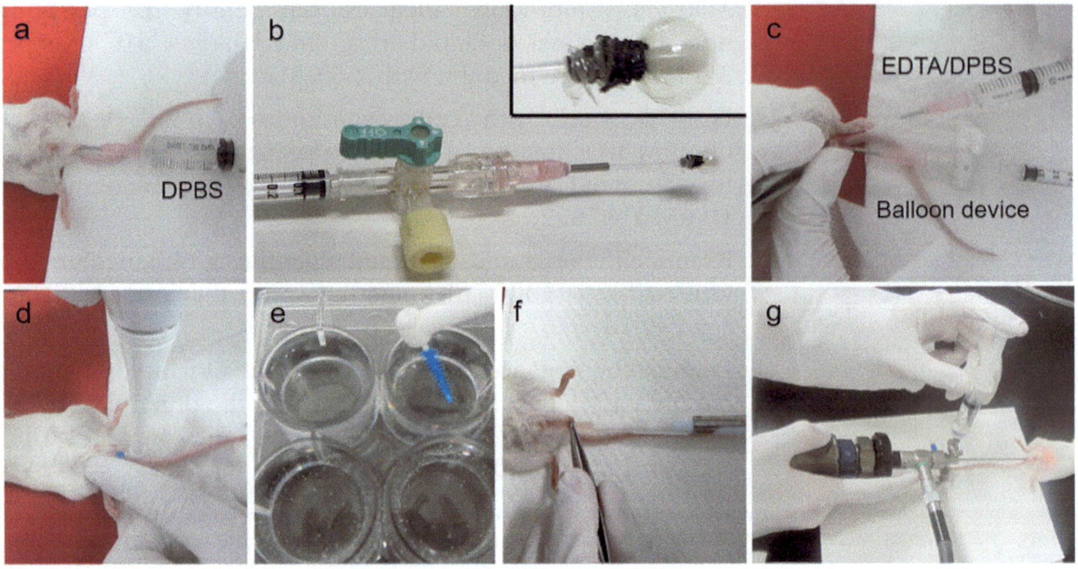

Fig. 2 Overview of mouse procedures. (**a**) Flushing the mouse colon using a thin catheter. (**b**) A handmade balloon device. (**c**) Filling EDTA/DPBS between the dilated balloon and anal verge covered with tweezers. (**d**) Epithelial abrasions using an electric toothbrush. (**e**) Observation of isolated crypts in a 24-well plate. Upper left is a negative control well. (**f**) Infusion of cell suspension into the anal. (**g**) Underwater colonoscopic observation for mice

4. Insert a handmade thin catheter (1 mm in diameter) equipped with a small balloon (Fig. 2b) via a transanal approach (Fig. 2c) (*see* **Note 9**).
5. Inflate the balloon with approximately 50–100 μL of air with a 1 mL syringe and lock it in the inflated position.
6. Infuse approximately 2 mL of prewarmed DPBS into the anus with a 5 mL syringe, and confirm the balloon inflation.
7. Infuse prewarmed EDTA/DPBS into the anus with a 2.5 mL syringe. To achieve topical exposure of the colonic mucosa to filled EDTA/DPBS, cover luminal space between the dilated balloon and anal verge with tweezers (Fig. 2c) (*see* **Note 17**).
8. After 2 min of exposure, wash the lumen with DPBS with a 5 mL syringe. After deflating the balloon, remove the catheter from the colon.
9. Create semicircumferential epithelial abrasions using an electric toothbrush with a soft interdental brush. Insert the head of the brush 1.5 cm into the colon, subsequently turn on the power and gently scratch the luminal surface in a half circle (Fig. 2d) (*see* **Note 18**).
10. Confirm the success of generation of colonic epithelial injury by observation of isolated crypts in a 24-well plate (Fig. 2e).

11. Insert the head of the brush 1.5 cm into the colon again, and gently scratch the luminal surface in a half circle with power remains off.
12. Gently repeat **steps 10** and **11** 2–5 times until sufficient epithelial removal can be confirmed (*see* **Note 18**).
13. Repeat **steps 4** and **5** and wash the lumen with approximately 2 mL of DPBS with a 5 mL syringe. After deflating the balloon, remove the catheter from the colon.
14. Put the mouse back into the cage and repeat **steps 4–14** for another mouse.
15. Keep the cell suspension on ice until transplantation and bring to mouse room.
16. After apparent bleeding stops within 2 h after colonic epithelial injury, anesthetize the epithelial-injured NOG mice. Confirm the absence of luminal contents within 3 cm from the anus again by using a thin catheter before organoid transplantation.
17. Using a 200 μL pipette and pipette tips, pipet the cell suspension briefly to prevent aggregation of cells.
18. Infuse 70 μL of cell suspension into the anus and pinch the anus with tweezers to prevent leakage of infused cells from the anus verge (Fig. 2f).
19. Temporary close the anus by attaching a bedding material to the anal verge with an adhesive to promote the retention of transplanted cells at the rectum. Put the mouse back into the cage.
20. Remove the bedding material from the anal verge after 3–6 h. Monitor the presence of stool carefully for a week until usual stool comes out. When they showed signs of irreversible bowel obstruction or became moribund, euthanize mice.

3.8 Endoscopic Observation

1. Endoscopic observation for confirming the success of xenograft is recommended ≥2 weeks after xenotransplantation (*see* **Note 19**). Set up the endoscopic device. Attach the light source cable, and camera head to the colonoscope covered with an endoscopic sheath.
2. Attach a 5 mL syringe filled with DPBS to the connection for air pump instead of air pump, and flush DPBS to the colonoscope (*see* **Note 19**).
3. Anesthetize the NOG mice by inhalation of 2–3% isoflurane in a plastic anesthetizing box. After confirming adequate levels of anesthesia, place the mouse in a supine position on a clean paper.

4. To empty the luminal contents within 3 cm from the anus, promote defecation by rubbing the mouse abdomen. Flush the mouse colon with approximately 2 mL of DPBS using a thin catheter with a 5 mL syringe.
5. Carefully insert the colonoscope into the rectum and fill the colon with a small amount of water (Fig. 2g) (*see* **Note 19**).
6. Carefully push the endoscope approximately 3 cm from the anus under visual control.
7. Record endoscopic movie during slow withdraw of the colonoscope. To obtain a clear view, infuse a small amount of water when necessary.
8. Remove the endoscope and attach an observation filter specific for GFP between the camera head and colonoscope.
9. Carefully push the endoscope approximately 3 cm from the anus again. Press the foot pedal to switch GFP observation mode, and record endoscopic movie during slow withdraw of the colonoscope. Put the mouse back into the cage.

3.9 Tissue Processing

1. Sacrifice the mouse and isolate the distal colon (2 cm from the anus) of each mouse.
2. Open the colon longitudinally on the opposite site of transplanted area using a scissors under stereomicroscopic observation with GFP fluorescence.
3. Stretch them flat on filter paper using ring tweezers and fix at least 3 h in 4% paraformaldehyde.
4. Cut the engrafted site of colon tissue samples using a knife under observation with GFP (Fig. 3a).
5. For paraffin sections, embed them in paraffin. For frozen sections, immerse them 15% sucrose in DPBS for 2 h, followed by 30% sucrose in DPBS overnight or until the tissue sinks to the bottom of the jar. Then, embed them in OCT compound (Fig. 3b, c) and freeze tissue block in liquid N_2. To avoid cracking the block, place the bottom third of the block into the liquid N_2 until all but the center of the OCT compound is frozen, and allow freezing to conclude on ice. Store frozen blocks at −80 °C.

4 Notes

1. Reconstituted basal culture medium can be stored at 4 °C for at least 2 months.
2. To prevent cells sticking to the sides of the tubes and pipettes, fill the tubes with 10% BSA/PBS, and then aspirate the solution by pipettes.

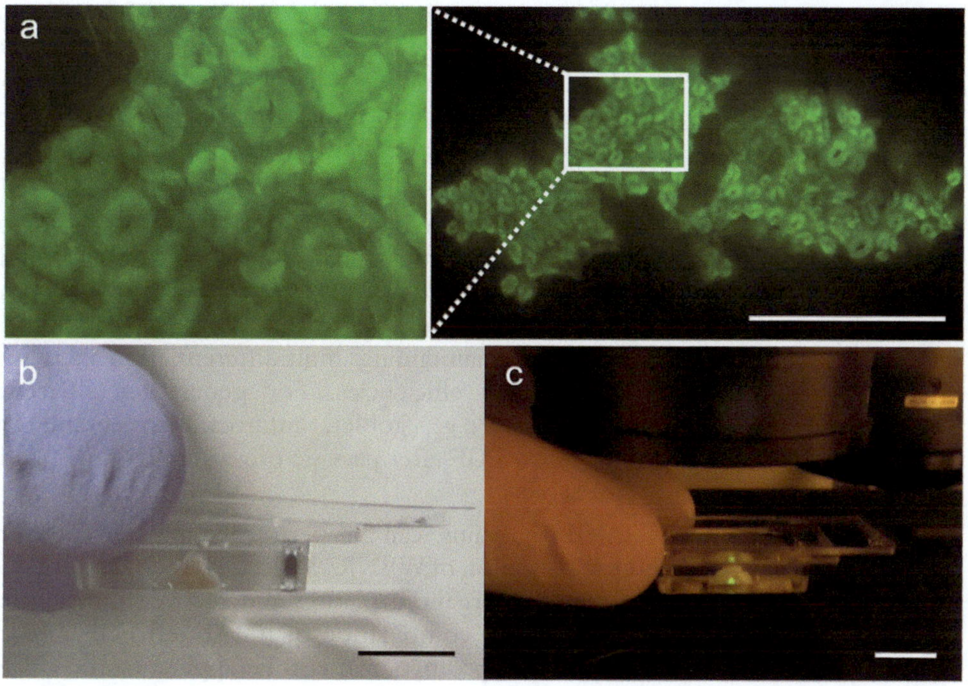

Fig. 3 Representative images of human colonic xenograft in NOG mice. (**a**) Xenotransplanted GFP-labeled colonic organoids reconstruct human colonic crypts. (**b**) Xenograft embed in the bottom of the cryo dish with OCT compound in prior to frozen sectioning. (**c**) Under fluorescent observation, GFP-labeled area can be detected. Insets show higher magnification. Scale bars, 1 mm (**a**) and 1 cm (**b**, **c**)

3. Samples should be obtained from patients with written informed consent under an approval of the appropriate ethical committee. Keep the samples be in ice-cold DPBS until crypt isolation. Although the samples can be stored up to 6 h on ice with minimal of loss of viability, we strongly recommend to proceed to the following steps as soon as possible.
4. An original 10 mL vial of Matrigel must be thawed on ice and aliquoted into precooled cryogenic tubes. Store the 1 mL aliquots at −20 °C and thaw on ice before use. They can be refrozen and thawed several times without substantial loss of culture efficiency.
5. The activity of recombinant Wnt3a from a commercial source is insufficient to culture human intestinal organoids. Afamin forms a stable complex with Wnt proteins and has higher biological activity without the addition of serum [3]. We confirmed lowering the concentration of Afamin-Wnt3a serum-free conditioned medium (J-ORMW301R) to 10% (v/v) does not affect the culture efficiency.
6. An economic way would be to alternatively use conditioned medium from mouse R-spondin1-Fc- [9] and Noggin-Fc- [10] producing HEK293T cell lines. Supplement the basal

culture medium with 10% (v/v) R-spondin1-conditioned medium and 10% (v/v) Noggin-conditioned medium. Divide into 2 mL aliquots and store at −20 °C for up to 6 months. When using a new batch of medium, the differences between the batches need to be tested by growing the organoids with the previous batch and the new batch at various concentrations.

7. Complete culture medium can be stored at 4 °C for up to 7 days. Through this protocol, conventional condition medium (WENRAS) is sufficient to perform experiments. However, refined condition medium (WENRAIF) has some advantages in maintaining multidifferentiation capacity and genome editing efficiencies. For generating differentiated secretory cells (e.g., goblet, enteroendocrine, and Paneth cells), remove EGF after passage onward (WNRAIF medium) [4] (*see* Table 1).

8. While GFP labeling can be performed by the conventional medium condition of WENRAS and CHIR99021 as previously described [7], organoid plating and recovery efficiencies are improved by the combination of IGF-1 and FGF-2 [4]. This electroporation protocol can also be applied to the CRISPR/Cas9-mediated genome engineering of organoids using designed pX330 donor vector as previously described [7]. Genome editing of the organoids could also be achieved by the delivery of the expression vectors by lentivirus infection [11].

9. Tightly bound the balloon from natural rubber latex textured condom with a thread at the tip of the flexible animal feeding needle [5, 6]. Adjust its size to the maximum inflation volume of approximately 100 μL. The balloon device can be reused while it works well. Before use, check the proper inflation and deflation of the balloon with no air leak. The method movie for preparation of balloon catheter device is available on the original report's web page [5].

10. Discard as much supernatant as possible to avoid dilution of Matrigel because diluted Matrigel breaks easily.

11. Use TrypLE Select after dilluting TrypLE Select Enzyme (10×) with DPBS (3×). While enzymatic passaging can be performed using TrypLE Express, TrypLE Select Enzyme dissociates the organoids more easily.

12. Organoids are passaged either as cell colonies or as single cells. For single-cell passage, extend the incubation time in a water bath to 20 min and dissociate the colonies into single cells by pipetting every 5 min. After single-cell dissociation, remove cell colonies from cell suspension with a 20-μm cell strainer. Passaging as cell colonies can also be achieved by mechanical dissociation using a fire-polished Pasteur pipette coated with 10% BSA/PBS.

13. Cryopreservation of excessive grown organoids may result in lower cell viability of cryopreserved organoids after thawing.
14. If the puromycin selection fails because the control does not die, try again after single-cell passage. Visual pickup under microscope or fluorescence- activated cell sorting is available to select GFP-high expressing organoids.
15. Approval of Institutional Animal Care and Use Committee and Safety Committee on Genetically Modified Organisms are needed. There is no need for fasting. NOG mice ($\geq$7 weeks old) weighting 20–25 g are desirable. Fasting the mice is not necessary. To promote defecation, slowly anesthetize in a plastic anesthetizing box.
16. Washing with a larger amount of DPBS may complicate the following procedures due to the outflow from the oral colon.
17. This step requires the most experience. Because the EDTA leakage in the oral side of colon may kill the mice, the infused EDTA/DPBS must be blocked by the inflated balloon. Gently infuse EDTA/PBS up to 100 μL since excessive pressure results in perforation of the colon, while maintaining modest pressure during the procedure is important for an efficient epithelia removal. When infused EDTA/DPBS leaked during this procedure, the leakage should be supplemented with an equal volume of EDTA/DPBS infusion.
18. Gently push the one side of the lower abdomen to attach the brush to the side of the colonic lumen. It is important to scratch the bottom of the crypt as well as the surface layer. Confirm successful epithelial removal by visual inspection of released crypts when washing the brush in water. To prevent excessive damage, only use electric vibration during the first scratch. Severe bleeding during the abrasion suggests the insufficient EDTA treatment.
19. Endoscopic observation is useful for confirming the success of xenotransplantation before sacrifice but is not essential. To prevent excessive insufflation and obtain clear view, we use water infusion instead of traditional air insufflation during the endoscopic procedure [5]. To reduce the load of the mice, be careful not to excessively infuse water.

Acknowledgments

This work was in part supported by the Research Center Network for Realization of Regenerative Medicine project from the Japan Agency for Medical Research and Development.

References

1. Sato T, Vries RG, Snippert HJ et al (2009) Single Lgr5 stem cells build crypt-villus structures in vitro without a mesenchymal niche. Nature 459:262–265
2. Sato T, Stange DE, Ferrante M et al (2011) Long-term expansion of epithelial organoids from human colon, adenoma, adenocarcinoma, and Barrett's epithelium. Gastroenterology 141:1762–1772
3. Mihara E, Hirai H, Yamamoto H et al (2016) Active and water-soluble form of lipidated Wnt protein is maintained by a serum glycoprotein afamin/alpha-albumin. Elife 5:e11621. https://doi.org/10.7554/eLife.11621
4. Fujii M, Matano M, Toshimitsu K et al (2018) Human intestinal organoids maintain self-renewal capacity and cellular diversity in niche-inspired culture condition. Cell Stem Cell 23:787–793.e6
5. Sugimoto S, Ohta Y, Fujii M et al (2018) Reconstruction of the human colon epithelium in vivo. Cell Stem Cell 22:171–176.e5
6. Fukuda M, Mizutani T, Mochizuki W et al (2014) Small intestinal stem cell identity is maintained with functional Paneth cells in heterotopically grafted epithelium onto the colon. Genes Dev 28:1752–1757
7. Fujii M, Matano M, Nanki K et al (2015) Efficient genetic engineering of human intestinal organoids using electroporation. Nat Protoc 10:1474–1485
8. Seino T, Kawasaki S, Shimokawa M et al (2018) Human pancreatic tumor Organoids reveal loss of stem cell niche factor dependence during disease progression. Cell Stem Cell 22:454–467.e6
9. Ootani A, Li X, Sangiorgi E et al (2009) Sustained in vitro intestinal epithelial culture within a Wnt-dependent stem cell niche. Nat Med 15:701–706
10. Farin HF, Van Es JH, Clevers H (2012) Redundant sources of Wnt regulate intestinal stem cells and promote formation of Paneth cells. Gastroenterology 143:1518–1529.e7
11. Koo BK, Stange DE, Sato T et al (2011) Controlled gene expression in primary Lgr5 organoid cultures. Nat Methods 9:81–83

Chapter 22

Advanced Colorectal Cancer Orthotopic Patient-Derived Xenograft Models for Cancer and Stem Cell Research

Irene Chicote, Juan Antonio Cámara, and Héctor G. Palmer

Abstract

In the recent years has being a great expansion of new preclinical models of colorectal cancer (CRC) based on patient-derived cells, from ex vivo 2D cell lines, toward 3D tumoroids or animal xenografts. These new technologies have been key to overcome historical limitations in CRC research such as precision medicine, pharmacogenomic screenings, or investigating mechanism of drug resistance. Here we describe a method to generate metastatic CRC in mice with patient-derived cells and the evaluation of drug response with computerized tomography. CRC at this advanced stage is the most frequent situation in patients enrolled in therapies with novel drugs that in some cases are designed to target metastatic cells. Therefore, these orthotopic models could be considered the best to recapitulate advance CRC and are therefore becoming instrumental to investigate the biology behind drug-response in metastatic disease.

Key words Colorectal cancer, PDX, Metastasis, Stem cells

1 Introduction

Patient-derived xenograft (PDX) models faithfully recapitulate original CRC at all relevant levels, from histology, molecular traits, and drug response [1]. The capacity of self-renewal and pluripotency of patient-derived cells demonstrate the existence of cancer stem cells in these models [2] (Fig. 1). Indeed, a minor population of colon cancer stem cells retain most of the tumor initiation potential present in PDXs as describe in patient samples [3–5].

These models are becoming the gold standard in cancer research particularly for the study to drug sensitivity or resistance. Most studies are based on experiments with tumors growing subcutaneously in immunodeficient mice. However, advanced CRC patients present metastatic lesion in liver, lungs or peritoneal carcinomatoses. The site of growth determines relevant aspects of tumor biology such as a distinctive vascularization, cross talk with immune system and most importantly the prognosis of the disease and patient's overall survival. As a consequence metastatic lesions may

Paloma Ordóñez-Morán (ed.), *Intestinal Stem Cells: Methods and Protocols*, Methods in Molecular Biology, vol. 2171,
https://doi.org/10.1007/978-1-0716-0747-3_22,

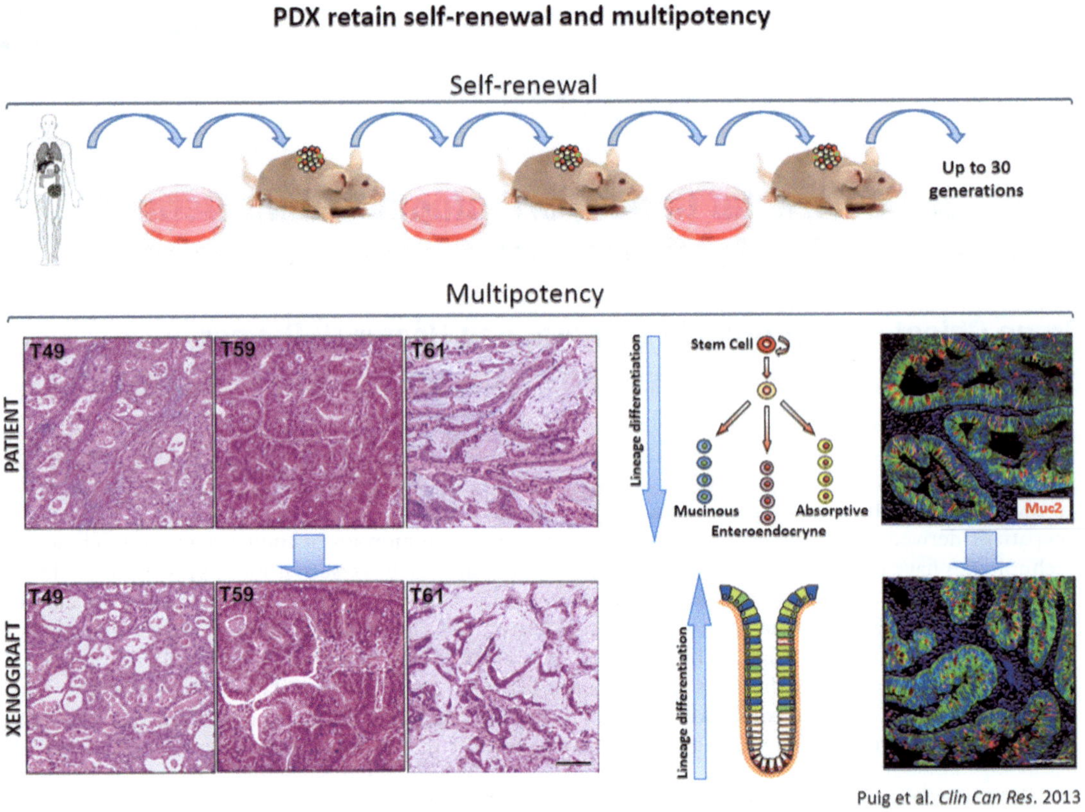

Fig. 1 The capacity of self-renewal and pluripotency of patient-derived cells demonstrate the existence of cancer stem cells in these models

not show equivalent drug responses than tumors growing subcutaneously. This can be particularly relevant when evaluating the response to drugs targeting molecules that are key for metastasis initiation or growth. Here we describe how to implant patient-derived cells in the cecum wall of immunodeficient mice, and how to evaluate intestinal tumor growth using microCT scan and liver or lung metastasis by histology at experiment end point. These orthotopic models permit to evaluate the activity of new drugs in a CRC advanced setting equivalent to clinically relevant scenarios. It is a particularly sensitive model to measure the response to drugs targeting metastasis.

2 Materials

2.1 Derivation of Patient Cells (for Product Concentrations See *Table 1)*

1. Forceps and surgical scissors (sterilize before use).
2. Phosphate-buffered saline (PBS), sterile.
3. Blade #24, sterile.
4. 29G U-100 insulin syringes
5. Filter strainer 100 μm.

6. Ammonium Chloride Solution RBC Lysis Buffer:
   NH_4Cl (ammonium chloride) 8.02 g. $NaHCO_3$ (sodium bicarbonate) 0.84 g. EDTA (disodium) 0.37 g. QS to 100 mL with Millipore water. Store at 4 °C.
7. Matrigel, Basement Membrane Matrix, 5 mL.
8. DNase I.
9. Collagenase.
10. DMEM/F12 Liq.
11. D-Glucose.
12. Apotransferrin.
13. Insulin.
14. Putrescin.
15. Sodium selenite (SS).
16. Progesterone.
17. Pen/Strep.
18. Fungizone.
19. Kanamycin and gentamycin.
20. Nystatin.
21. B27 supplement.
22. Heparin sodium salt (HSS).
23. Nonessential amino acids.
24. Sodium pyruvate.
25. L-Glutamine.
26. Epidermal growth factor (EGF).
27. Fibroblast growth factor-2 (FGF2).
28. CoCSCM 6Ab medium (Table 1).

2.2 Orthotopic Injection in Cecum

1. 8-week-old NOD.CB17-Prkdcscid/NcrCrl mice.
2. 29G needles (insulin syringes).
3. 1×10^6 patient-derived cells in 50 μL of PBS per injection.
4. Phosphate-buffered saline (PBS), sterile.
5. Betadine.
6. 70% alcohol
7. Buprenorphine.
8. Isoflurane.
9. Suture PROLENE 5-0.
10. 29G U-100 insulin syringes
11. Forceps and surgical scissors (sterilize before use).
12. Heating pad.

Table 1
Colon cancer stem cell media

	Stock concentration	Final concentration	Dil factor	Volume	Stock preparation
Digestion medium	(prepare to use)				
CoCSCM			1×	5 mL	Commercial
DNAse I	8 kU/mL	0.08 kU/mL	100×	50 μL	10 kU/1.25 mL NaCl 150 mM
Collagenase	150 mg/mL	1.5 mg/mL	100×	50 μL	500 mg/3.3 mL Tris 50 mM ClCa 0.36 mM pH 7 a 37 °C
Growth factors (GF) MIX 10× (200 mL)	(make 10 mL aliquots and store to −20 °C)				
DMEM/F12 Liq	300 mg/mL (30%)	60 mg/mL (6%)	1×	128 mL	Commercial
D-(+)-glucose	10 mg/mL	1 mg/mL	5×	40 mL	30 g in 100 mL H_2O MQ + filter 0.22 μm
Apo-transferrin	5 mg/mL	0.25 mg/mL	10×	20 mL	Prepare in sterile H_2O MQ
Insulin	9.6 mg/mL	96 μg/mL	20×	10 mL	Commercial
Putrescine	520 μg/mL	52 ng/mL	100×	2 mL	Prepare in sterile H_2O MQ+ filter 0.22 μm
Sodium selenite	630 μg/mL (2 mM)	63 ng/mL	10,000×	20 μL	Prepare in sterile H_2O MQ
Progesterone			10,000×	20 μL	Prepare in EtOH absolute
Final volume				200 mL	
CoCSCM 6Ab without EGF, FGF2 and growth factors (final volume 450 mL)					
DMEM/F12 Liq			1×	402 mL	Commercial
Pen/strep	10,000 U/ 10000 μg/mL	100 U/ 100 μg/mL	100×	5 mL	Commercial
Fungizone	250 μg/mL	10 μg/mL	25×	20 mL	Commercial
Kanamycin	1000 μg/mL	10 μg/mL	100×	5 mL	Commercial

(continued)

Table 1 (continued)

	Stock concentration	Final concentration	Dil factor	Volume	Stock preparation
Gentamycin	50 mg/mL	50 μg/mL	1000×	500 μL	Commercial
Nystatin	5 mg/mL	5 μg/mL	1000×	500 μL	Prepare in DMSO
B27 supplement	250×	1×	250×	2 mL	Commercial
Heparin sodium salt	40 mg/mL	4 μg/mL	10,000×	50 μL	Prepare in sterile H_2O MQ
Nonessential amino acids	100%	1%	100×	5 mL	Commercial
Sodium pyruvate	100%	1%	100×	5 mL	Commercial
L-Glutamine	200 mM	2 mM	100×	5 mL	Commercial
CoCSCM 6Ab complete	(prepare FRESH) store to 4 °C (max. 1 month)				
Growth factors mix (GF)	10×	1×	10×	10 mL	
EGF	500 μg/mL	20 ng/mL	25,000×	4 μL	Prepare in sterile H_2O MQ
FGF basic	100 μg/mL	10 ng/mL	10,000×	10 μL	Prepare in sterile H_2O MQ
CoCSCM without EGF, FGF2 and GF				90 mL	Commercial
Final volume				100 mL	

2.3 Micro-CT Scanning for Orthotopic Tumor Evaluation

Imaging studies acquired with microCT were performed with a Perkin Elmer's Quantum FX microCT Imaging system (Perkin Elmer. 940 Winter St. Waltham, Massachusetts. EEUU). This piece of equipment is specifically designed for lab animal imaging studies (http://www.perkinelmer.com/es/product/quantum-gx-instrument-120-240-cls140083).

3 Methods

3.1 Derivation of Patient Cells

3.1.1 Tumor Extraction

Cells are obtained from patients by surgery or biopsy or from a tumor xenograft growing subcutaneously in mice. In the case of a mouse xenograft, tumor is removed using a sterile technique. Extract the tumor from the mice (free from the skin), carefully removing as much excess of surrounding tissue. In all cases, store the harvested tumors in PBS on ice until the digestion procedure (*see* **Note 1**).

3.1.2 Cell Preparation

1. All the procedure should be performed in a biological cabinet at room temperature (RT).
2. In a 10 cm Petri dish with 1 mL of complete CoCSCM 6Ab medium (Table 1) (to make mincing easier), dissect the tumor with a scalpel blade until you get a homogeneous sample and place into a 15 mL conical tube (*see* **Note 2**).
3. Add up to 5 mL of complete CoCSCM 6Ab medium (no more than 3 mL of disaggregated tissue in the same tube) (*see* **Note 3**).
4. Add 50 μL of DNase I and 50 μL of collagenase. Vortex the sample.
5. Incubate the tube 1 h at 37 °C in the cell culture incubator in an inclined position. Pipet the sample every 15 min with a 5 mL pipette.
6. Dilute the 5 mL digested mixture with complete CoCSCM 6Ab medium at a 1:1 ratio.
7. Filter the mixture with a 100-μm cell strainer into another sterile 50 mL conical tube.
8. Centrifuge the filtered cells at 500 × *g*, 10 min at room temperature.
9. Remove supernatant.
10. Resuspend in 3 mL of RCB Lysis Buffer.
11. Incubate for 10 min at RT.
12. Add 3 mL of complete CoCSCM medium and centrifuge at 500 × *g*, 10 min at RT.
13. Cell counting: Resuspend the pellet with 5–10 mL of complete CoCSCM 6Ab medium (depending on how big the pellet is).
14. Make sure you have a homogeneous cell suspension.
15. The tumor cell suspension should be loaded into the syringe prior to extraction of the cecum and kept on ice (*see* **Note 4**).

3.2 Orthotopic Injection in Cecum

1. Remove hair from the surgical area.
2. Anesthetize the mouse using 2% isoflurane in an anesthesia chamber and placed in supine position in the anesthesia nose.
3. The surgical site is draped in a sterile fashion.
4. The abdomen is then disinfected for the procedure.
5. Grab the skin with forceps and make an approximately 1 cm longitudinal incision over the lower abdomen. Gently free the skin from the peritoneum.
6. Grab the peritoneum lifting it straight upward and make a small incision. Extend the peritoneal incision.
7. The cecum is identified and exteriorized (*see* **Note 5**).

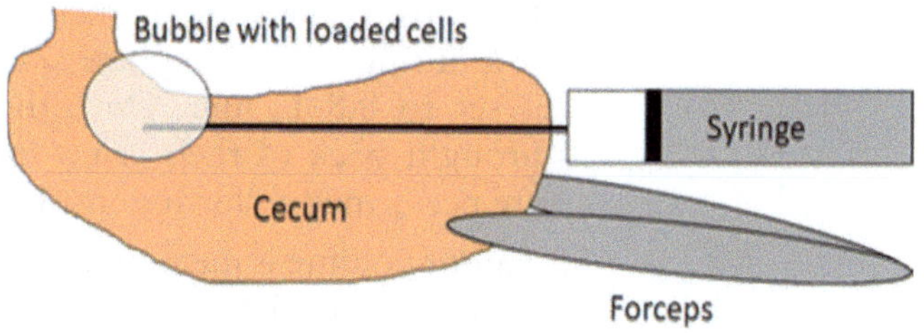

Fig. 2 Diagram showing the best position for cell injection in the mouse cecum

8. The cecum is isolated from the rest of the mouse using a precut, sterile gauze.
9. Use saline solution to keep the cecum moist during the entire procedure.
10. Grab the cecum blind ending pouch and insert the needle into cecal wall (Fig. 2). The insertion should be superficial avoiding capillaries and vessels until the injection area. Make sure that bubbles are removed from the cell suspension (*see* **Note 4**).
11. Inject slowly the cell suspension. The needle should not penetrate caecum lumen because cell suspension would be eliminated from the body through intestinal peristalsis (bleeding may occur from superficial vessels) (*see* **Note 6**).
12. After the injection, slowly retract the needle from the cecum and place gentle pressure on the needle insertion site with a cotton-tipped applicator for several seconds (*see* **Note 7**).
13. Clean with saline solution and discard the gauze.
14. The cecum is returned into the abdomen.
15. Using 5-0 suture, close the peritoneal layer.
16. Using 5-0 suture, close the skin.
17. Apply postoperative antibiotic and analgesics. Place and keep the mice on a heating pad until they recover.

3.3 Micro CT Scanning for Orthotopic Tumor Evaluation

1. Animals should be controlled daily. Tumor growth is monitored by microCT starting 2 weeks after cell injection.
2. Knowing the timing for the contrast agent to arrive to the cecum, an oral dosage of contrast agent is administered via oral gavage. After waiting for this timepoint, an intraperitoneal injection of contrast agent is administered and CT scan is acquired.
3. The parameters defined for the acquisition of the images are FOV30 mm, acquisition time 26 s, current voltage 90 kV, and current amperage 200 μA. Reconstruction of the studies is performed with the microCT software, based on Feldkamp's method.

4. Previous to the scan, animals are anesthetized with isoflurane (5% during induction phase, 2% during maintenance). Air flow is set to 0.8 L/min. Once the scan is finished, animals are brought back to their cages for recovery. All the procedures are performed following institutional ethics committee.
5. Tumor volume is measured in three different axes and ellipsoid volume is calculated using the following formula: VOL = 4/3PI × (semiaxis X × semiaxis Y × semiaxis Z).

4 Notes

1. Tumors should be digested for preparing cell suspension as soon as possible after removal from their original location in patients' lesion or subcutis in mice. Cell viability drops significantly 24 h after tissue removal compromising an effective implantation in recipient mice.
2. Tumor tissue has to be minced intensively and digested tissue repetitive pipetted for an optimal single cell preparation. This is crucial for an accurate counting of cells prior injection in recipient mice.
3. The cocktail of six antibiotics is essential to remove bacteria present in the original tumor tissue sample. This is particularly relevant in the case of pieces of primary colon or rectal tumors resected from patients that are naturally contaminated with bacteria. Injecting cancer cells contaminated with bacteria in recipient immunodeficient mice can be lethal for the animals.
4. Eliminating air bubbles for syringes loaded with the cell preparation is important to avoid excessive injection volume in the cecum wall, potential tissue break and sample loss.
5. Be extremely careful when exteriorizing cecum since excessive manipulation of internal organs can be fatal for animals.
6. The injection of cells in the cecum is the most delicate part of the whole protocol. This should be done with a bright light focused on the injection site and using a magnifying loupe. Needle should be positioned parallel to the mouse cecum surface. Cecum should be immobilized with forceps but applying soft pressure to avoid tissue damage, perforation or hemorrhages. When cells are implanted properly, a bubble of white material (pellet of cells) will be visualized. When this does not occur it is mostly due to a perforation of the cecum. In this case cells mostly end in the cecum lumen where they will be eliminated by the intestinal tract.

7. Pressuring with a cotton-tipped applicator when removing the needle form the cecum is also essential to avoid the scape of injected cells and bleeding. Breaking capillary or small blood vessels frequently occurs without major consequences.

Acknowledgments

We acknowledge Cellex Foundation, CIBERONC network, and Instituto de Salud Carlos III for their support.

References

1. Byrne AT, Alferez DG, Amant F, Annibali D, Arribas J, Biankin AV et al (2017) Interrogating open issues in cancer precision medicine with patient-derived xenografts (ReviewResearch support, non-U.S. Gov't research support, N.I. H., extramural). Nat Rev Cancer 17 (4):254–268. https://doi.org/10.1038/nrc.2016.140
2. Puig I, Chicote I, Tenbaum SP, Arques O, Herance JR, Gispert JD et al (2013) A personalized preclinical model to evaluate the metastatic potential of patient-derived colon cancer initiating cells (research support, non-U.S. Gov't). Clin Cancer Res 19(24):6787–6801. https://doi.org/10.1158/1078-0432.CCR-12-1740
3. Kreso A, van Galen P, Pedley NM, Lima-Fernandes E, Frelin C, Davis T et al (2014) Self-renewal as a therapeutic target in human colorectal cancer (research support, non-U.S. Gov't). Nat Med 20(1):29–36. https://doi.org/10.1038/nm.3418
4. O'Brien CA, Pollett A, Gallinger S, Dick JE (2007) A human colon cancer cell capable of initiating tumour growth in immunodeficient mice (research support, non-U.S. Gov't). Nature 445(7123):106–110. https://doi.org/10.1038/nature05372
5. Puig I, Tenbaum SP, Chicote I, Arques O, Martinez-Quintanilla J, Cuesta-Borras E et al (2018) TET2 controls chemoresistant slow-cycling cancer cell survival and tumor recurrence. J Clin Invest 128(9):3887–3905. https://doi.org/10.1172/JCI96393

Chapter 23

Modeling Colorectal Cancer Progression Through Orthotopic Implantation of Organoids

Felipe de Sousa e Melo, Jonathan M. Harnoss, Noelyn Kljavin, Ryan Scott, Catherine Sohn, Kevin G. Leong, and Frederic J. de Sauvage

Abstract

Colorectal cancer (CRC) related death has often been attributed to the presence of metastatic disseminated disease. A concise understanding of the molecular mechanism(s) that drive metastatic progression is therefore needed but has thus far been hampered by the limited number of CRC mouse models that progress toward this disease stage. In addition, preclinical evaluation of therapeutic modalities aimed at managing metastatic disease also rests on the availability of relevant in vivo models that faithfully recapitulate the key molecular features of metastatic human CRC. To overcome these limitations, we have recently developed methodologies that enable the study of CRC progression at relevant orthotopic sites. Here, we provide a detailed methodology that describes the injection of CRC derived cell lines and organoids directly into the colorectal mucosa. This results in the growth of a single tumor mass within the colon, that can spontaneously metastasize to the liver. Furthermore, we also present a surgical procedure to directly inject cells into the portal venous circulation to induce CRC tumor growth in the liver without the requirement of a primary tumor.

Key words Colorectal cancer, Metastasis, Mouse models

1 Introduction

Colorectal cancer (CRC) is a leading cause of cancer related death [1] and is largely thought to progress through the acquisition of specific genetic alterations, including functional loss of the tumor suppressors *APC*, *TP53*, and *SMAD4* as well as activating mutations in *KRAS* [2]. Despite extensive biological, molecular, and clinical knowledge, CRC remains a high unmet medical need. This is especially the case once the tumor has disseminated beyond its primary site. Indeed, patients that present with metastatic disease, liver metastases in particular, have an overall dismal prognosis [1, 3, 4]. Hence, a better understanding of the processes that drive CRC

Felipe de Sousa e Melo and Jonathan M. Harnoss contributed equally to this work.

Paloma Ordóñez-Morán (ed.), *Intestinal Stem Cells: Methods and Protocols*, Methods in Molecular Biology, vol. 2171, https://doi.org/10.1007/978-1-0716-0747-3_23,

progression, metastatic dissemination, and maintenance are key to enable the development of novel and effective therapies. To do so, preclinical in vivo models that faithfully recapitulate the human disease are critically needed. Despite the availability of xenograft, chemically induced, and genetically engineered mouse models of CRC, these models fail to recapitulate full disease progression, including dissemination to the liver, the most relevant site of metastasis of human CRC. This is evident with the Apc^{min} mouse model for example, wherein tumors do not progress beyond the stage of adenoma and fail to develop distant metastasis, with animals succumbing due to high intestinal tumor burden [5]. To circumvent this issue, Cre-lox technology can be used to spatiotemporally control recombination of multiple alleles in the gut. This approach has been successfully employed to control primary tumor burden and to generate models that progress to more advanced stages of disease including metastatic dissemination [6–8]. Nevertheless, the long latency and low penetrance of metastasis in these models has made it challenging to harness them for therapeutic intervention in the metastatic setting. Several colon orthotopic transplantation techniques have been described to date [5, 6, 9–12], including the injection of cancer cell suspensions directly into serosal wall of the cecum, or the surgical implantation of intact tumor fragments onto the serosal side of the cecal wall [10]. One limitation of cecum implantation is the fact that tumors are implanted on the serosal side of the cecal wall, thus bypassing the requirement for primary tumors to invade through the mucosa to the serosa—a critical step in the development of metastasis. Importantly, a major caveat of all or some of these established techniques is the development of widespread peritoneal carcinomatosis, which is likely a consequence of seeding and shedding of tumor cells into the peritoneal space, rather than bona fide metastatic dissemination.

To overcome these limitations, we have recently developed two orthotopic implantation methodologies that form the basis of this protocol [9]. The first approach relies on the induction of a rectal prolapse that exteriorizes the lumen of the host colon, thus rendering the colonic mucosal surface amenable to cell injection [5, 9]. A variety of murine or human cell lines or primary cultures can be implanted using this technique. Importantly, this implantation procedure does not breach the colon wall, as tumors are implanted directly and exclusively on the mucosal surface of the colon. Furthermore, depending on the cell line used for implantation, spontaneous metastatic dissemination to the liver and/or lungs can occur. This methodology enables the study of each-and-every step of the metastatic process as well as the testing of therapeutic modalities relevant to the adjuvant setting. Finally, this method does not rely on any complicated surgical procedures and thus is cost-effective as it does not require specialized equipment for its implementation. As an example, this method was recently leveraged to reveal the requirement of stem cells during colon cancer dissemination and metastatic colonization [9].

The second approach consists of infusion of tumor cells directly into the portal vein, leading to robust tumor growth in the liver only [9, 13]. This method, however, requires a more complex surgical procedure. Of note, the portal venous injection approach bypasses the initial steps of the metastatic cascade—tumor cell extravasation and survival in the bloodstream. A key advantage of this method, however, is that it allows one to study the growth of colon cancers in the liver environment for a prolonged period of time, since the growth of the primary tumor does not contribute to comorbidity.

2 Materials

2.1 General Equipment

1. LED cold light source for routine, lower magnification applications.
2. Halstead mosquito 5″ forceps.
3. Glass Hamilton Syringe 50 μL Gastight.
4. TSK sterile hypodermic needle, 33G × 1/2″ (0.24 × 13 mm).
5. 1.0 mL tuberculin syringe.
6. Animal feeding needle, Reusable AFN 22G × 1.5″, 1.25 mm-Straight.
7. K&H small animal heated pad.
8. Dry sterilizer.
9. Rodent nonrebreathing circuit with 9 mm nose cone.
10. Magic Touch Ice Pans 4 L.

2.2 Endoscopic Equipment

1. STORZ ENDOSKOPE.
2. Electronic endoflator.
3. Image 1 hub.
4. SCB D Light AF.
5. AIDA Control II.
6. SCB Tricam SL II.
7. Point Setter Mondel.
8. STORZ Office Kart.

2.3 Surgical Equipment

1. Rectal temperature probe.
2. Electric razor.
3. Sterile surgical gloves.
4. Nair hair removal cream.

5. Surgical drape, Steri-Drape™, Towel drape.
6. Disinfection: Povidone–iodine swab sticks, alcohol pads.
7. Hemostyptic calcium sodium alginate dressing.
8. Nonwoven gauze pads.
9. Surgical platform including heating pad.
10. Cotton-tipped applicators.
11. Suture: Vicryl 6-0, 5-0 Monocryl Plus.
12. Retractors: Magnetic Fixator Retraction System, Bowman Retractor.
13. Surgical instruments: Straight scissors, Dumont forceps, Halsted-Mosquito hemostat, Castroviejo needle holder.
14. Sterile towel drape.
15. Criterion PF latex surgical gloves.
16. Surgical facemask Critical Cover CoolOne.
17. Conical 50 mL tubes.
18. Puritan medical 6″ sterile standard cotton swab with wooden handle.

2.4 Reagents

1. Water (sterile).
2. Buprenorphine hydrochloride injectable, dose at 0.05–0.1 mg/kg body weight.
3. Advanced DMED/F12.
4. 100× GlutaMAX.
5. Matrigel, growth factor reduced (GFR), phenol red-free.
6. PBS.
7. Penicillin–streptomycin.
8. Zoetis IsoFlo 250 mL isoflurane gas anesthesia.
9. Buprenorphine slow-release pain medication.
10. Sterile eye ointment—Puralube.

2.5 Cell Lines and Organoids

A number of human colorectal cancer cell line can be used for these studies. Most human cell lines can be purchased through vendors including the American Type Culture Collection (ATCC) and European Collection of Authenticated Cell Cultures (ECACC). Of note, if human CRC are used, the mouse strain which serves as the host must be immunocompromised to prevent immune rejection of the cell line. We have obtained high tumor take rates using NOD.Cg-PrkdcscidIl2rgtm1Wjl/SzJ (NSG) animals.

In addition, the presented methods are suitable for intestinal and colonic organoids. However, the establishment of organoid culture systems for mouse and human intestinal organoids are well established and will not be described here. The reader is referred to several key methodologies describing organoid cell isolation, expansion, and preparation for injection [9, 14, 15]. We have used small intestinal organoids derived from $Apc^{min/+}$;$Kras^{LSL\text{-}G12D/+}$;$Villin^{Cre}$ (*AKV*) or variants that were engineered to harbor mutations in Tp53 (AKVP) and SMAD4 (AKVPS) using CRISPR/Cas9 [9, 16, 17].

2.6 Animals

NSG mice. All animal procedures in this study are approved by the Genentech Institutional Animal Care and Use Committee (IACUC) and are performed at an AAALAC international accredited facility. Animals ranging from 8 to 16 weeks have been used for our studies using the presented methods. Alternatively, when using mouse derived organoids, the immunocompetent host animals matching the strain from which organoids have been initially derived from can be used if a syngeneic immunocompetent system is favored.

3 Methods

3.1 Rectal Prolapse Orthotopic Injection Procedure (Fig. 1)

1. Autoclave or hot bead-sterilize hemostat and forceps.
2. Set up heating pad and microscope and ensure an anesthesia extension with a nose cone is available (Fig. 1b).
3. Animals are anesthetized by inhalation of isoflurane in an induction chamber using 2.0% (vol/vol) isoflurane delivered in 100% oxygen.
4. Use a 1 mL syringe with a 25-gauge needle to subcutaneously inject 100 μL of 0.03 mg/mL buprenorphine hydrochloride (dose per mouse 0.05–0.1 mg/kg body weight subcutaneously (SC) every 8–12 h).
5. While animals are anesthetized, load cells into Hamilton syringe through back end with a pipette (see cell loading).
6. Place mouse in supine position (ventral side up) into nose cone on heating pad (*see* **Note 1**).
7. Confirm mouse is adequately anesthetized by gently pinching hind leg and confirming lack of withdrawal reflex.
8. Administer eye ointment topically to both eyes using a sterile cotton-tipped applicator to protect the cornea from desiccation.
9. Insert sterile blunt gavage needle slowly into rectum to remove residual stool (Fig. 1c) (*see* **Note 2**).
10. Insert sterile hemostat approximately ~0.5 cm into rectum (Fig. 1d).

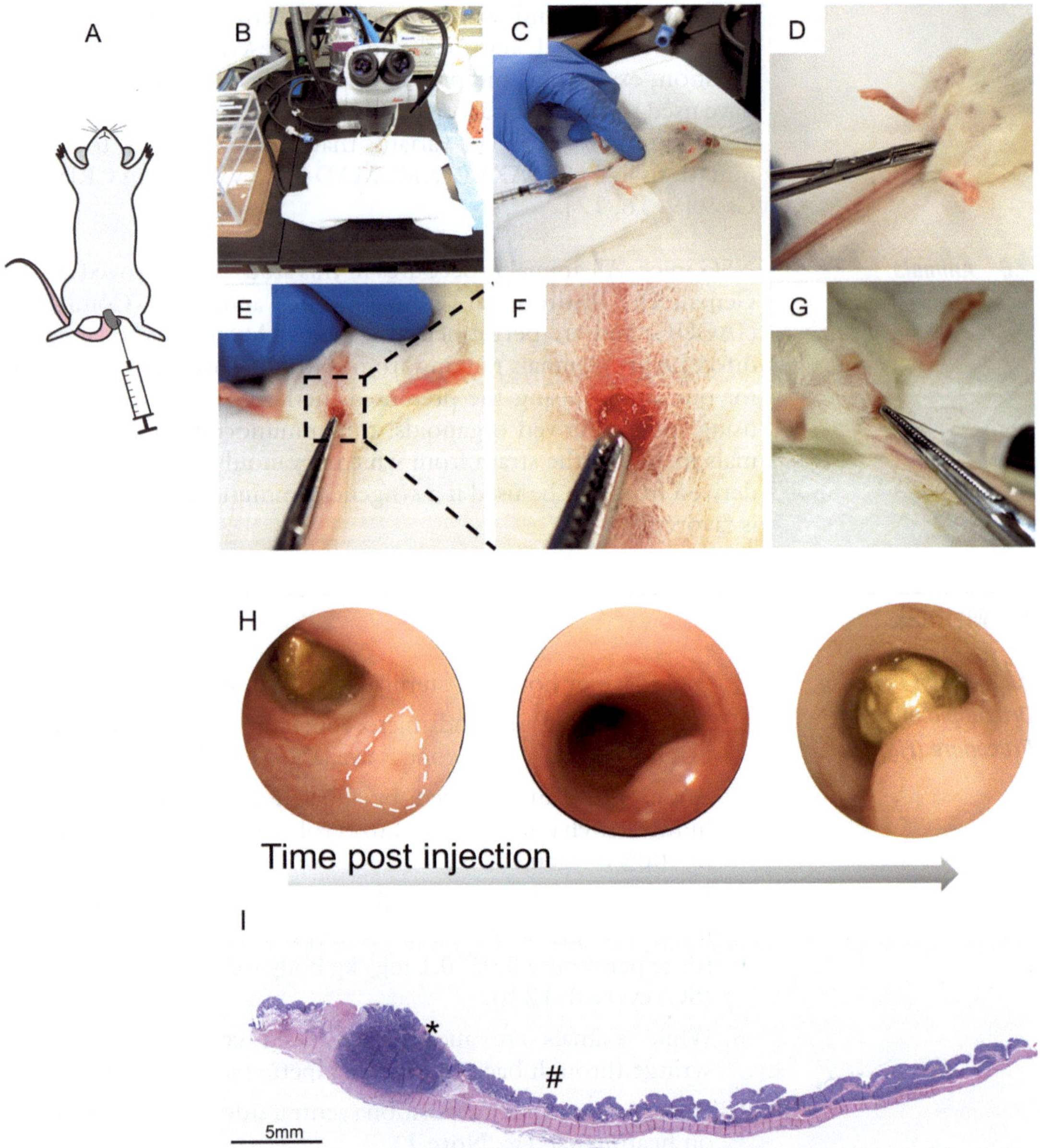

Fig. 1 Orthotopic colonic implantation methodology and anticipated results. (**a**) Schematic representation of orthotopic rectal prolapse induction. (**b**) Microscope and heating pad with an anesthesia extension with a nose cone available. Correct positioning of the animal is displayed. (**c**) Residual stool can be removed using a blunt sterile gavage needle. (**d**) Hemostat insertion and opening into the rectum. (**e**) Prolapse induction by slowly retracting the hemostat to expose the rectal mucosa (**f**). (**g**) Injection of cells into the portion of the submucosa that is pinched by the hemostat. (**h**) Representative colonoscopic still images of implanted colon tumor derived organoids. Time points depicted are from left to right: 3 weeks, 4 weeks and 5 weeks postinjection. (**i**) Representative hematoxylin and eosin section of a colonic tumor at 5 weeks post implantation. $*$ indicates tumor tissue, # indicates adjacent normal tissue

11. Incline the hemostat and gently press down the tip of the hemostat while opening it (*see* **Note 3**).
12. Gently take hold of the colon wall with the hemostat and close the hemostat to the first ratchet.
13. Slowly retract to exteriorize rectal region (Fig. 1e, f) (*see* **Note 4**).
14. Visualize colon mucosa that is pinched by the hemostat under dissecting scope or magnifier glass (*see* **Note 5**).
15. Rest both elbows on the table, if needed adjust the height of your chair. Take the Hamilton syringe + 33G needle in your dominant hand, placed in-between the index and the middle finger, with the thumb on the plunger.
16. Inject 10 μL of cells into the portion of the submucosa that is pinched by the hemostat (Fig. 1g) (*see* **Note 6**).
17. Stop any leakage or bleeding with a sterile cotton-tipped swab.
18. Gently push colon back into body and release the hemostat.
19. Return the mouse to its cage and carefully monitor recovery from anesthesia.
20. After procedure recovery, perform a general health check on mouse prior to the end of day for any bleeding or scab formation, which may indicate procedure complication.

3.1.1 Anticipated Results

Tumor growth in the colon will depend on the cell line injected and the animal strain. We have observed the highest tumor take with NSG animals and recommend their use for initial experiments. Most established human CRC cell lines will show evidence of growth within 3–4 weeks in the colon (Fig. 1h, i). Some lines will spontaneously metastasize to the liver. The earliest time point where we have seen metastasis for the most aggressive cell lines is 3 weeks post implantation. If metastatic dissemination has occurred, burden in the liver will be evident around 8 weeks post-implantation. In addition, potential clinical signs of metastasis and endpoints, including hunched posture and lethargy can indicate the presence of metastatic burden.

If available, endoscopic imaging can be used to score rectal tumor take and also monitor tumor growth longitudinally (Fig. 1h). Below, we describe the endoscopic imaging procedure.

3.1.2 Optional: Endoscopic Imaging

1. Set up isoflurane anesthesia machine according to the manufacturer's instructions.
2. Place a sterile sheet on table.
3. Cover a surgical platform board with another sterile sheet.
4. Place platform at an angle on a table with one side slightly elevated and facing operator.

5. Place a multiple (up to 5) mouse anesthesia nose cone set-up on a covered surgical platform with a waste gas scavenger vacuum tube positioned above it.
6. Prepare and place to the right side of the surgical platform, a gavage needle with 1 mL syringe filled with sterile water, Kim-wipes, and sterile cotton-tipped applicators.
7. Assemble telescope and camera according to the manufacturer's instructions.
8. Remove cover from monitor and cover from the camera.
9. Place the camera in telescope and turn left to tighten, slightly off center.
10. Attach camera to arm.
11. Fit the telescope to endoscope and turn knob to lock it in.
12. Attach light source top port.
13. Attach air source to right port.
14. Close left port on sheath.
15. On endoscope cart, turn Master switch on first followed by all other switches.
16. Open CO_2 tank and press CO_2 button until it stays on to start flow of CO_2.
17. Place up to 5 mice into an anesthesia induction chamber set on a heated animal pad.
18. Induce anesthesia at 2.5% isoflurane delivered in 100% O_2 at a flow rate of 1.5–2 L/min.
19. When mice are anesthetized, remove from the chamber and apply sterile eye lubricant topically to each eye.
20. Position each mouse in the supine position with its nose in a nose cone to maintain an adequate anesthesia level. Confirm each mouse is fully anesthetized by gently pinching the rear toes and observing no withdrawal reflex. This is to ensure no movement during the procedure which can result in damage to the colon.
21. Using the gavage needle placed on a 1 mL syringe, remove feces from colon by gently injecting sterile, warm (room temperature) water into the colon. This can be done several times to clean the colon of any feces. It is important that the colon is clear to insure a clear picture.
22. Hold the fur of the mouse just above the anus in one hand and the scope in the other, then carefully insert the endoscope into the anus.

23. Insufflation pressure should be maintained at approximately 14 mm/Hg pressure and 14 L/min flow in order to keep the colon adequately inflated.
24. Gently push the endoscope forward about 3 cm at which point the colon makes a curve and is not accessible to the rigid endoscope.
25. At this point, a video can also be recorded by pressing the gray video button (left) on the endoscope.
26. Slowly withdraw the endoscope while filming and maintaining insufflation pressure similar to entry.
27. Stop recording when endoscope reaches the rectum.
28. Remove animal from anesthesia nose cone, then return to its home cage for recovery, and monitor until fully awake and ambulatory.

3.2 Portal Venous Injection Procedure (Fig. 2)

1. The surgery must be performed in a sterile working environment and an appropriate sterile field must be maintained. An isoflurane anesthesia machine will be used to induce and maintain anesthesia, and an external waste scavenger will be used to minimize gas exposure to the surgeon. All scissors, forceps, and hemostats must be sterilized prior to surgery. This can be done using a hot bead sterilizer, or having multiple sterilized surgical kits. Autoclave the kit together with some surgical drapes and cotton swabs. Ensure the microscope is thoroughly cleaned with 70% ethanol before proceeding with surgery.
2. Using ~2.0% (vol/vol) isoflurane, anesthetize a mouse in an induction chamber until loss of righting reflex. Remove the mouse from the induction chamber and place on a heating pad in the supine position with its nose in the facemask to maintain anesthesia.
3. Use a 1 mL syringe with a 25-gauge needle to subcutaneously inject 100 μL of 0.03 mg/mL buprenorphine hydrochloride.
4. Confirm mouse is adequately anesthetized into a surgical plane of anesthesia by gently pinching toe to confirm absence of reflexes.
5. Shave abdomen and chest of the mouse with an electric razor (*see* **Note 7**).
6. Monitor animal breathing and reflexes regularly throughout the procedure to ensure adequate anesthesia.
7. Administer eye ointment topically to both eyes using sterile cotton-tipped applicator to protect the cornea from desiccation.
8. Secure all four extremities on a heating pad using tape.

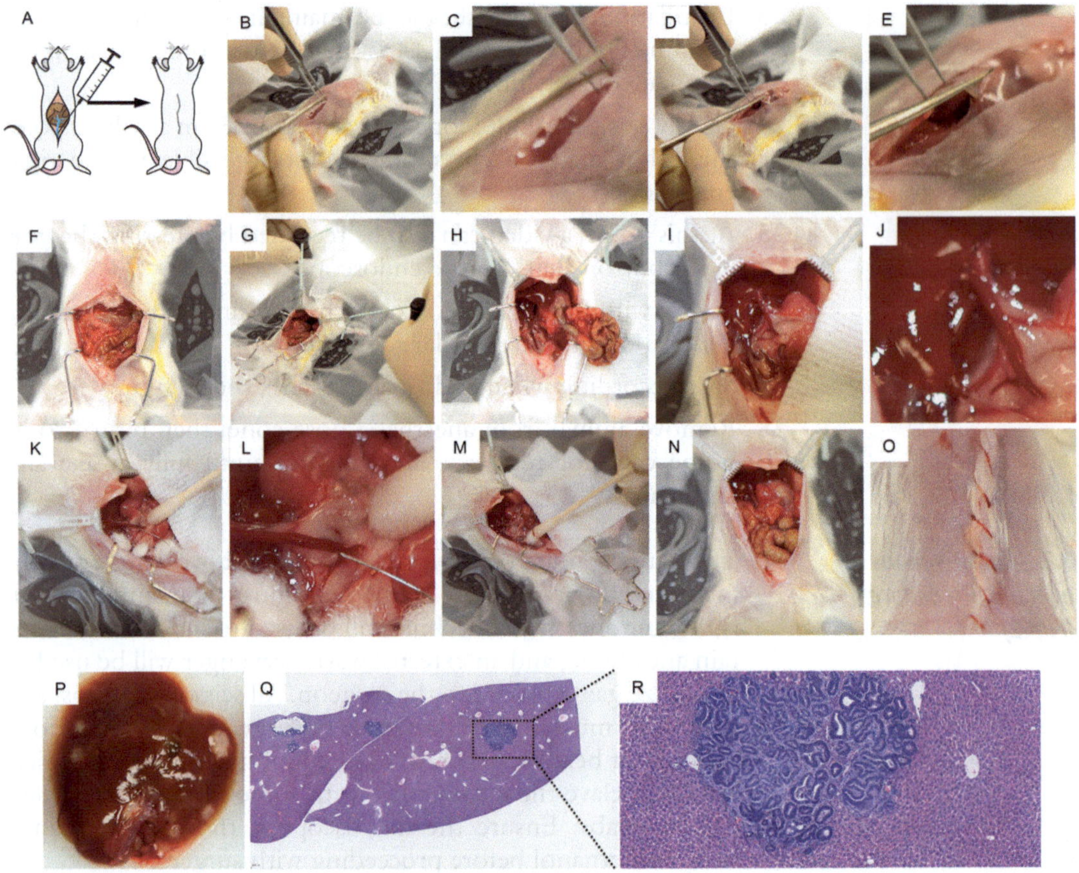

Fig. 2 Portal venous injection methodology and anticipated results. (**a**) Schematic representation of the portal venous injection method. (**b–o**) stepwise image description of the methodologies as described in details in **steps 19–55** in Subheading 3. (**p**) Gross metastatic nodules visible as white spots on a liver from an animal euthanized 5 weeks post cell infusion. (**q**) Representative hematoxylin and eosin of a cross section of a liver containing metastatic nodules. (**r**) Enlarged view of metastatic nodule in **q**. Note the presence of glandular structure of the colon derived liver metastasis

9. Decrease isoflurane to 0.5–1% (vol/vol) with continuous air or oxygen flow of 0.5–1 L/min for maintenance of anesthesia (*see* **Note 8**).
10. Apply hair removal cream to remove hair in the shaved abdominal and chest area.
11. Subsequently remove hair removal cream after 2–3 min using moist gauze soaked in sterile 0.9% saline solution. Repeat this step if necessary (*see* **Note 9**).
12. Disinfect the abdominal incision site with betadine and 70% (vol/vol) ethanol following surgical antiseptic guidelines.
13. Place sterile surgical adhesive towel drapes around the surgical site.

14. Make a vertical midline incision through the skin extending from the xiphoid process to the pubis using a straight scissors (Fig. 2b, c).
15. Lift the abdominal wall with straight forceps and carefully incise the linea alba to separate the rectus abdominis muscles and open the abdominal cavity. Make sure to avoid damaging the abdominal organs during the incision (Fig. 2d, e).
16. Carefully insert a Bowman retractor into the incision (Fig. 2f) (*see* **Note 10**).
17. Carefully insert two small blunt retractors underneath the costal arches using a magnetic fixator retraction system (Fig. 2g) (*see* **Note 11**).
18. Place a sterile nonwoven gauze pad drenched in warmed sterile 0.9% saline (IV injectable grade) or similar sterile lavage solution on the left side of the mouse, drape out the cecum using two sterile cotton swabs drenched in PBS and then gently transfer the entire intestine out of the abdominal cavity to the left side of the mouse onto the sterile gauze. Cover the intestine with sterile gauze drenched in PBS to keep the intestine hydrated (Fig. 2h) (*see* **Note 12**).
19. Using sterile cotton swabs drenched in sterile PBS carefully push the stomach cranial so that it rotates about 45° counterclockwise. Thus, following the stomach, the duodenum also rotates into the same direction exposing the portal vein. Flip the right, median, and left lateral liver lobes upward, so that they are attached to the abdominal cavity and the liver hilum is exposed (*see* **Note 13**).
20. Cover all exposed areas inside the abdominal cavity but the portal vein with sterile gauze pads drenched in PBS to keep the organs hydrated (Fig. 2i).
21. Form 3 to 4 5 × 5 mm sized round balls of hemostyptic calcium sodium alginate dressing (CSAD) and place them on the right side of the portal vein on the sterile gauze (Fig. 2j, k).
22. Take a 1-mL syringe needle with a 33-gauge needle and fill the syringe with the cells in media. Carefully remove all air bubbles and reduce the volume in the syringe to 100 μL.
23. Use a dry sterile gauze pad, place it on the table and clean the tip of the needle from any cells (*see* **Note 14**).
24. Use a hemostat to carefully bend the needle twice 5–8 mm apart at 45°, bevel up (*see* **Note 15**).
25. Take the syringe in your left hand, placed in-between the index and the middle finger, the thumb on the plunger. A sterile cotton swab drenched in sterile PBS is in the right hand. Rest both elbows on the table, if needed adjust the height of your chair (*see* **Note 16**).

26. Check again that the 3–4 CSAD balls are in close proximity to the portal vein.
27. While looking through a stereomicroscope using a 25–40× magnification, carefully push the duodenum to the left side using the sterile cotton swab in the right hand. Thus, the portal vein will be stretched out. Exchange the cotton stick with fine forceps and use the forceps to take hold of the needle of the syringe for stabilization. Guided by the forceps, carefully poke the needle of the syringe through the vessel wall of the portal vein about 1–2 mm upstream of the splenic vein, and carefully continue to insert ca. 3 mm of the needle. If done properly, the tip of the needle should be visible through the vessel wall (Fig. 2l) (*see* **Note 17**).
28. Slowly inject the cells by pressing the plunger over a period of 30–60 s (*see* **Note 18**).
29. While maintaining the tip of the needle inside of the portal vein, carefully release the grip of the forceps on the needle and use the forceps to place one of the CSAD balls directly on top of the injection side.
30. Exchange the forceps with a sterile cotton swab drenched in PBS, and gently push on the CSAD ball with the needle underneath. Use the left hand to rapidly pull out the needle while simultaneously using the right hand to push the CSAD ball directly onto the injection site (Fig. 2m).
31. Using fine forceps in the left hand, place two additional CSAD balls on top of the first, and apply pressure for 2–4 min using a sterile cotton swab drenched in PBS in the right hand (*see* **Note 19**).
32. Gently release the pressure on the CSAD balls and remove the first layer of CSAD balls using fine forceps. It is important not to pull too hard at any time. If the CSAD balls become stuck to adjacent organs or to each other, add 300–400 μL of warm (37 °C) sterile PBS and wait another 2–4 min. The balls should detach easily. Repeat this step until all CSAD balls have been removed (*see* **Note 20**).
33. Carefully remove all gauze pads and rotate the stomach back and transfer the intestine into their original positions using sterile cotton swabs drenched in PBS (Fig. 2n).
34. Remove all retractors. Inject 2 mL of warm (37 °C) sterile saline in the abdominal cavity. Make sure there is no intra-abdominal bleeding.
35. Close the abdominal incision in two layers, using a simple continuous or running pattern with 6-0 vicryl absorbable suture to oppose the rectus abdominis muscles and 5-0 nonabsorbable ethilon to close the skin (Fig. 2o).

36. Shut off isoflurane and increase oxygen flow to induce recovery of the animal. Remove adhesive tape used for limb fixation, turn mouse over onto its normal ventral resting position, and remove eye ointment with a gauze pad. Monitor the animal until it is fully awake and ambulatory. Place the mouse in the home cage on a heating pad to recover. Observe the mouse every 1–2 h for the first 4–6 h to ensure recovery, and does not show any signs of pain before returning the cage to the housing location.
37. Animals will need to be monitored on the first post-operative day and regularly throughout the study to ensure sutures are still correctly in place. Since this is a major surgery, use appropriate analgesics (e.g., buprenorphine) and antibiotics (e.g., cefazolin) as required by your institutional veterinarian.

3.2.1 Anticipated Results

As in the colon, tumor growth in the liver will depend on the cell line injected and the animal strain. We have observed the highest tumor take with NSG animals and recommend their use for initial experiments. Most established human CRC lines will show evidence of growth within 5–6 weeks in the liver. The earliest time point where we have seen growth for the most aggressive lines is 3 weeks post implantation. We recommend euthanizing animals every week starting 4 weeks post implantation and check for liver metastasis by gross inspection. Metastatic nodules will be easily visible as white spots on the liver (Fig. 2p–r).

If available, bioluminescence imaging and/or microCT scanning can be used to score tumor take as well as to monitor tumor growth longitudinally.

3.3 Cell Preparation and Syringe Loading

We described the procedure using the AKVPSL organoids generated previously in de Sousa e Melo et al. [9]

1. Prepare a single cell suspension of organoids as previously described.
2. For orthotopic injections, resuspend cells in 100% Matrigel to obtain a concentration of 5000 cells per microliter. Keep Matrigel on ice at all times to prevent polymerization. Remove plunger from the Hamilton syringe.
3. Slowly attach and seal the 33G needle to the Hamilton syringe.
4. Use a pipette to take up 100 μL of Matrigel and insert the pipette tip at the back of the syringe and into the barrel.
5. Slowly load the Matrigel into the Hamilton syringe until the hub is filled and the solution begins to exit the needle bevel. Ensure that no air bubbles are trapped in the barrel.
6. Keep syringe on ice to prevent Matrigel polymerization.

7. For portal venous injections, resuspend cells in advanced DMEM/F12 at a concentration of 500 cells per microliter in a regular 1 mL syringe with an attached 33G needle.

4 Notes

1. Ensure animals are immobilized on the heating pad, with the anus should be visible when looking through the microscope.
2. Perform this step very slowly to avoid any rupture or perforation of the colorectal mucosa.
3. If the hemostat is correctly positioned, a tail twitch reflex may occur.
4. Only retract a few millimeters. While doing so, a small amount of resistance can be encountered. If too much resistance is encountered, do not pull. Release the hemostat and reclamp the mucosa.
5. Carefully monitor the pinched mucosa area. If too little tissue is clamped, the mucosa can rupture. If too much is clamped, the rectum will be difficult to retract.
6. Because Matrigel is used, it will solidify rapidly at body temperature. Hence, a small plug of Matrigel lodged under the mucosa should be evident if the injection is successful.
7. It is critical to remove as much hair as possible to prevent contamination of the surgical wound.
8. Caution: Do not overheat the mouse as this can cause burn injuries and also enhances the depth of the anesthesia, which can cause respiratory failure. If available, a rectal probe is recommended to monitor body temperature.
9. Caution: follow product instructions, hair removal cream can cause irritation and injuries to the skin.
10. Caution: avoid any organs to be trapped and incarcerated underneath the jaws of the retractor, this can cause damage to the organs.
11. Caution: check mouse's breathing and make sure that breathing is normal.
12. Caution: During the organ transfer avoid damaging the intestine. Do not use sharp tools to hold the intestine.
13. Caution: Rotate the stomach and flip the liver gently and avoid damaging any adjacent organs. Do not use sharp tools.
14. Caution: do not hold the gauze pad in your hands to avoid needle injuries.
15. Caution: do not do this with your fingers to avoid needle injuries.

16. Caution: it is critical that your forearms are rested to avoid any movements in the hands.
17. Caution: during insertion of the needle into the portal venous vessel wall, the surgeon must refrain from any movements. Stabilizing the needle with the forceps is an absolute essential.
18. If the injection is successful, turbulence within the blood stream caused by the injection should be readily visible through the portal venous vessel wall.
19. Caution: This step is the most critical step of the procedure. It is important to work calmly but steadily with precision. Perforating the back of the portal vein, damaging the hepatic artery, or rupturing the vessel wall when pulling out the needle can compromise the surgery, leading to hemorrhage and premature death of the experimental animal.
20. Caution: it is important to work slowly and carefully. Do not forcibly pull off the CSAD balls, but rather wait until they detach by themselves to avoid accidently removing the coagulation plug that has been deposited on the injection site. If the injection site starts to bleed, apply another CSAD ball and repeat this step until bleeding has stopped. Hemostasis is absolutely essential, and do not proceed without having achieved this.

Acknowledgments

The authors would like to acknowledge Genentech's lab animal staff for their invaluable help in implementing all the described methodologies.

Conflicts of Interest: All authors are/were employees of Genentech or own/owned shares from Roche.

References

1. Siegel R, Desantis C, Jemal A (2014) Colorectal cancer statistics, 2014. CA Cancer J Clin 64 (2):104–117. https://doi.org/10.3322/caac.21220
2. Fearon ER, Vogelstein B (1990) A genetic model for colorectal tumorigenesis. Cell 61 (5):759–767
3. Jorissen RN, Gibbs P, Christie M, Prakash S, Lipton L, Desai J, Kerr D, Aaltonen LA, Arango D, Kruhoffer M, Orntoft TF, Andersen CL, Gruidl M, Kamath VP, Eschrich S, Yeatman TJ, Sieber OM (2009) Metastasis-associated gene expression changes predict poor outcomes in patients with dukes stage B and C colorectal cancer. Clin Cancer Res 15 (24):7642–7651. https://doi.org/10.1158/1078-0432.ccr-09-1431
4. Sheffer M, Bacolod MD, Zuk O, Giardina SF, Pincas H, Barany F, Paty PB, Gerald WL, Notterman DA, Domany E (2009) Association of survival and disease progression with chromosomal instability: a genomic exploration of colorectal cancer. Proc Natl Acad Sci U S A 106(17):7131–7136. https://doi.org/10.1073/pnas.0902232106
5. Enquist IB, Good Z, Jubb AM, Fuh G, Wang X, Junttila MR, Jackson EL, Leong KG (2014) Lymph node-independent liver metastasis in a model of metastatic colorectal cancer.

Nat Commun 5:3530. https://doi.org/10.1038/ncomms4530
6. Heijstek MW, Kranenburg O, Borel Rinkes IH (2005) Mouse models of colorectal cancer and liver metastases. Dig Surg 22(1–2):16–25. https://doi.org/10.1159/000085342
7. Boutin AT, Liao WT, Wang M, Hwang SS, Karpinets TV, Cheung H, Chu GC, Jiang S, Hu J, Chang K, Vilar E, Song X, Zhang J, Kopetz S, Futreal A, Wang YA, Kwong LN, DePinho RA (2017) Oncogenic Kras drives invasion and maintains metastases in colorectal cancer. Genes Dev 31(4):370–382. https://doi.org/10.1101/gad.293449.116
8. Tauriello DVF, Palomo-Ponce S, Stork D, Berenguer-Llergo A, Badia-Ramentol J, Iglesias M, Sevillano M, Ibiza S, Canellas A, Hernando-Momblona X, Byrom D, Matarin JA, Calon A, Rivas EI, Nebreda AR, Riera A, Attolini CS, Batlle E (2018) TGFbeta drives immune evasion in genetically reconstituted colon cancer metastasis. Nature 554 (7693):538–543. https://doi.org/10.1038/nature25492
9. de Sousa e Melo F, Kurtova AV, Harnoss JM, Kljavin N, Hoeck JD, Hung J, Anderson JE, Storm EE, Modrusan Z, Koeppen H, Dijkgraaf GJ, Piskol R, de Sauvage FJ (2017) A distinct role for Lgr5(+) stem cells in primary and metastatic colon cancer. Nature 543 (7647):676–680. https://doi.org/10.1038/nature21713
10. Fumagalli A, Drost J, Suijkerbuijk SJ, van Boxtel R, de Ligt J, Offerhaus GJ, Begthel H, Beerling E, Tan EH, Sansom OJ, Cuppen E, Clevers H, van Rheenen J (2017) Genetic dissection of colorectal cancer progression by orthotopic transplantation of engineered cancer organoids. Proc Natl Acad Sci U S A 114 (12):E2357–E2364. https://doi.org/10.1073/pnas.1701219114
11. O'Rourke KP, Loizou E, Livshits G, Schatoff EM, Baslan T, Manchado E, Simon J, Romesser PB, Leach B, Han T, Pauli C, Beltran H, Rubin MA, Dow LE, Lowe SW (2017) Transplantation of engineered organoids enables rapid generation of metastatic mouse models of colorectal cancer. Nat Biotechnol 35 (6):577–582. https://doi.org/10.1038/nbt.3837
12. Roper J, Tammela T, Cetinbas NM, Akkad A, Roghanian A, Rickelt S, Almeqdadi M, Wu K, Oberli MA, Sanchez-Rivera FJ, Park YK, Liang X, Eng G, Taylor MS, Azimi R, Kedrin D, Neupane R, Beyaz S, Sicinska ET, Suarez Y, Yoo J, Chen L, Zukerberg L, Katajisto P, Deshpande V, Bass AJ, Tsichlis PN, Lees J, Langer R, Hynes RO, Chen J, Bhutkar A, Jacks T, Yilmaz OH (2017) In vivo genome editing and organoid transplantation models of colorectal cancer and metastasis. Nat Biotechnol 35(6):569–576. https://doi.org/10.1038/nbt.3836
13. Thalheimer A, Otto C, Bueter M, Illert B, Gattenlohner S, Gasser M, Meyer D, Fein M, Germer CT, Waaga-Gasser AM (2009) The intraportal injection model: a practical animal model for hepatic metastases and tumor cell dissemination in human colon cancer. BMC Cancer 9:29. https://doi.org/10.1186/1471-2407-9-29
14. Fumagalli A, Suijkerbuijk SJE, Begthel H, Beerling E, Oost KC, Snippert HJ, van Rheenen J, Drost J (2018) A surgical orthotopic organoid transplantation approach in mice to visualize and study colorectal cancer progression. Nat Protoc 13(2):235–247. https://doi.org/10.1038/nprot.2017.137
15. Sato T, Clevers H (2013) Primary mouse small intestinal epithelial cell cultures. Methods Mol Biol 945:319–328. https://doi.org/10.1007/978-1-62703-125-7_19
16. Drost J, van Jaarsveld RH, Ponsioen B, Zimberlin C, van Boxtel R, Buijs A, Sachs N, Overmeer RM, Offerhaus GJ, Begthel H, Korving J, van de Wetering M, Schwank G, Logtenberg M, Cuppen E, Snippert HJ, Medema JP, Kops GJ, Clevers H (2015) Sequential cancer mutations in cultured human intestinal stem cells. Nature 521 (7550):43–47. https://doi.org/10.1038/nature14415
17. Matano M, Date S, Shimokawa M, Takano A, Fujii M, Ohta Y, Watanabe T, Kanai T, Sato T (2015) Modeling colorectal cancer using CRISPR-Cas9-mediated engineering of human intestinal organoids. Nat Med 21 (3):256–262. https://doi.org/10.1038/nm.3802

Index

Paloma Ordóñez-Morán (ed.), *Intestinal Stem Cells: Methods and Protocols*, Methods in Molecular Biology, vol. 2171, https://doi.org/10.1007/978-1-0716-0747-3, © Springer Science+Business Media, LLC, part of Springer Nature 2020

V

W

X

MIX
Papier aus verantwortungsvollen Quellen
Paper from responsible sources
FSC® C105338

If you have any concerns about our products,
you can contact us on
ProductSafety@springernature.com

In case Publisher is established outside the EU,
the EU authorized representative is:
Springer Nature Customer Service Center GmbH
Europaplatz 3, 69115 Heidelberg, Germany

Printed by Libri Plureos GmbH
in Hamburg, Germany